Routledge Revivals

Safe Drinking Water

Originally published in 1978, this volume addresses the scientific, economic, and administrative aspects of the public policy problem raised by the United States' Safe Drinking Water Act of 1974. In this collection of papers, the authors examine the various problems presented by drinking public water in the United States, the United Kingdom, and throughout Europe. This is an ideal title for students interested in environmental studies and public policy reform.

Safe Drinking Water

Current and Future Problems

Edited by
Clifford S. Russell

RFF PRESS
RESOURCES FOR THE FUTURE

First published in 1978
by Resources for the Future, Inc.

This edition first published in 2016 by Routledge
2 Park Square, Milton Park, Abingdon, Oxon, OX14 4RN
and by Routledge
711 Third Avenue, New York, NY 10017

Routledge is an imprint of the Taylor & Francis Group, an informa business

Publisher's Note

The publisher has gone to great lengths to ensure the quality of this reprint but points out that some imperfections in the original copies may be apparent.

Disclaimer

The publisher has made every effort to trace copyright holders and welcomes correspondence from those they have been unable to contact.

A Library of Congress record exists under LC control number: 78019840

ISBN 13: 978-1-138-96258-3 (hbk)
ISBN 13: 978-1-315-65932-9 (ebk)
ISBN 13: 978-1-138-96262-0 (pbk)

Safe drinking water: Current and future problems

Proceedings of a national conference in Washington, D.C.

CLIFFORD S. RUSSELL, editor

RESEARCH PAPER R-12

RESOURCES FOR THE FUTURE / WASHINGTON, D.C.

RESOURCES FOR THE FUTURE, INC.
1755 Massachusetts Avenue, N.W., Washington, D.C. 20036

Resources for the Future is a nonprofit organization for research and education in the development, conservation, and use of natural resources and the improvement of the quality of the environment. It was established in 1952 with the cooperation of the Ford Foundation. Grants for research are accepted from government and private sources only if they meet the conditions of a policy established by the Board of Directors of Resources for the Future. The policy states that RFF shall be solely responsible for the conduct of the research and free to make the research results available to the public. Part of the work of Resources for the Future is carried out by its resident staff; part is supported by grants to universities and other nonprofit organizations. Unless otherwise stated, interpretations and conclusions in RFF publications are those of the authors; the organization takes responsibility for the selection of significant subjects for study, the competence of the researchers, and their freedom of inquiry.

Library of Congress Catalog Card Number 78-19840

ISBN 0-8018-2181-9

Manufactured in the United States of America

Published September 1978. $12.00

TABLE OF CONTENTS

LIST OF TABLES

Page

LIST OF FIGURES

PREFACE

The United States Environmental Protection Agency is now in the process of setting the standards called for in the Safe Drinking Water Act of 1974. This summer (1978) there will be oversight hearings on the Act itself. This seemed, therefore, an appropriate time for a conference on drinking water. The heart of the conference was the series of papers found in this volume addressing scientific, economic, and administrative aspects of the public policy problem raised by our new knowledge in the drinking water area. RFF's goal was to have these papers, along with the formal and informal discussion, delineate the existing state of this knowledge (and the nature of our ignorance) and thus form a useful and balanced basis for informed policy choice.

This is the second time in recent history that safe drinking water has become a major issue of public health. In the nineteenth century, concern focused on the spread of communicable diseases, and we may find useful perspective in reminding ourselves that many of the issues with which we are now struggling were also issues in that first round. For example, how much evidence linking disease and water is "enough"? How much does prevention cost and what is its most effective form? How much preventive action should be taken? Who should be responsible? Although these same questions trouble us today, and useful comparisons can be drawn from history, there are indeed differences in today's version of the drinking water problem.

One hundred years ago we were coming to grips with cholera and typhus and were remarkably successful in bringing these diseases under control. Now our concern has shifted to cancer, lead poisoning, and heart disease. These conditions have latency periods far longer than the old waterborne epidemic diseases. Thus, it appears that we are unlikely to develop clear, overwhelmingly convincing evidence comparable to that available to decision makers in the earlier setting, even if there exist direct causal links between drinking water and cancer, or other long-term latent diseases. People move, or die of other diseases. Records are not kept or are kept poorly. There is a question of the statistical power of epidemiology to uncover hidden effects, if present. We hoped, therefore, to learn something about the uses and limits--the state of the art--of epidemiological investigation applied to the potential long-term hazards of drinking water. Similarly, we wished to assess the nature of our uncertainty, or the state of our knowledge, with respect to toxicology and extrapolation from laboratory experiments to normal human life.

Since we, as a society, must make decisions now, we must make them under uncertainty. We are forced to weigh one uncertain cause against another, and it is important that we weigh them in a symmetrical manner. An epidemiological study can fail to find an effect when it is present or can identify an "effect" when none in fact exists. What are the costs associated with these two types of errors? For a given statistical technique, what the probabilities? The same weighing of uncertainties can be applied to the costs of treatment, which can be over- or underestimated. A policy which says that we should take no preventive action until it is proved beyond a reasonable doubt that carcinogens at the parts per billion level are substantially raising cancer rates is just as biased, though from the opposite direction, as a policy which says that we should use activated carbon filtration unless it is proved beyond a reasonable doubt that carcinogens are not increasing cancer rates. A balanced approach attempts to assess the costs of wrong decisions either way and to minimize the total expected cost.

Our problems do not end, however, with scientific or engineering uncertainties. Even if these were miraculously resolved, there would remain formidable regulatory problems. The public policy picture is clouded by the accretion of historical, legal, and administrative structure, and even our recent efforts to set new national policies have created difficulties of their own, as I think this volume will show. For example, we have been involved in drinking water regulation at every major jurisdictional level in our federal system; and within the federal government there is overlapping legislation and the appearance, at least, of conflicting mandates. In some cases, legislative intent and the standard of proof required for precautionary action are in question.

In addition, of course, attempts to clarify the situation and perhaps to make changes in it must reckon with the evolving complications of administrative law and standard setting, with the financial strains which may be imposed on particular public or private entities, and with the level and sophistication of public perception of the issues.

The papers, discussions, and panel transcripts contained in this volume take on this range of concerns and, I believe, accomplish RFF's aims. Every reader will find something to reject, but I hope that he or she will also find arguments against which to measure previously accepted positions--perhaps even persuasive reasoning for new positions.

Clifford S. Russell
June 1978

ACKNOWLEDGMENTS

Organizing a conference on a broad topic that involves as diverse a set of participants and interested audience is inevitably the work of many individuals. On the heart of the conference, the program, I received valuable advice from a number of sources. I had an informal advisory committee consisting of: Clifford Atkinson, American Water Works Association; Jeffrey Fletcher, National League of Cities; Robert Harris, Environmental Defense Fund; Jerry Hutton, Foremost Foods; Victor Kimm, U.S. Environmental Protection Agency; Emil Mrak, University of California at Davis. Of these people I must single out Robert Harris and Jerry Hutton for their continuing and conscientious efforts to answer my questions, react to my suggestions, and make contacts for me. In addition, I called upon any number of other people for general and specific advice, including: John Boland, Blair Bower, J. Clarence III, David Kinnersley, Daniel Okun, Hans Landsberg, Charles Malone, Talbot Page, Paul Portney, Gerard Rohlich, Patricia Rosenfield, Theodore Schad, Jeffrey Schwartz, and Pat Tobin.

On the management side, the conference was the product of the care, skill, and imagination of associates Sherrill Malloy (through November 1977) and Margaret Parr (beginning in November 1977). At the time of the conference itself, we received additional valuable assistance from Audrey Lord-Hausman.

Others at RFF who contributed to whatever success we had, include Jo Hinkel, Herb Morton, Shirley Woodford, Emery Castle, Marion Clawson, and Diana Tasciotti. We are also grateful to Nick Gavin of the Brookings Institution and the conference staff people at the L'Enfant Plaza Hotel, Michael Plott and Gunter Karsch for their help in making conference arrangments.

Two people deserve special mention. Talbot Page of the RFF staff had been thinking of a drinking water conference for some time but was prevented from taking charge of this one principally by his absence from Washington during the academic year 1977-78. He had many useful thoughts to pass on and helped with the funding proposal and preface to this volume. David Ringler of Foremost Foods was, at every stage of the operation, an invaluable guide, critic, and confidant. His knowledge of whom to involve on the federal government side made all the difference in the "relevance" of the sessions; and his professional experience with press relations kept us on the right track, if never on quite as fast a portion of that track as he knows.

Beyond all this, of course, the value of the conference rests on the contributions of the participants. Before giving them the floor, I should add that I am grateful for their intellectual efforts as well as their attempts to stay within my very tight deadlines. And finally, the proceedings volume has been vastly improved as a result of comments by RFF's official reviewers: Riley Housewright of the National Academy of Sciences and Daniel Okun of the University of North Carolina.

The responsibility for the product rests, of course, with me and the authors.

C.S.R.

INTRODUCTION OF PAUL G. ROGERS

Charles J. Hitch

If there is such a thing as extraterrestrial intelligence--and Hollywood and millions of moviegoers testify at least to our wish for intergalactic company--it is probably a safe bet that the men from Mars, or wherever, have us filed under the heading of "water." Viewed from space, water seems to characterize the earth: some two-thirds of the planet's surface is covered by water, and this remarkable combination of hydrogen and oxygen is responsible for the chemistry of all forms of life, save possibly those inside a visiting spaceship. Indeed, if we are not alone and there have been close encounters of whatever kind, an analysis of our particular life form would reveal to the visitors that homo sapiens is about 70 percent water.

Without water we would not exist. Places where it is not found are referred to as Godforsaken, and regardless of the strength of one's religious convictions, that seems an apt description. From earliest times, we and our ancestors have depended on water as a highway, a sewer, a pathway to discovery, a means to empire, an irrigator of crops--in short, as a social as well as a chemical necessity. Chemistry, however, remains the bottom line: whatever else we do with water, we must also drink it.

So it is surprising, even shocking, that safe drinking water is so rare at this stage of human history. Roughly 2 billion people in the developing nations are completely without safe drinking water. Less than 15 percent of the entire world's population has access to public water supply, either piped into the home or pumped through a public spigot within reasonable walking distance. Even in such a relatively advanced nation as Mexico we are warned, with good reason, to beware of Montezuma's revenge.

And the definition of "safe" is elusive. As recently as the last century, mosquito larvae, toads, leeches, and eels abounded in the public water supplies in Europe. It did not take much to screen out contaminants of this size, of course, but then it was found that what you could not see was often worse than what you could. Typhoid fever, dysentery, cholera, and other diseases from bacterial contamination of drinking water were common.

Chlorination was a public health breakthrough early in this century and made possible major reductions in the incidence of enteric, waterborne bacterial diseases in the United States. Ironically, as we now know, the cure itself is suspect. Moreover, the water problem is not confined to chloroform but includes heavy metals, exotic chemicals, pesticides, and other concomitants of our vastly increased capacity to produce things.

Since the need for safe drinking water is so universal, its provision is a major responsibility of government, and in the House of Representatives that responsibility has come to rest on Paul Rogers of Florida. It is his partly because he is chairman of the Subcommittee on Health and the Environment of the Commerce Committee, but that would be true even if he were a figurehead, an accident of seniority. He is not, of course, but is a purposeful, vigorous activist who has transformed his subcommittee, and himself as chairman, into one of the major influences on health policy in the country. Moreover, he is a student of the field. He does his technical homework, and he knows what he is talking about--no small feat in this complex area.

Mr. Rogers represents the Eleventh Congressional District in Florida, the northern tip of what is called the gold coast. He was first elected in 1954 and is only the second person to hold the seat. The first was his father, Dwight Rogers, and the son has added luster to the family tradition. He practiced law before his election to Congress and, while at the University of Florida, he was a national debate champion, so be forewarned should you want to question him later.

Ladies and gentlemen, it is my honor and privilege to introduce to you the chief architect of the Safe Drinking Water Act of 1974, the Honorable Paul G. Rogers.

ADDRESS BY PAUL G. ROGERS

Thank you for that introduction and for affording me the opportunity to address this distinguished group tonight. It is always rewarding to see the private sector anticipate future problems and involve itself in an issue before it reaches the crisis stage. I think that this conference indicates your willingness to do precisely that--address a pressing national issue now, before we are forced to take action in a crisis setting.

In studying the program, I would commend Resources for the Future, along with the many cosponsors, for ensuring that the full range of topics relating to safe drinking water will receive attention: science and health, economics and law, administration, commerce, and the various international aspects of the issue.

This gathering of individuals is not only impressive, it is also vitally important, for I believe that it represents one of the keys to solving the problems which we presently face in providing safe drinking water. By examining every facet of the issue, and bringing together individuals with expertise ranging from the academic to the industrial, from the scientific to the governmental, you are certain to receive significant contributions. This type of public evaluation and contribution serves two very important purposes--first, it clearly defines the problems and with some good fortune, it begins the solution process. Second, the visibility of such a meeting will further emphasize the importance of giving this matter high priority.

I, of course, have a special interest in the Safe Drinking Water Act. The committee on which I serve worked on this legislation for several years before we were able to secure its passage. One of the major reasons we had such problems was that the previous administrations did not see the need for such legislation. To turn a phrase, we found ourselves dead in the water because the Office of Management and Budget was carrying the water for the Nixon and Ford Administrations.

Between 1971 and 1973, the issue lacked the type of visibility which ensures congressional action. Without a focal point, without a smoking pistol, there was no compelling force for the Congress to institute a new law governing drinking water. But then came surveys of the drinking water in New Orleans, and the Ohio River Valley. Suddenly, with the discovery of possible carcinogens in drinking water, we had the visibility and sense of urgency which we needed.

Up to that time, the people of this nation took their drinking water for granted. True, a move from one part of the country to another might cause one to wonder why the water tasted or appeared different. But the general public was confident that the water which came out of the tap was safe. The New Orleans situation challenged that complacency and brought needed legislation. Passage of the law did not end the issue; it gave us the tools to begin the solution to these problems.

Unfortunately, for most Americans, the problem of quantity often seems more pressing than the issue of quality of drinking water. The recent drought in the western part of the United States, specifically California, is the most recent example. However, many people do not realize that only 10 percent of the water this nation needs for its daily life is used for drinking purposes. The rest goes to agriculture and industry. It is the way in which this 90 percent is used which often creates the problems for that 10 percent.

As an inadvertent by-product of our industrial growth, the commercial application of our scientific and technological progress, we are dumping a Niagara Falls of chemical compounds into our streams and rivers. Pollution of every description--pesticides from farms and orchards, municipal wastes, urban runoff--finds its way into the waterways of our nation, and thus into drinking water. Despite natural dilution, we know that trace amounts of these contaminants elude even the best of conventional treatment and are consumed daily by millions of Americans. Of course, this problem is not unique to the United States; industrial nations around the globe are facing the same issue, and few have taken the steps that we have to correct the problem.

There is one further important factor. While the New Orleans situation was the catalyst for the passage of the original legislation, it also made another important contribution--it delineated the existing state of the art with respect to our ability to detect contaminants in our water supply.

This new knowledge and increased awareness of possible dangers were also frustrating, when we learned how far we still had to go. The study turned up nearly 100 elements. Some were known carcinogens, others were suspect. When we asked for the findings, we assumed that the scientific experts would also be able to provide information as to how these elements should be controlled, and what the consequences would be if we failed to control them adequately.

This information, however, was not forthcoming. In many cases we did not know where the elements originated, and in almost all cases we were missing a threshold level which would indicate the danger to human health. In short, we were finding potentially dangerous elements, but we did not know how dangerous they were. The problem is that we still do not know, and this is the major obstacle which we must overcome if we are to move ahead and establish standards to protect the public.

Since 1975 we have spent $20 million on research. This will aid us to better understand pollutants and to arrive at the stage where regulations will be possible. In the meantime, the Environmental Protection Agency was mandated to issue interim primary regulations based on the 1962 public health standards. At the same time the National Academy of Sciences began its study to give us guidance for the future. We are still not ready to publish the final regulations, and we have only recently seen the proposed regulations on organics published.

It is meetings such as this that give me hope that as we move forward, those parties concerned with the subject of drinking water will contribute. I strongly feel that time and effort can be saved and that reliable information can be developed if the private sector works with the government at this crucial stage, instead of simply reacting to federal efforts at a later date. I believe that you will do this, for it is in your best interest to share your expertise and provide us with your comments.

We must also look to the administration and ensure that they join us in our efforts. I believe that the tenor of things to come was evident in Doug Costle's statement of January 28, 1978, when he noted that the aim of the Environmental Protection Agency would be to prevent the pollutants from entering the environment, rather than trying to take them out once they got there. This seems to be the most effective and most logical way of accomplishing our goals and I commend Mr. Costle for his position. Additionally, I think that we can read another sign of commitment in EPA's proposed budget. There will be an additional $14.5 million for the drinking water division and 213 positions have been added. This reflects an important reassessment of priorities, not only within EPA, but also within the administration.

As you know, it was the intent of the committee and the Congress that this program be mainly administered by state and local authorities. We hoped that the federal involvement would be limited to developing the data and giving guidance, and we provided for the training of local people and made available technical assistance. As of now, there are thirty-five states which have qualified for primary enforcement authority over their programs, and this clearly indicates that the states are willing to take on the responsibility for ensuring safe drinking water in their regions.

As to the mechanics of the law, some deadlines for compliance have been extended and others have yet to be met. We are still waiting for the final promulgation of the underground injection control regulations, very important items in many states. Additional attention is being given to existing filtration systems. Methods now universally used have served us well, but they may represent one of the major problems of the future. Chlorine treatment has saved millions from death and disease. Now we know that there may be a substantial risk because of the reaction of chlorine and organics. EPA has moved to bring new technology in the form of activated charcoal, but this may be just the beginning of changes which need to be made.

As our population grows, it is obvious that we will use and reuse our drinking water resources. This will demand higher levels of treatment. Paints, pipelines, chemicals, and equipment used for more than 100 years are now suspect. Given present evidence of leaching and corrosion, it is logical that those components should be given high consideration in regard to new construction.

I understand that the existing regulatory responsibility in this area falls in a gray area between the EPA and the

Food and Drug Administration. In this regard, I am pleased that EPA, FDA, and the Occupational Safety and Health Administration and the Consumer Product Safety Commission have formed a series of task forces to eliminate duplication in areas such as toxicological testing, data gathering, and risk assessment. I fully appreciate the new ground which is being developed here, but I would add that I hope we do not see undue delays or interagency chauvinism. Recent reports of better communication are encouraging.

In conclusion, I think that in the near future we will look back at the revelations of the past four years and the progress which we make over the next several years, and evaluate them as the second great step in man's scientific involvement with drinking water. The first step was the realization that disease was often waterborne. Drinking water systems and the use of chlorine were milestones in correcting this. I feel that we are on the verge of the second generation. We have a technology which can identify additional, and potentially more dangerous, pollutants which may have more subtle consequences over a longer period of time. It is our responsibility to see to it that we enter the second generation of protecting drinking water with the best data available and that we do so as soon as possible.

INTRODUCTION TO THE CONFERENCE

Charles Warren*

On February 28, 1978, President Carter submitted the Annual Report of the Council on Environmental Quality (CEQ) to the Congress. In that report we highlighted examples of marked improvement in various water bodies around the United States. Our press release accompanying the report began: "For the first time since World War II, Pearl Harbor is open to the public for swimming, fishing, and boating."

That was Tuesday. Wednesday morning I walked into my office and found on my chair a cable from the commander of Pearl Harbor. That facility, read the cable, was definitely not open to the public.

After some embarrassed and rapid investigation with the sources from which we had obtained our illustrations of water-body improvements, we found that the thrust of our statement was accurate, but not its detail. That is, the water in Pearl Harbor is now safe for swimming, fishing, and boating--but the place, for military reasons, is off limits. Thus the only hazard in using the water is that you might get shot by a Navy sentry.

This tale illustrates--in an admittedly roundabout fashion--one of the problems to be addressed by this conference:

---

*Chairman, Council on Environmental Quality.

the contamination of drinking water by organic chemicals in trace amounts, and by the possible by-products of our purification efforts. If the older forms of drinking-water contamination don't get you, some of the newer ones might. And this problem, in a broader context, reflects the frustrating, back-and-forth progress of some of our efforts to address environmental problems of all kinds. In several areas, as we learn better measurement techniques and develop more sensitive methods for identifying the potential for human and ecological damage, we are finding that our victories in one direction are sometimes offset by costs in another. Thus, retarding spoilage in food through the use of chemicals may pose dangers for consumers, as may the use of synthetic hormones, such as DES, in cattle feeds with the aim of stimulating faster growth. And this same DES, when used to prevent miscarriages in human pregnancies, seems to have increased the risk of cancer for the resulting daughters and of genital track abnormalities for the sons.

In so many matters involving environmental protection and human health, we are faced with troublesome questions that involve trying to balance costs and benefits. Some of the characteristics of this problem include:

1. The precision of the costs of regulatory action on the one hand, and the imprecision of environmental costs on the other. In 1976, when the federal government investigated the possible impact of fluorocarbons on the ozone shield, the

chemical industry could tell us exactly how much a fluorocarbon ban would cost in terms of jobs and sales lost. On the other side, scientists could not tell us how rapidly the ozone shield would be eroded by continued use of aerosols, the extent of the health effects that might result--nor, indeed, could they even tell us for certain _if_ the ozone shield was being affected.

2. The problem of cost evaluation is heightened by the difficulty of appraising benefits. Frequently we have only a poor understanding of the mechanisms involved in ecological problems, or we only understand after the damage has been caused. Thus, the construction of canals through the Great Lakes--conferring on us a transportation benefit--also introduced into the Lakes a predator that nearly destroyed commercial fishing. Our efforts to balance costs and benefits are further complicated by the difficulty of assigning value to certain benefits, such as the avoidance of disease or the worth of human life.

3. Especially with carcinogenic chemicals--or chemicals that _may_ cause cancer--our appraisals must take into account the latency problem. Definite proof of carcinogenicity in humans may not show up for fifteen to forty years after exposure; thus, by the time conclusive evidence is in hand, millions may have been irreversibly exposed.

4. Finally, we have the problem of additive or even synergistic effects. If we are receiving tiny doses of chem-

ical X from several different sources, we may not judge correctly if, in our decision making, we assume that the effects are simply additive. The last microgram may cause more damage than the first. Or, as has been determined from a study of asbestos workers, a carcinogen's effects may be *multiplied* by the effects of cigarette smoking.

In addition, we are finding that certain combinations of chemicals may interact in the bodies of man and other living things in ways we would never anticipate from looking at such chemicals in isolation. Thus, if one agent is present in drinking water, and another in poultry, a single-source evaluation may overlook the possibility of combined effect. By the same token, we may encounter situations in which we spend massive amounts to reduce contamination of one kind close to zero, while leaving unregulated a second, more significant source.

More and more often in recent years, the public has tended to accept a conservative solution to such problems--that is, to accept greater current obvious costs than our certain knowledge of benefits can support. Even that perspective is far from giving us the answer to some central questions, however: How conservative should we be? Is *no* chance of future damage acceptable? Are certain types of possible damage, such as future cancer incidence, so repellent to us that we are willing to go to abnormal lengths to avoid them?

These ultimate judgments are not something this conference can decide, though I hope it will succeed in clarifying the pertinent evidence on costs and benefits for drinking-water policy.

Such a clarification is particularly important for what I call "environmental credibility." In our highly industrialized, chemical-laden society, it seems that not a month goes by without a revelation of some new hazard. One possible public reaction can be a kind of despair--a sense that, when we get around to analyzing every substance, all of them will prove harmful in one way or another. A further consequence of that despair might be a growing rejection of environmental regulation.

We see one symptom of that disillusionment in public comment on the debates over carcinogens and the Delaney clause. Quite frequently I have seen it asserted that a large enough dose of any chemical will cause cancer. Similarly, the high doses given to animals in chemical-testing have received a dangerous amount of ridicule. We must strive, in our approaches to drinking-water policy and to environmental regulation of every sort, to educate the public about the scientific and policy issues underlying decisions--to emphasize, for example, that cancer-causing agents are relatively rare, and that we are in no danger of being forced back into a dark age of preindustrial hardship by regulation. Further, we must convey more effectively the difficulty of making prudent social policy in the <u>absence</u> of definitive scientific evidence. The prestige of science with the general public often leads people to assume that scientists have answers to every important question. That is not the case, of course, and policy makers must often frame regulations despite ignorance or uncertainty. Unless we make this point again and again, our efforts to safeguard the public may be eroded when--as is likely to happen--some suspect chemicals prove to be harmless.

These public policy problems are accompanied by institutional issues, particularly in the case of drinking water. Where and how should fundamental decisions on costs, benefits, and risks be made? Should they be made by the states individually or the federal government? By legislators or agency administrators? How should evidence on questions of "fact" be gathered and evaluated? Are there objective sources for such determinations, or is the adversary process of claim and counterclaim most likely to lead us to the truth?

As the past history of governmental regulation illustrates, we will always have to contend with charges of governmental paternalism--a theoretical problem that can quickly become a practical one if any regulation leads to widespread noncompliance. Is the federal government always the proper jurisdiction for addressing environmental problems? Will state decisions and enforcement prove lax out of a desire to encourage industrial immigration? Or might the states be too harsh in an effort to export environmental problems? Would industry prefer the convenience of a single set of national rules, or the potential flexibility of fifty different sets? And a question especially pertinent to drinking water: How do we reconcile national directives with traditions of local responsibility?

It is my good fortune to be able to raise these questions instead of having to answer them. They recur again and again in a dozen environmental contexts, and I think I shall find it extremely helpful to hear such questions debated in this critical area.

Part I

Technical Background: Data and Experience

Chapter 1

DRINKING WATER POLICY PROBLEMS
BACKGROUND OF THE CURRENT SITUATION

C. C. Johnson

Introduction

More than sixty years ago Congress recognized that an adequate supply of readily available, safe drinking water is an essential requirement for healthy citizens. Under authority provided by Congress, the Surgeon General of the U.S. Public Health Service promulgated, in 1914, national drinking water standards. A federal/state program for ensuring the safety of drinking water served on carriers in interstate commerce was initiated under federal supervision and every state and territory instituted a program to ensure the safety of public drinking water supplies.

The success of that national effort is a testament to the persons in the water supply industry and the government who joined together in this early effort. With rare exceptions, epidemics traceable to "waterborne diseases" are no longer a part of our way of life. People expect to travel anywhere in the United States and drink water without fear of getting sick. In other times, and even today in other countries, that accomplishment would be considered an idealistic dream. The U.S. waterworks industry, under the unifying controls of federal and state health regulations, made that dream a reality in this country. Three ingredients contributed to this success: an understanding of the problem, a willingness to overcome the problem, and a recognized need to protect the public's health.

In those early days the recommended criteria, standards, and design practices had as their aim the prevention of communicable diseases. However, the standards designed to cope with the bacteria-caused diseases of the 1920s provide an inadequate safeguard in the technological-megalopolis environment of the 1970s. As a result, new legislation--the Safe Drinking Water Act of 1974--was enacted by Congress. This in turn requires a new set of policy determinations to assure the achievement of the objectives set forth in the act. So, as we look into drinking water policy problems today we are likely to find that they are inextricably associated with the waste disposal practices we endorse and the attitudes we take as to what constitutes safe water. In an effort to provide some background for the current situation related to the provision of safe drinking water, a more detailed discussion of each of these two subject areas follows.

Waste Disposal Practices Related to Water Supply

There are those who will argue that America's biggest problem, and indeed maybe its biggest enemy, is its ever-growing and as yet unmanaged production and disposal of waste products. The impact of waste production on the quality of the nation's water supplies is one of the manifestations of this all-pervasive problem of waste disposal.

A brief review of the tremendous volumes of waste produced in this country may help to place the problem of disposal in perspective. Major U.S. industries treat about 5,000 billion (10^9) gallons per year of wastewater before discharging it to the environment (U.S. Environmental Protection Agency, 1977a). Of this volume, about 1,700 billion gallons are pumped to oxidation ponds or lagoons for treatment or as a step in the treatment process. Another 5,000 billion gallons per year of municipal wastewater are discharged to the environment, with an estimated 730 billion gallons of this amount discharged to the land. In the United States, municipal sludge production amounts to about 5 million dry tons/yr. Industrial sludge production is believed to be many times this amount. EPA (USEPA, 1977b) estimated annual solid waste production at 136.1 million tons in 1975. In addition, we must deal with the millions of tons of gaseous waste that are produced annually; the untold hundreds of million tons of mine tailing disposed of each year; and the tremendous volumes (more than 24 million barrels per day) of oil field brines produced each day. The discharge to the environment of such large volumes of waste with varying concentrations of toxic and hazardous substances must have a detrimental effect on the quality of our nation's water supplies. Perhaps it is time to remind ourselves that our water resources, more than any other, illustrate the interaction of all parts of the environment and particularly the recycling process that characterizes every resource of the ecosystem. Everything that man injects into the biosphere--chemical, biological or physical--can ultimately find its way into the earth's water, and these contaminants must be removed, by nature or by man, before that water is again potable.

In looking at the waste disposal practices in this country and the impact that they have or can have on the quality of our water supply sources, some persons are inclined to raise the question of who should pay for making water safe for human consumption. In earlier days a concept prevailed that the public's requirements for domestic water supply were guaranteed top priority in the hierarchy of uses. Some voices now suggest that water polluters should be required to discharge wastewater meeting even more stringent quality standards than required at present in order to assure water supply sources that require minimum treatment for the production of safe water. A National Academy of Sciences panel (1972) defined water quality recommendations as "those limits, characteristics and concentrations of substances in raw waters that will allow the production of a safe, clear, potable, esthetically pleasing and acceptable public water supply after treatment." In the absence of water quality standards that can be used to support the objectives of this definition, the quality of a public water supply is essentially governed by the chance that discharges to the source will not reduce its quality below the level acceptable for producing safe drinking water.

Because of the proclivity for waste production in this country, and as a result of the rather indiscriminate waste disposal practices that prevail, some basic concepts regarding water supply protection have been destroyed, and some new ones, which affect established policies in the water supply field, have surfaced.

The Public Health Service Drinking Water Standards (USPHS, 1962) required that water supplies be obtained from a "protected source." Adequate protection by natural means was defined as involving one or more of the following natural processes which produce water consistently meeting the requirements of the standard: dilution, storage, sedimentation, aeration, and exposure to sunlight. In addition there are the associated physical and biological processes which tend to accomplish natural purification in surface waters. In the case of groundwaters, the natural purification of water is achieved by infiltration through soil, percolation through underlying material, and storage below the groundwater table. The standard then required that the water supply be obtained from the most desirable source feasible, and that efforts be made to prevent or control pollution of the water supply source. If the source was not adequately protected by natural means, the supply was to be adequately protected by treatment. It is generally agreed that these drinking water standards were designed to protect against bacteriological contamination and that this definition of a "protected source" provides insufficient protection to today's public.

Recently a committee of the National Academy of Sciences, commenting in an advisory capacity to a government agency, voiced its concern on the treatment and use of water for human consumption from an unprotected source (National Academy of Sciences, 1977a). The potential source of supply was an estuary that receives the discharges of a regional wastewater treatment plant. The NAS committee noted that there are questions as to what characteristics of treatment plant output should be measured and what standards they should be compared with if raw water is taken from the estuary. Further, it is generally conceded that many communities in heavily industrialized regions of the country are currently utilizing water sources of much poorer quality than the one under evaluation in the NAS comment.

Therefore, the question must be raised: On what basis should the treated water from these other sources be evaluated? It must be recognized that it is almost impossible for nature's processes to neutralize the impact of the toxic substances that assault our surface and underground water sources.

Well-meaning but poorly informed persons would have us transfer the discharge of waste products from surface streams, which are already contaminated, to the land and thus, subsequently to the groundwater, which is relatively very clean. This would be an excellent idea if complete reliance could be placed on the natural action of most soils to remove the potentially harmful contaminants that seem to permeate today's waste stream. The problem is that we cannot rely on soils for this. In nearly every instance where land treatment of wastewater has been considered successful, the evaluation has not included measurement of the increase in organic contamination of the underlying groundwater.

One preliminary study by the EPA's Ground Water Research Branch (USEPA, 1977c) provides information on the detection of low levels of chloroform (1-13 μg/liter), trichloroethylene, (2-10 μg/liter), and benzene (8 μg/liter) in the effluent flow from the Muskegon land treatment system. In EPA's judgment, no significant levels of arsenic, beryllium, cadmium, cyanide, mercury, or other toxic metals were found. It is noted, however, that they were present in detectable quantities. The report concluded that "the presence of low levels of organics in the final effluent indicates the need for definitive information concerning the movement and fate of organic pollutants in the subsurface." While the report recognized in these preliminary study results the ability of the Muskegon treatment system to reduce to very low or nondetectable levels all of the identified organic pollutants, two important issues are raised. One is whether the policy of nondegradation, so vocally argued in the recent debates on the Clean Air Act Amendments, should apply to our water resources, particularly where the health significance of the contaminants, at the concentrations expected, is unknown. The second is whether a waste treatment system should be permitted to introduce a contaminant into an otherwise safe water supply at the same time that a national policy is being proposed which calls for the use of best available technology to remove to the maximum extent possible that same contaminant from water before it is consumed? It is important to note that this latter policy, which proposes to require a technology-based treatment technique, is not governed by any maximum acceptable contaminant level for any synthetic organic pollutant or group of such pollutants, but rather by the potential for contamination and the need for protecting the public's health (USEPA, 1978).

Land treatment of wastewater, by national policy, is to be given preferential consideration over other methods of wastewater treatment (Costle, 1977). It is most difficult to understand why this policy enjoys such a high priority when one reviews EPA's report to Congress on "Waste Disposal Practices and Their Effects on Ground Water" (1977a). The report notes that half the population of the United States is served by groundwater and that its use is increasing at a rate of 25 percent per decade. It also found that groundwater has been contaminated on a local basis in all parts of the nation and on a regional basis in some heavily populated and industrialized areas, with the contamination in some instances precluding the development of groundwater for drinking. It is interesting to find that almost every instance of groundwater contamination has been discovered only after drinking water has been affected. Most important, EPA has told the Congress that soil attenuation alone will not be sufficient to prevent groundwater contamination from a waste disposal source. It stated that land disposal of wastes is not environmentally feasible in many areas and that a national strategy of groundwater protection will require a better understanding of the environmental, legal, technical, and economic complexities of dealing with the resource.

Obviously the writers of EPA's land treatment policy and the authors of the reference report to Congress had different objectives in mind. Nevertheless, their concerns appear to be in conflict and cannot help but confuse that portion of the public sector concerned with disposal of wastewater and protecting water supply sources. As a matter of policy it should be recognized that contamination of groundwater aquifers can occur easily and rapidly. Once one is contaminated, it may take years and cost millions to purge the aquifer of the contaminant, if in fact this can be done at all. Unless the public is willing to treat the water from these underground sources as it does that from polluted surface streams, every effort must be made to prevent underground water sources from becoming contaminated.

The impression should not be left that no waste materials can or should be placed on, in, or under the ground surface. Given the proper hydrogeological conditions and using appropriately designed facilities, there are situations where selected wastes can be disposed of into the ground without appreciably modifying the quality of the potable groundwater. Disposal of brine from petroleum exploration and development is one example. Another situation may require the use of a nonpotable water to enhance the quality of an unusable supply or to protect the quality of a potable supply, as in the case of agricultural production or salt water intrusions, when the required results warrant the implied risk. Further, a decision could be made that an underground aquifer's most beneficial use is for the disposal of waste products. In each of these situations, the choice is deliberate. It is made with full knowledge of its subsequent effects, either actual or potential.

Attitudes Relative to Safe Drinking Water

In the previous paragraphs considerable space has been given to talking about waste disposal practices and their impact on water supply sources. Because of this impact, many believe that for the present, and for some time to come, a discouragingly large number of our water supply sources probably require the best treatment that technology can provide and that the public is willing to pay for. Such a statement presupposes that there is general understanding and agreement upon what constitutes safe drinking water. Therein lies one of the major drinking water policy problems. While many definitions have surfaced over the years, the one that counts at this point must reflect the voice of the people as revealed in the intent of the Congress in its passage of the Safe Drinking Water Act (U.S. Congress, 1974a). For this, reference is made to the committee report (U.S. Congress, 1974b) that supports the act. The stated purpose of the legislation is to assure "that water supply systems serving the public must meet minimum national standards for protection of public health." It "requires the Administrator of the Environmental Protection Agency to prescribe national primary drinking water regulations for contaminants which may adversely affect the public health"; further, the report is explicit in its explanation of the use of the term may. It says "because of the essentially preventive purpose of the legislation, the vast number of contaminants which may need to be regulated, and the limited amount of knowledge presently available on the health effects of various contaminants in drinking water, the committee did not intend to require conclusive proof that any contaminant will cause adverse health effects as a condition for regulation of a suspect contaminant."

Thus the stage was set for all the controversy that currently surrounds the development of interim primary and revised primary drinking water standards. How, where, and when does the government establish drinking water regulations for contaminants that may adversely affect the public health? It has been established that most surface streams will require treatment to some level. Also, experience has demonstrated that the nation's groundwater resources are vulnerable to contamination to some degree and when this contamination occurs they also may have to be treated to some level. In the absence of all the answers, indeed in the face of a great deal of ignorance, what level of safety should drinking water standards provide? What risks should the public be asked to take? Who should make these decisions and on what basis should they be made? These questions are inherent in the national debate that is taking place in the United States at this time on what to do about organic contamination of the nation's drinking water.

A brief recounting of the significant events that surround this debate will help to sharpen the focus of the present situation. In 1974 the EPA reported on the discovery of minute quantities (parts per billion) of 66 organic substances in the New Orleans water supply. Six of these substances were suspected carcinogens. In the same year the Environmental Defense Fund released a report that indicated a correlation exists between the quality of the New Orleans water supply and increased cancer mortality in white males (Harris, 1974). This report is generally credited with providing the impetus that led to the passage of the 1974 Safe Drinking Water Act, though a statistical review of the EDF report by Tulane University scientists commented that the study had many shortcomings (DeRouen and Diem, 1975).* The EPA, in an effort to determine the national implications of the findings of organic contaminants in the New Orleans water supply, launched a nationwide survey of the presence of suspected carcinogens in water supplies. The results of this preliminary assessment confirmed the presence of organic compounds in very small quantities (parts per billion) in all 80 drinking water supplies surveyed (USEPA, 1975). Chloroform was one of these compounds and a study by the

*Ed. Note: See the paper below by Schneiderman for a fuller discussion of this dispute and particularly some of the more recent findings of DeRouen and Diem.

National Cancer Institute (1976) concluded that chloroform administered in very large doses caused cancer in rats and mice. Because of these events, and mindful of the intent of the Congress when it passed the Safe Drinking Water Act, a question is logically raised as to why the Interim Primary Drinking Water Standards did not include a maximum contaminant level for chloroform and other organics. This question was raised in a very poignant way by EDF in its suit (1976) against EPA filed in the U.S. Court of Appeals. In the meantime the National Academy of Sciences was completing its study on "Drinking Water and Health" which was released in late spring of 1977. Such a study by the Academy or other similar body was required by the Safe Drinking Water Act. The results of this two-year effort were to provide the background information necessary for determining what maximum contaminant levels (MCL) in drinking water are required for protection of health. Finally, after considering public comments on a preliminary proposal for regulating organic contaminants in drinking water, and influenced by the findings of the NAS report, the EPA has recently proposed an amendment to the Interim Primary Drinking Water Regulations (USEPA, 1978) that would regulate trihalomethanes and other synthetic organics in drinking water.

It is not the purpose of this paper to answer the question of why an MCL on organics was not included in the original Interim Primary Drinking Water Regulations; rather, it is to raise the question of what constitutes the rationale for creating any water supply standard. Is this done on the basis of the findings of one study? What would be the reactions if the levels of contaminants in this instance were parts per trillion, rather than parts per billion or some other concentration? Does any detectable quantity of a substance that is considered carcinogenic under differing test conditions unrelated to human exposure dictate the setting of an MCL for water supplies and other purposes? What is the recognized and utilized rationale for setting standards? In the area of standard setting, is it possible to establish a logical sequence of thought and events that would dictate a course of action for accomplishing desired objectives? Is it possible to develop a document that would, among other things, define the kinds of substances considered to adversely affect the public health; define the considerations used in evaluating the significance of their presence; define the parameters for assessing the applicable risk factors for public exposure; define when the federal presence is required and when local jurisdictions are expected to prevail in correction of problem situations?

Is it necessary to repeat this process, with varying degrees of consistency, for every instance of a new discovery of a contaminant in the water environment? Discussions in meetings of the National Drinking Water Advisory Council (1977a) suggest it is possible to develop a set of criteria and guides that will enable the regulatory agencies to know when they should act, and the public to know why they are acting. At the very least, it is believed such a document would focus public attention and discussion on the effects of the process rather than on the determination of acceptable levels of exposure, which is beyond the scientific capability of most persons in the public sector. Also, it would help the public to decide in a more intelligent fashion whether they are willing to pay the costs involved in correction, or protection against the environmental harm in question. Such an approach requires clear statements of policy on the part of the regulating body, sometimes in very controversial areas. Without such statements, ambiguity, contradictions, and conflict often occur among agencies and levels of government, and even within a single agency. Further, a clear statement of policy enhances the opportunity for uniform approaches to standard setting by environmental regulatory programs that affect all media: air, water, and land.

Every day the need for a rationale for setting standards is made clearer to persons associated with the protection of water supplies. In the wake of the drought that plagued much of the nation last year, and under the continuing pressure to conserve energy and to reuse and recycle resources, the possibility of water reuse poses even greater threats to the already questioned quality of our nation's water supply. The state of California is planning to double the volume of wastewater that is currently recycled, to an additional 400,000 acre-feet (equivalent to 355 mgd) of water each year by 1982 (National Drinking Water Advisory Council, 1977b). The California State Department of Health is developing standards that are designed to provide a proper degree of public health protection.

Water reuse and recycling is not limited to California. In fact, it is practiced every day in every state in the nation. What rationale will California use to protect the health of its people? Should each of the fifty states develop its own criteria for water reuse? It seems fairly clear to me that the answer should be no! Yet the only visible program at the national level that has a potential for controlling the quality of treated water for reuse and recycling in the environment is embraced by the National Pollutant Discharge Elimination System program under the Federal Water Pollution Control Act Amendments (FWPCA)(U.S. Congress, 1972). The program calls for the issuance of permits for the discharge of pollutants or a combination of pollutants. Under the act, the EPA is required to prepare or develop comprehensive programs for preventing, reducing, or eliminating the pollution of the navigable waters and groundwaters and improving the sanitary conditions of surface and underground waters. In developing such programs, EPA is to give due regard to the withdrawal of such waters for public water supplies. Even a casual review of the permit program from the perspective of health protection and withdrawal of water for public water supplies would raise serious questions as to the rationale used to establish the limitations in the permits which allow the discharge of wastewater to streams.

Another limiting factor in relying upon the Water Pollution Control Act to provide the focus for federal action in the water reuse area is inherent in the act itself. The basic and essentially sole interest of "water pollution control" is to control the pollution of surface waters. In fact, when EPA's preferred method of land treatment of wastewater is used, it is questionable whether a discharge permit is legally required under the act unless runoff is collected and discharged to a surface stream. Under these circumstances, there is no enforceable provision at the federal level to prevent or control the contamination of the nation's groundwater resources resulting from wastewater discharges. Even if there were enforceable provisions, the basis for determining when groundwater is contaminated does not exist. There are no standards that govern the quality of our groundwater supplies prior to their use as public water supply sources.

Today the United States uses water in excess of 300 billion gallons per day. The supply of fresh water for these purposes is relatively fixed. In many places during part of the year surface stream flow is less than the withdrawal rate. In some places stream beds are dry a portion of the time. Water in quantities necessary for dilution following discharge of wastewater more often than not is not available. When the flow is sufficient, then the concentrations of people and industry often produce wastewater that overloads the dilution capacity of the streams. For years we have been reading reports of declining groundwater tables in the Southwest and elsewhere. Groundwater recharge takes place purposefully and accidentally in many forms.

It follows, whether our concern is for surface or groundwater, that we have been and will continue using much of the same water supply over and over again. The unanswered questions facing all of us are: Under what conditions should this be permitted to occur? What is the rationale upon which standards and actions should be based? Within this context, a few words should be addressed to the relationship of wastewater effluents to water supply influents.

The most advanced wastewater treatment (AWT) technology, when properly designed and operated, can produce an effluent that is superior in every respect to the quality of water in some receiving streams. Yet in one instance this capability was not considered sufficient to permit the location of an AWT plant above a water supply intake. Literally, this was the decision of the EPA, which ruled on the construction of the Dickerson, Maryland AWT Plant (USEPA, 1976). In that situation the final assessment of the administrator's executive panel was "though it would be preferable to discharge below the metropolitan area's water intakes, the health risks associated with the discharge from the proposed project do not, by themselves, warrant rejection of the proposed facility and denial of the grant application." Ultimately the grant was denied. It is in the public interest that EPA publish the specific criteria that guided the government's conclusions relative to the protection of the public's health. The basis for EPA's evaluation can be of enormous benefit to the communities that utilize surface sources which receive treated sewage effluent and also serve as water supply sources.

Whenever it is difficult to establish some specific limiting standard, people tend to resort to technology-based standards to compensate for the lack of knowledge, lack of enforcement resources, or other inconveniences. This is the basis on which regulations implementing the Water Pollution Control Act are written, as indeed the act itself prescribes technology as the vehicle by which its objectives would be obtained. Now, in part, where synthetic organics are concerned, regulation of technology is presented as the means by which the goal of the Safe Drinking Water Act can be attained.

The water pollution program, with which we have had some experience, is a good example to use in commenting on this practice. The stated objective of the FWPCA is "to restore and maintain the chemical, physical, and biological integrity of the Nation's waters." To achieve this objective, among other things, "it is the national policy that the discharge of toxic pollutants in toxic amounts be prohibited." The nation is putting forth considerable effort and dollars to require the use of "best practicable control technology" and "best available technology" without a real knowledge or basic agreement of what constitutes a "toxic amount," particularly where people are concerned. What, then, is the impact of control technology on the reduction of toxic substances to "less than toxic amounts?"

Many agree that technology-based standards probably are a good first step in controlling the quantity of contaminants discharged to the environment. This is particularly true if the basis for their requirements follows a rational design for setting standards. Such standards as presently conceived cannot, and should not, be relied upon to protect the public health against the long-term consequences of trace elements in drinking water; nor has the EPA proposed this. In all probability it is not known if the technology available is sufficient for the required treatment. Neither is it agreed how much effort and resources should be applied to generate the required technology. Research must be relied upon to produce the answers that help us understand what we are doing to our environment and what that environment is doing to us. Adequate support for this research simply has not been forthcoming. However, this in no way excuses any regulatory agency from articulating to the public the rationale against which its operable standards are set and the limitations that attend the results that can be expected. History does not reveal that this philosophy permeates many of the environmental control programs in existence today.

A final concern associated with standard setting that affects the degree of protection afforded the public involves the application of an economic feasibility test. When a decision has been made that a contaminant poses a serious threat to public health and that regulation is required to protect the public, what role should economic feasibility play in deciding who gets the benefit of that protection, and when? Some specific guidance in such instances is provided by the Congressional Committee Report (1974). The report seems to say that economic feasibility is not a basis for denying equal protection to the public in the development and promulgation of primary drinking water standards. While limited time exemptions can be granted to those systems unable to comply promptly, Congress suggests the time should be used to develop and execute plans that will enable full compliance. There seems to me to be a basic inconsistency in a policy that suggests to one group it is all right to continue to drink water that poses a serious health threat since it cannot afford better water, and says to a second group that it is required to reduce this serious health threat by improving the quality of the water because it is economically feasible to do so. A uniform approach to setting standards would preclude these kinds of decisions, which on their face are not supportive of good preventive public health practices, the primary purpose for which the act was passed.

Summary

A discussion has been presented of the impact of wastewater disposal practices on U.S. water supplies and the attitudes we hold on what constitutes safe drinking water. I believe a working appreciation of these subject areas and their interrelationship is paramount to resolving the drinking water policy problems of the present and the future. The absence of a published rationale that follows a uniform and consistent approach to standard setting produces economic and health protection inequities for the public. Also, single-view approaches to establishing policy and setting standards tend to promote correction of one environmental problem at the expense of another environmental entity. The failure of many government authorities at all levels to fully recognize the relationship between water resources and the total environment is a serious problem associated with protection of drinking water supplies.

Very recent activities in EPA's Office of Water and Hazardous Materials offer a glimmer of hope that better coordination of programs in that office (water and solid waste particularly) is taking place. Perhaps the deliberations of this conference will provide the impetus required to assure the continued development of a strong national water supply program that creates a recognition of its necessity, enables a maximum application of current knowledge, stimulates a broadened support for required research, and provides the protection to the public health envisioned in the Safe Drinking Water Act.

References

Costle, Douglas M. 1977. EPA Policy on Land Treatment of Municipal Wastewater (Memorandum of October 3, 1977 to Assistant Administrators, and Regional Administrators)(Washington, D.C.).

DeRouen, T. A., and J. E. Diem. 1975. "The New Orleans Drinking Water Controversy: A Statistical Perspective," American Journal of Public Health vol. 65, p. 1060.

Environmental Defense Fund. 1976. "Petition for Review of Regulations Promulgated by the Administrator of the United States Environmental Protection Agency" (Brief for the Petitioner EDF, in the United States Court of Appeals For the District of Columbia Circuit).

Harris, R. H. 1974. "Implications of Cancer-Causing Substances in Mississippi River Water" (Environmental Defense Fund, Washington, D.C.).

National Academy of Sciences/National Research Council. 1972. Water Quality Criteria of 1972. A Report of the Committee on Water Quality Criteria, Environmental Studies Board, National Academy of Sciences, for U.S. Environmental Protection Agency (EPA R3-73-033) Washington, D.C. (NTIS PB 236 191).

National Academy of Sciences. 1977a. "Comments on the Proposed Testing and Evaluation Program for the Experimental Estuary Water Treatment Plant Project" (A Letter Report of the Committee to Review the Potomac Estuary Experimental Water Treatment Plant Project, Assembly of Engineering, National Research Council) (Washington, D.C.)

______. 1977b. "Drinking Water and Health" (A Report of the Safe Drinking Water Committee, Advisory Center on Toxicology, Assembly of Life Sciences, National Research Council) (Washington, D.C.).

National Cancer Institute. 1976. "Report on the Carcinogenesis Bioassay of Chloroform" (Division of Cancer Cause and Prevention)(Washington, D.C.).

National Drinking Water Advisory Council, Transcripts of Meetings. 1977a. (Environmental Protection Agency, Washington, D.C.).

______. Transcripts of Meeting, December 15. 1977b. (Environmental Protection Agency, Washington, D.C.).

U.S. Congress. 1972. "P.L. 92-500-Federal Water Pollution Control Act Amendments of 1972" (Washington, GPO).

______. 1974a. "Public Law 93-523, Safe Drinking Water Act" (Washington, GPO).

______. 1974b. "Report No. 93-1185, Safe Drinking Water Act" (Washington, GPO).

U.S. Environmental Protection Agency. 1974. "New Orleans Area Water Supply Study" (Draft Report) (Cincinnati, Ohio).

______. 1975. "Preliminary Assessments of Suspected Carcinogens in Drinking Water" (Interim Report to Congress) (Washington, D.C.).

______. 1976. "The Proposed Montgomery County Advanced Wastewater Treatment Plant" (Executive Panel's Report to the Administrator).

______. 1977a. "Waste Disposal Practices and Their Effects on Ground Water" (The Report to Congress) (Washington, D.C.).

______. 1977b. "Resource Recovery and Waste Reduction" (Fourth Report to Congress) (Washington, D.C.).

______. 1977c. "Preliminary Survey of Toxic Pollutants at the Muskegon Wastewater Management Systems" (ADA, Oklahoma).

______. 1978. "Control of Organic Chemical Contaminants in Drinking Water" (Amendments to the Interim Primary Drinking Water Regulations) (Washington, D.C.).

U.S. Public Health Service Drinking Water Standards. 1962. (Department of Health, Education and Welfare, Washington, D.C.).

Chapter 2

DRINKING WATER AND HEALTH*

Gerard A. Rohlich

The Safe Drinking Water Act (P.L. 93-523) directed the administrator of the EPA to arrange with the National Academy of Sciences or other appropriate organizations to conduct a study to determine the adverse effects on health attributable to contaminants in drinking water. The report of the study has been completed and is now available from the academy. The report is rather lengthy (more than 900 pages), perhaps too lengthy, but reflects the large number of topics and substances that it was necessary to consider. Simply stated, the report attempts to evaluate the public health implications of various contaminants found in drinking water. For many contaminants this cannot be done because the most current knowledge is too often inadequate and incomplete. A number of the subjects considered are complex and characterized by uncertainty, conflicting

---

*This paper is based on, and in large part excerpted, with minor changes, from the report entitled "Drinking Water and Health" prepared by the Safe Drinking Water Committee of the Advisory Center on Toxicology, Assembly of Life Sciences, National Research Council, National Academy of Sciences, Washington, D.C. 1977. Gerard A. Rohlich is a professor in the Department of Civil Engineering, the University of Texas at Austin.

opinions, and inconclusive or incomplete data, necessitating detailed consideration from several points of view before a balanced judgment can be made. What is presented in this report must therefore be considered as an initial contribution to an effort that should receive continuing study and reevaluation.

The committee posed several questions:

1. What reason is there for concern about the contaminant? What risks are associated with its presence in water?

2. What sources are there other than water?

3. What contaminants need to be controlled?

4. Are there special places or persons at higher than average risk?

5. Are there essential health requirements for this substance?

6. In view of the data at hand, can one say that this is a substance which causes temporary ill effects? Permanent ill effects? Reversible effects? What additional data are needed?

7. In view of these effects and their reversibility, or lack of it, is it possible to set "no-adverse-health-effects" levels?

8. If some substances have special health benefits, what concentrations will maximize these benefits while keeping any associated health risk at an acceptably low level?

Many of the contaminants considered occur in nature. Some are natural waste products of men or animals. Some are man-made, inadvertently finding their way into water. A "natural" substance was not considered safer than one that was man-made.

The Safe Drinking Water Committee was responsible for the general direction of the study and assigned specific areas to subcommittees, each of which was chaired by a member of the parent committee. The main areas considered were:

1. Safety and risk assessment
2. Microbiology of drinking water
3. Solid particles in suspension
4. Inorganic solutes
5. Organic solutes
6. Radioactivity in drinking water

Considerable effort was directed to the question of assessing risk arising from the ingestion of substances in drinking water. Pollutants in water have different effects. At one extreme they may cause the water to have a disagreeable taste or odor, quickly perceived by the user. At the other extreme, a waterborne carcinogen may have effects on human health that go undetected, particularly if it produces only a small increase in the incidence of a common cancer.

Within the limitations of the available information, discussion is provided on the occurrence, chemical characteristics, metabolism, health effects, and analytical methods for detection of the substances included in the report. The discussion of each substance is followed by a section on conclusions and recommendations.

The application of modern methods of analysis has greatly expanded and diversified our knowledge of the occurrence of trace amounts of impurities in water, and is continuing to do so much more rapidly than the rate at which information about their toxicity or adverse health effects is accumulating. This necessitated detailed scrutiny of a large and diverse segment of the scientific literature. Furthermore, the central effort of the study, assessing the long-term biological effects of different substances that are present in trace amounts, made severe demands on the committee to apply the contemporary knowledge of toxicology and epidemiology to quantitative estimation of public health risks in terms that would be useful in framing regulations.

In recognition of these limitations and taking only health effects into account, the scope of the committee's work was defined on the following bases:

1. Where there were sufficient data for projecting a human dose-response relationship with some degree of precision, a projection was to be made and its qualifications made explicit.

2. Where data for contaminants were sufficient to conclude that at certain levels the risks of adverse health effects were specifiable and slight, those contamination levels were to be proposed.

3. In the case of contaminants for which there is no scientific basis or methodology for recommending such levels, the procedure was to describe the available data and their

significance in terms of known or anticipated adverse health effects.

The report is concerned only with water that is used for drinking. Although all contaminants may cause problems when present in water used in health care facilities, the health hazards associated with nondrinking uses such as in humidifiers, kidney dialysis units, laundries, or other special uses were not considered.

It is not possible in this brief summary to do full justice to the efforts of the many people who generously contributed their time in the review and evaluation of the literature. I can only attempt to highlight some of the considerations and conclusions of the subcommittees.

Microbiology

The chapter on microbiology of drinking water deals with a variety of bacteria, viruses, protozoa, and helminths. Methods of detecting these contaminants in drinking water are reviewed, together with the determination of permissible levels. Because the control of waterborne epidemics still depends largely upon the control of infectious enteric diseases, considerable attention is devoted in the report to the validity and health significance of microbiological standards. Good engineering and public health practices emphasize the need for using raw water of the highest possible quality. Bateriological testing, or the imposition of bacteriological standards, are adjuncts to, not substitutes for, good quality raw water, proper water treatment, and integrity of the distribution system.

Effective water treatment systems in the United States have had a major impact on the reduction of waterborne infectious diseases during this century. However, waterborne diseases continue to occur. In 1975, twenty-four waterborne disease outbreaks involving 10,879 cases were reported to the Center for Disease Control (CDC) in Atlanta. No etiologic agent was found for the two largest outbreaks (Sewickly, Pa.--5,000 cases--and Sellersburg, Ind.--4,000 cases). Seventeen of the twenty-four outbreaks and about 90 percent of the cases reported to CDC were designated as "acute gastrointestinal illness."

The reported numbers of outbreaks and illnesses probably represent only a portion of the true totals. The number and kinds of reported outbreaks and of some etiologies may depend upon the interest or capabilities of particular health agencies or individuals. Many small outbreaks are not reported to state health departments.

The current coliform standard is of value in protecting against frank outbreaks of bacterial diseases, but it may not protect against low levels of virulent pathogens. It is clear that further protection could be achieved by analyzing larger samples for coliforms. More frequent sampling throughout the distribution systems would provide additional safeguards. In the final analysis, testing for coliform, while not perfect bacteriologically, is still the most reliable indicator of the possible presence of fecal contamination, and therefore of the pathogens that may be present in water.

The subcommittee gave consideration to the standard plate count and concluded that although it is not a substitute for total coliform measurements, it is a valuable procedure for evaluating bacterial quality of drinking water.

The report stresses the need for a research program to increase the value of the relatively simple bacteriological tests in controlling the sanitary quality of drinking water, with emphasis on the following:

1. Epidemiological studies of water quality and health, with application of more sensitive methods for detecting pathogens in drinking water and better reporting of outbreaks of waterborne disease;

2. Development of membrane-filtration methods to allow testing of larger samples and to reduce interference by overgrowth and disinfectants;

3. Improvement of procedures for making total plate counts and study of the utility of such tests for assessing the health hazards of drinking water;

4. Research on more rapid and sensitive methods for detecting pathogens and the use of such methods for monitoring the quality of water;

Viruses differ fundamentally from other microorganisms that may occur in water. They are transmitted as submicroscopic, inert particles that are unable to replicate or adapt to environmental conditions outside a living host.

The presence of infective virus in drinking water is a potential hazard to the public health, but there is no valid basis on which a no-effect concentration of viral contamination in finished drinking water might be established.

Microbiological indicators may have a limited correlation with viral contamination. The coliform count and standard plate count are the best indicators available today for routine use in evaluating the presence in water of intestinal pathogens, including viruses. Use of such indicators and chlorination afford more protection to public health than any available alternative.

Protozoa. The principal pathogenic parasites that may escape our sanitary barriers in public water supplies are the protozoa *Entamoeba histolytica* and *Giardia lamblia*. The cysts of these organisms are not completely destroyed by the usual chlorination. Most cysts, however, will be removed by sedimentation and filtration through sand.

Giardiasis is emerging as a major waterborne disease. The largest outbreak of giardiasis in the United States occurred in Rome, New York during 1974-75. All the U.S. outbreaks involving municipal systems, except for a large outbreak at Aspen, Colorado in 1965-66, have been associated with surface water sources where disinfection was the only treatment. At Aspen it appears that wells serving as a source of water for part of the community were contaminated by old broken sewer lines that passed close to them.

One of the major deficiencies of customary methods for evaluating the bacteriological quality of water is that the water is distributed and used before test results are known. Successful regulation of the microbiological quality of drinking water therefore depends on the use of raw water supplies of relatively invariant high quality. It is unlikely that sudden contamination will be detected promptly enough to protect the public and, in addition, corrective treatments may be inadequate.

Nevertheless, it is essential that present methods of microbiological testing be continued, in order to validate the effectiveness of disinfection, and for detecting defects within the system.

Solid Particles in Suspension

Materials suspended in drinking water include finely divided inorganic and organic solids ranging in size from colloidal dimensions to over 100 μm. Such particles may also have other substances and microoorganisms attached to them. Other particles, such as the asbestos minerals, may have the potential to affect human health directly when ingested. Many kinds of particles, though apparently harmless in themselves, may indirectly affect the quality of water by acting as vehicles for concentration, transport, and release of other pollutants.

Water treatment can often be effective in removing most of the suspended particles, but conventional methods of detection of particulate material by turbidity measurements have serious deficiencies.

Pollutant concentrations are higher in sediments than in the waters with which they are associated. It should be recognized that the consequences of pollutant adsorption by clays may be very important in natural systems and may affect drinking water quality. If a clay-pollutant complex survives water treatment and enters the drinking water system, it would then be available for ingestion by humans. In the adsorbed state the pollutant may not be toxic, but the possibility exists that the pollutant might be released from the clay in the alimentary tract and thus exert toxic effects. Whether or not such a process might take place would depend on the complex in question. Generalities are not possible since information is completely lacking in this area.

Particles of asbestos and other fibrous minerals occur in raw water, usually in a range of sizes from fractions of a micrometer to a few micrometers. Generally there are fewer than 10 million fibers per liter, but waters are found with from less than 10,000 to more than 100 million fibers per liter. Some of the highest counts have been found near some cities.

Certain mineral fibers found in water are suspected of being harmful upon ingestion. The available data on asbestos ingested through drinking water do not suggest an immediate hazard to health, but additional research in animal experiments and epidemiological studies is required to determine the degree of hazard. Studies of cancer death rates in Duluth, Minnesota, where the water supply has been contaminated with mineral fiber, have so far not revealed any increase of gastrointestinal cancer with time, in comparison with death rates in other areas. However, contamination of the Duluth drinking water began less than twenty years ago, and since many cancers have long latency periods, these negative findings do not exclude the possibility that an increase may appear within the next five to fifteen years.

Although a high turbidity measurement may indicate a potential adverse health effect, a low turbidity measurement does not guarantee that a water is potable. Turbidity measurements do not indicate the type, number, or mass of particles in a water supply. Where particulates in water are suspected of being harmful, the particulate content should be identified and counted by more specific techniques. Such techniques may include biological, organic, inorganic, and fibrous particulate surveys. Turbidity measurements are valuable for process control. However, the results obtained with present instruments, procedures, and units of measurement are not well correlated with particle concentrations and size distribution. The test must be refined to facilitate its use for these and other purposes.

Inorganic Solutes

The Interim Primary Drinking Water Regulations list maximum allowable concentrations for six metallic elements--barium, cadmium, chromium, lead, mercury, and silver. Ten additional metals were reviewed in the study--beryllium, cobalt, copper, magnesium, manganese, molybdenum, nickel, tin, vanadium, and zinc. Sodium, which is also a metallic constituent, is considered in a separate section because the problems it poses are quite distinct from those associated with the other metallic substances.

Eight of these selected metals are known to be essential to human nutrition: chromium, cobalt, copper, magnesium, manganese, molybdenum, tin, and zinc. Nickel and vanadium probably are also essential and it is possible that barium can be beneficial under certain conditions. The toxic metals lead, mercury, and cadmium are believed not to be essential to humans. Beryllium and silver also are not known to be essential. Elements that are beneficial in small quantities very often exhibit toxic properties when ingested in excessive amounts or concentrations.

Chemical weathering, soil leaching, runoff, mining, and industrial and municipal discharges are of major importance as sources of most inorganic materials. Other important sources of trace metals in drinking water are water treatment processes and pickup of metallic ions during storage and distribution. Although a large fraction of the United States continues to receive water from ground sources or from impounded upland sources without treatment other than disinfection, most large surface supplies are subjected to treatment that includes coagulation, sedimentation, filtration, and disinfection. Should trace metals occur in the raw water supply, these normal water-treatment processes have either no effect or an uncertain one on removing the usual low level concentrations of these metals. Moreover, probable trace metal impurities in the technical-grade chemicals used may introduce additional concentrations into the treated water.

Control of the corrosive properties of the finished water is important to prevent increases in trace metal concentrations during storage and distribution. A wide variety of materials are used in the pumps, pipe fittings, and reservoirs of distribution systems and home plumbing. Reactions, particularly of soft, low-pH waters, with materials of the distribution system very often have produced at the tap concentrations of iron, copper, zinc, lead, and cadmium that are much greater than those in the raw or treated waters at the plant.

All of the trace metals studied are known to exhibit toxic effects at some level of intake. Many of these, however, are at levels greater than the maximum concentrations found in drinking water. To include such materials in primary drinking-water standards with a requirement for mandatory surveillance does not confer any health benefit.

Adverse health effects associated with trace metals depend upon the total intake from all sources--water, air, and food. As a general rule, concentrations of trace metals in foodstuffs greatly exceed those found in drinking waters. Because the diet of most of the U.S. population is increasingly varied and comes from diverse geographical sources as a result of modern food distribution practices, the dietary intake of trace metals exhibits relatively small variation throughout the United States. This factor is helpful in evaluation of maximum no-adverse-health-effect concentrations for drinking water.

In addition to the trace metals, the effects on health of sodium, arsenic, selenium, fluoride, nitrate, and sulfate are also considered in the report. A section on water hardness and health reviews the literature on this subject.

It is of interest to note that the intake of sodium in adults in the United States often exceeds body needs by an order of magnitude. Sodium intake from drinking water represents less than 10 percent of the normal adult total intake. Although there is evidence linking excessive sodium intake to hypertension, the evidence for man is indirect. It is not possible to specify a "no-observed-adverse-health-effect" concentration in water for sodium since the usual food intake is already greater than a desirable level. It is desirable to know the sodium ion content in the water supply when prescribing diets for special risk groups on restricted sodium intake, and in some cases it may be necessary to prescribe bottled water low in sodium. The report points out that at least 40 percent of the population would benefit if sodium ion intake were maintained at less than 2,000 mg/day. The contribution of water, containing not more than 100 mg/liter, to the desired total intake would be less than 10 percent for a daily consumption of 2 liters.

Summarizing the health effects of the twenty-two inorganic solutes reviewed in the report could result in a misleading oversimplification, and the reader is referred to the report for the detailed analysis made by the committee. Generalizations might be equally misleading, but most of the materials considered were judged to have little or no threat to human health either because of low concentration in drinking water or minimal potential toxicity, or both. The health issue is complicated because of potential toxicity at higher concentrations than usually encountered in drinking water and the nutritional value at low levels. Lead appears to have the greatest potential toxicity, and the present standard may not provide an adequate margin of safety for infants and young children. In light of the data presented in the report, reexamination of current standards for arsenic and selenium is justified. Although the causal relationship between the incidence of cardiovascular disease and water hardness is not clear, the evidence indicates the existence of an inverse correlation between the two.

Organic Solutes

Only a small fraction of the total organic matter present in water has been identified. Organics include a volatile fraction of about 10 percent, and nonvolatile of about 90 percent. Although approximately 90 percent of the volatile organics have been identified, only 5-10 percent of the nonvolatile organic compounds (which comprise 90 percent of the total organic material in water) have been identified.

The report considered 74 nonpesticide organic compounds of the approximately 309 volatile compounds so far identified in drinking water, and 55 pesticides. Some of the pesticides studied have not yet been detected in drinking water, but were included because they are or have been used in large quantities. A compound was selected for consideration if any of the following criteria applied:

1. Experimental evidence of toxicity in man or animals: carcinogenicity, mutagenicity, and teratogenicity.

2. Identified in drinking water at relatively high concentrations.
3. Molecular structure closely related to that of another compound of known toxicity.
4. Pesticide in heavy use that could result in contamination of drinking water supplies.
5. Listed in the Safe Drinking Water Act or National Interim Primary Drinking Water Regulations.

In its consideration of safety factors and risk assessment of chemical contaminants, the committee felt that predictions that are risk-related provide a more meaningful first approximation than safety-related predictions. The concept of safety factors and use of such factors for pollutants whose toxic effects are reversible and nonprogressive involves empirical calculations based on past history of use and concentrations that appeared safe. The highest apparently safe dose or concentration is divided by a "safety factor" that varies by one or more magnitudes depending on the investigator's assessment of the adequacy of the data. Opinions differ on whether or not the "safety factor" approach can be used with pollutants that cause chronic, irreversible, or progressive disease in laboratory animals. Supporters of the "safety factor" concept and threshold values argue that body defenses can protect at doses less than the threshold values. Those who argue that "safety factors" are inadequate suggest that even one or a very few events have a finite probability of causing disease.

After much discussion, the committee adopted a "non-threshold" approach for estimating risk from pollutants that have been shown to be carcinogenic in laboratory animals, believing that it is not possible to guarantee a risk-free society since there are subgroups that are more susceptible than the normal population. The establishment of non-threshold risk estimates does not imply that water containing the contaminant should not be used, for even greater risks to health might arise from such a prescription. It is important that a net risk be estimated and society must attempt to balance the risks and benefits.

Most of the experimental results on which the current knowledge of toxicity depends are based on doses and dose rates for man and animals that are much larger than those that correspond to the usual concentrations of harmful materials in drinking water. Consequently there is great uncertainty in estimating the magnitude of the health risk arising from the ingestion of pollutants. The theoretical and experimental bases for extrapolating estimates of risk to low levels of dose are reviewed in the report and the principles used to guide the study are presented in detail.

A more traditional approach, combining the maximum dose producing no observed adverse effects with an uncertainty (risk) factor to calculate an acceptable daily intake

(ADI),[1] was used for agents not considered suspected carcinogens, and for which there were adequate toxicity data from prolonged ingestion studies in man or animals. The ADI represents an empirically derived value that attempts to reflect a particular combination of knowledge and uncertainty concerning the relative risk of a chemical. The uncertainty factors used to calculate ADI values represent a level of confidence judged to be justified on the basis of the animal and human toxicity data available. The calculation of the ADI value is based on the amount of a chemical that is ingested. It cannot be applied solely to drinking water because of possible intake of the chemical from other sources such as food and air.

Because the experimental data on the effects of many substances are inconsistent, "no-observed-adverse-health-effect" levels cannot be firmly specified for all organic contaminants. Most of the materials considered have not been studied sufficiently to establish their carcinogenic potential with certainty. The risk assessments do not take into account interactions such as additive toxicity,

---

[1]The committee considered several alternative terms, other than ADI, but concluded that the introduction of a substitute for ADI might well lead to confusion. The term "Acceptable Daily Intake" is used throughout the discussion because of its adoption by international organizations.

synergism, and antagonism. What ultimately may be most important is the interaction of these compounds with each other and with other material in contributing to the total body burden resulting from multiple sources of contaminant exposure. For these reasons the ADI is intended to be used only as a guide for assessing toxicity from chronic exposure.

With our present limited knowledge concerning structure-activity relationships for carcinogenic and other toxic effects, we cannot consistently and accurately extrapolate these properties from one compound to another. Nevertheless, in certain instances (for example, the substitution of bromine for chlorine in a halogenated methane) it is presumed that the relationship is sufficiently strong to justify the suspicion that the related compounds may be similarly toxic.

The potential for existing concentrations of organic pesticides and other organic contaminants in drinking water to adversely affect health cannot be estimated with certainty at this time. The key issue is whether or not certain organic chemicals found in very low concentrations can cause or increase the rate of cancer development in man. Even though several of these chemicals have demonstrated carcinogenicity in laboratory animals, the extrapolation of such results to man remains difficult for a number of reasons. First, the bioassays which have been used to establish carcinogenicity are conducted at doses which are hundreds to thousands of times greater than the levels at which these organic chemicals occur in water. The risks at these low levels must be obtained by extrapolation from higher doses. There is no hard evidence that low level oral exposure to any of these chemicals produces cancer. An argument has been made that the dose levels used to establish carcinogenicity are so high that they overwhelm normal detoxification or repair mechanisms or both, and produce cancer by some mechanism that does not operate under low dose conditions. Second, risk assessments based on extrapolations which fail to consider species differences in sensitivity, tissue susceptibility, kinetics, pathology, or biotransformation pathways may be inappropriate. This kind of information is not currently available.

In light of such uncertainties, a cautious approach must be adopted when dealing with potentially harmful chemicals. Even more uncertainty exists when one considers the possibility that some of these chemicals may also be mutagenic or teratogenic. The methodologies used to establish these effects are even less applicable to man than cancer bioassays.

Virtually no toxicity data are available for many of the organic compounds identified in drinking water. Ideally, all of these agents (as well as any new ones) should be subjected to an extensive battery of toxicity tests, including chronic bioassay. In practice, there is a need to determine those agents for which the generation of data is most pressing. Several criteria are important for the development of an order of priority for testing. The main factors identified in the assignment of priorities are:

1. The relative concentrations of the compounds and the number of people likely to be exposed, as well as the identity of defined subpopulations exposed.

2. The number of water systems in which they occur.

3. The existence of positive responses to *in vitro* mutagenic screening systems.

4. The existence of positive responses to *in vitro* carcinogen prescreens (mammalian cell transformations).

5. Similarity of chemical structure of the test compound to those of other compounds having defined toxic properties (i.e., structure-activity relationships).

6. Relationship of dose from water to total body burden.

A number of assays using bacteria and yeast have shown promise in yielding high correlations between mutagenic activity and known carcinogenic activity for certain classes of materials. These may prove to be useful in establishing a first level screen for potential carcinogens.

Conclusions

Table 2-1 lists the organic contaminants for which positive data on carcinogenesis exist. For these compounds, where adequate (lifetime) feeding studies were available, a statistical extrapolation of risk was performed. The numbers of table 2-1 are upper 95 percent confidence estimates of cancer risk to man from a lifetime of exposure to

Table 2-1. Categories of Known or Suspected Organic Chemical Carcinogens Found in Drinking Water

Compound	Highest observed concentrations in finished water (μg/liter)	Upper 95% confidence estimate of lifetime cancer risk per μg/liter[a]
Human carcinogen		
Vinyl chloride	10	5.1×10^{-7}
Suspected human carcinogens		
Benzene	10	I.D.[b]
Benzo[a]pyrene	D[c]	I.D.
Animal carcinogens		
Dieldrin	8	2.6×10^{-4}
Kepone	N.D.[d]	4.4×10^{-4}
Heptachlor	D	4.5×10^{-5}
Chlordane	0.1	1.8×10^{-5}
DDT	D.	1.2×10^{-5}
Lindane (γ-BHC)	0.01	9.3×10^{-6}
α-BHC	D	6.5×10^{-6}
β-BHC	D	4.2×10^{-6}
PCB (Aroclor 1260)	3	3.1×10^{-6}
ETU	N.D.	2.2×10^{-6}
Chloroform	366	3.7×10^{-7}
Carbon tetrachloride	5	1.5×10^{-7}
PCNB	N.D.	1.4×10^{-7}
Trichloroethylene	0.5	1.3×10^{-7}
Diphenylhydrazine	1	I.D.
Aldrin	D	I.D.
Suspected animal carcinogens		
bis(2-chloroethyl)ether	0.42	1.2×10^{-6}
Endrin	0.08	I.D.
Heptachlor epoxide	D	I.D.

[a] * See report for details.
[b] I.D. = Insufficient data to permit a statistical extrapolation of risk.
[c] D = Detected but not quantified.
[d] N.D. = Not detected.

a particular compound. These estimates have been corrected for interspecific differences (that is, between the experimental animal and man on the basis of relative surface area).

Data for mutagenesis in bacteria, or other test systems were also examined. Available data are summarized as follows: (1) benzo[a]pyrene, chlorodibromomethane, captan, and folpet have been found to be mutagenic; (2) bromoform and vinyl chloride, weakly mutagenic; (3) carbon tetrachloride, bromobenzene, nicotine, DDE, dieldrin, carbaryl and trifluraline, nonmutagenic.

Data on teratogenic potential exist for twenty-four of the compounds under study. Hexachlorophene, nicotine, the phthalate esters, 2,4-D, 2,4,5-T, and folpet have been shown to be teratogens, while benzene, benzo[a]pyrene, carbon tetrachloride, PCB's, captan, carbaryl, chlordan, DDT, Kepone, malathion, methyl parathion, mirex, paraquat, and parathion have been reported to be nonteratogenic. Nowhere is the paucity of toxicological data more evident than in the data on teratogenesis.

ADI values were calculated for forty-five compounds not considered to be known or suspected carcinogens, and for which there were adequate toxicity data from prolonged ingestion studies in man or animals. The report presents details on these compounds, including calculations that suggest a no-anticipated-adverse-effect level for these chemicals in drinking water using two hypothetical exposures.

The health effects of many compounds of interest could not be assessed because toxicological information about them was inadequate or unavailable. These compounds are listed in tables 2-2 and 2-3, together with their reported occurrence in drinking water in the United States.

Radioactivity in Drinking Water

Everyone is exposed to some natural radiation that comes from both cosmic rays and terrestrial sources. Although there are large geographic variations in the amount of natural background radiation, the average background dose in the United States is about 100 mrem/year. A small proportion of this unavoidable background radiation comes from drinking water that contains radionuclides.

By far the largest contribution to the radioactivity in drinking water comes from potassium-40, which is present as a constant percentage of total potassium. Only a small percentage of the total potassium-40 body burden, however, comes from drinking water. The total body dose from other possible radioactive contaminants of water constitutes a small percentage of the background radiation to which the population is exposed. Although the amounts of individual radioactive contaminants fluctuate from place to place, calculations made for a hypothetical water supply that might be typical for the United States have shown that a total soft tissue dose of only 0.24 mrem/year would be contributed

Table 2-2. Organic Contaminants of Drinking Water, for Which There Are Insufficient Data on Chronic Toxicity

	Highest concentration in finished water (μg/liter)
Acetaldehyde	0.1
Acrolein	b
Bromobenzene	D[a]
Bromoform	D
Carbon disulfide	D
Chloral	5.0
Chlorobenzene	5.6
Cyanogen chloride	0.1
1,2-Dichloroethane	21.0
2,4-Dichlorophenol	36.0
2,4-Dimethylphenol	D
ϵ-Caprolactam	D
Hexachloroethane	4.4
O-Methoxyphenol	D
Methyl chloride	D
Methylene chloride	7.0
Phenylacetic acid	4.0
Phthalic anhydride	D
Propylbenzene	<5.0
t-Butyl alcohol	0.01
Tetrachloroethane	4.0
Tetrachloroethylene	<5.0
Toluene	11.0
Trichlorobenzene	1.0
1,1,2-Trichloroethane	D
Nicotine	3.0
Methomyl	b
Cyanazine	D
Xylene	<5.0

[a] D = detected but not quantified.
[b] Not available.

Table 2-3. Organic Contaminants of Drinking Water for Which Information on Chronic Toxicity Is Lacking

Compound	Highest concentration in finished water (μg/liter)	Highest concentration in raw water (μg/liter)
1,2-bis(chloroethoxy)ethane	0.03	
Bis(2-chloroisopropyl)ether	1.58	
Bromochlorobenzenes	D[a]	
Bromodichloromethane	116	11
Butyl bromide	D	
Chloroethyl methyl ether	D	
Chlorodibromomethane	100	1.4
Chlorohydroxybenzophenone	D	
Chloromethyl ethyl ether	D	
Chloropropene	D	
Crotonaldehyde	5.0	
Dibromobenzene	D	
Dibromodichloroethane	0.63	
1,3-Dichlorobenzene	<3.0	
Dichlorodifluoroethane	D	
Dichloroiodomethane	0.5	
1,1-Dichloro-2-hexano	1.0	
1,2-Dichloropropane	<1.0	
1,3-Dichloropropene	<1.0	
1,2-Demethoxybenzene	D	
4,6-Dinitro-2-aminophenol	D	
Dioctyladipate	20.0	
Hexachloro-1,3-butadiene	0.07	
Isodecane	5.0	
Metachloronitrobenzene	D	
Methylstearate	D	
Nonane	4.0	
Octyl chloride	D	
Pentachlorophenyl methyl ether	0.1	
1,1,3,3-Tetrachloroacetone	1.0	
2,4,6-Trichlorophenol	D	
Trimethylbenzene	6.1	

[a] D = detected but not quantified.

by all the radionuclides found in the water. Even with rather wide fluctuations in the concentrations, the total contribution of the radionuclides will remain very small.

However, bone-seeking radionuclides--such as strontium-90, radium-226, and radium-228--account for a somewhat larger proportion of the total bone dose. This is particularly true for the two isotopes of radium because they emit high-linear-energy-transfer (LET) radiation, and because certain restricted localities have been found to have rather high concentrations of radium in drinking water. Nevertheless, in the hypothetical typical water supply, less than 10 percent of the annual background dose comes from such radiation. It has also been estimated that the total population exposed to levels of radium greater than 3 pCi/liter is about a million people. About 120,000 people are exposed to radium at levels greater than 9 pCi/liter.

Risk estimates were made for three kinds of adverse health effects that radiation could produce: developmental and teratogenic effects, genetic effects, and somatic (chiefly carcinogenic) effects.

The radiation associated with most water supplies is such a small proportion of the normal background to which all human beings are exposed that it is difficult, if not impossible, to measure any adverse health effects with certainty. In a few water supplies, however, radium can reach concentrations that pose a higher risk of bone cancer for the people exposed.

Concluding Comment

The high quality of drinking water supplies in the United States is recognized throughout the world, but it is necessary that we continue to direct our efforts toward maintaining and improving the quality of public water supplies in this country. A continuing research program is needed to develop a better understanding of the mechanisms by which water pollutants produce toxic effects. There is a need to develop rapid, inexpensive, accurate screening tests. Although much of this research is underway, the committee is convinced that more must be done. We are seeking and in fact applying solutions to problems that many professionals view as yet neither fully understood nor defined. We shall have to continue to do this, while recognizing that we must be able to retrace our steps as additional knowledge is acquired.

Discussion by Carol H. Tate

My comments on the NAS report are intended to put its findings in perspective. General comments on the NAS study are followed by specific comments on contaminant groups with an attempt to use the NAS findings to compare the relative risks of the various groups. Finally, the discussion concludes with suggestions on the usefulness of the NAS study in making forthcoming decisions on drinking water regulations.

General Comments

The NAS report, Drinking Water and Health, represents a major advance in the water quality field. In response to the congressional mandate of the Safe Drinking Water Act (U.S. Congress, 1974), the NAS has made an initial effort to consider all drinking water contaminants in terms of their effect on health. Although the NAS indicated that present scientific information was inadequate to specify for all contaminants the "maximum contaminant levels which should be recommended . . . in order to protect the health of persons from any known or anticipated adverse health effects" as required by the act, the academy did provide a useful service in systematically evaluating contaminants on a common basis--their impact on health.

Although it may come as a surprise to some, drinking water regulations have not been based entirely on health considerations, even though setting drinking water standards at a level to protect users' health appears to be eminently logical. For various reasons, including lack of solid information, product quality standards for drinking water have historically been set on the bases of contaminant background levels, analytical detection limits, technological feasibility of treatment processes, aesthetic considerations, health effects, and combinations of the above (Tate and Trussell, 1977). Since the NAS report summarizes the current state of knowledge of the health effects of contaminants in drinking water, it represents a major step toward the goal of establishing all drinking water regulations on the basis of health effects.

Before proceeding further, it is important to note that the academy chose to take the honest stance that it could not provide Congress with all answers on maximum contaminant levels which would cause no known or anticipated adverse health effects. With the recent passage of the amendments to the Safe Drinking Water Act (U.S. Congress, 1977), Congress apparently recognized the need to pursue defining the relationships between drinking water and health by requiring the NAS to update the report every two years.

One hopes that with the present study and continuing updates, the EPA will be able to move closer to the goal of setting all drinking water regulations on the basis of health effects, while recognizing, of course, the relevant costs and technological feasibility of proposed regulations.

Contaminant Groups

The remainder of this discussion will summarize and compare findings from the five sections of the academy report: (1) microbiological contaminants, (2) particulates, (3) radioactivity, (4) inorganics, and (5) organics. This division was probably essential to enable completion of the comprehensive study within the allocated time period. However, it should be recognized that this approach also has shortcomings. Perhaps the key drawback in dividing the contaminants into groups is the limits it places on evaluating synergistic or antagonistic effects of more than one contaminant. A second problem relates to difficulty in establishing consistent evaluation criteria among the subcommittees for the groups. Therefore, it is suggested that some attention be paid to trying to overcome these possible disadvantages in the scheduled NAS update reports. Perhaps an overview subcommittee would help the NAS in looking at contaminant interactions, in setting more consistent evaluation methodology, and in putting all sections of future reports in perspective.

Microbiological Contaminants. The NAS report pointed out several factors related to microbiological contaminants in drinking water that are relevant for decision makers today and indicate a need for continued vigilance on the part of water suppliers, engineers, regulatory agencies, and others.

First, the study indicated that there continue to be outbreaks of waterborne disease in the United States. This situation occurs in spite of the fact that microbiological contaminants were the first major health-related water quality problem attacked by the public health and sanitary engineering professions in this country; and in spite of the widespread practice of disinfection and filtration. Since contamination of water sources, treatment deficiencies, and degradation within distribution systems appear to be key factors in these outbreaks, the proper maintenance of water supply systems should not be overlooked, even today.

Second, the NAS study allows us to review the problems of microbiological contaminants in relation to other water quality characteristics. There were approximately 10,000 cases of illness resulting from twenty-four outbreaks of waterborne diseases reported to the Center for Disease Control in 1975, but no reported deaths. Bacteria, protozoa, and viruses were listed among the most common causative agents in recent years, but in some years only 50 percent of the etiologic agents were identified. When it is realized that many cases of waterborne disease are probably not reported or are not diagnosed as due to a waterborne agent, it can be concluded that there is still considerable cause for concern about microbiological contaminants.

Third, the NAS pointed out that there are still unknowns surrounding virus contamination. Virus testing is complex; a single virus may be capable of infecting an individual; and the link between an infected individual and a water route of transmission could be easily missed. For all these reasons the relationship of viruses to waterborne diseases is not now quantifiable. When this fact is coupled with a potential for some viruses to act as carcinogenic agents after a long latency period, it is evident that more study is needed to clearly define any potential role of viruses in water supplies and disease transmission.

Fourth, the NAS described current thinking on the coliform bacteria test and its debatable validity as an indicator of microbiological quality. Perhaps of equal importance is the fact that there is a time delay of one to two days or longer from the time a sample is collected to the time coliform results are known. In the interest of continued public health protection, it can be concluded that improvements are necessary in this area. Development of a faster and perhaps more valid microbiological test, and, in the meantime, supplementing coliform tests with a disinfectant residual test to monitor quality in a more timely manner are suggested.

Other comments that should be emphasized include:

- Degradation in the distribution system could alter microbiological contaminant levels, so monitoring in the distribution system is important.
- The existence of a high quality source is important if microbiological contamination is to be minimized. This point cannot be overemphasized, particularly when it is realized that in the future good quality sources are likely to diminish in number. Therefore, the need for protecting existing good quality sources should be recognized by water, wastewater, and other planning agencies and compared with alternatives (such as treatment costs associated with increasing development or pollutant discharges to these sources) before decisions are made that could lead to source degradation.
- Finally, the potential for microbiological contamination must be balanced against the desire to reduce chlorinated organics in drinking water. It would be unwise to trade one potential public health problem for another.

Particulates. Particulates or solid particles in suspension were correctly identified in the NAS study as contaminants about which little is known but which, nevertheless, could indirectly affect drinking water quality and health. Precise information on their health effects is not available. Principally, the concerns over particulates are as follows. First, particulate clays tend to adsorb heavy metals, organic materials, bacteria, and viruses, all of which could have health-related effects. Second, it has been demonstrated that particulate material can protect microorganisms from the action of a disinfectant. Third, some particulates, such as asbestos fibers, may in themselves be harmful.

The NAS indicated there is a need for methods to measure particulate material in water supplies. This author and others (Tate and Trussell, 1976; Beard and Tanaka, 1976) have routinely used automatic particle counting equipment for several years to determine particle number concentration, size distribution, and calculated volume in water supplies. Therefore, there are techniques readily available at least to begin to gather more information on the presence and characteristics of particulates in water.

A useful effort suggested by the NAS and seconded by this author would be a survey of suspended particulate matter in several drinking water supplies. Because a wide variety of raw water sources and treatment methods exist today, it would be most useful if such a survey were nationwide in scope. One recent survey by this author and others (James M. Montgomery, Consulting Engineers, Inc., 1976) of twenty-five cities in the southwestern state represents a start in this direction.

Unfortunately, as the NAS pointed out, little is known about the characteristics of particulates in water supplies. This author agrees with the NAS suggestion that analytical methods be developed to determine biological, organic, inorganic, and fibrous (e.g. asbestos) fractions and components of particulates.

The relationship between particulates and turbidity is important since the National Interim Primary Drinking Water Regulations (EPA, 1975) include maximum contaminant levels for turbidity--which is an indirect or surrogate measurement of particulates--but do not include maximum contaminant levels for direct measures of particulates. While studies have shown that particle counts generally increase as turbidity increases (Tate and Trussell, 1976), the spread in data is such that turbidity cannot be used to accurately predict particle counts. Conversely, theoretical and experimental data on particulate measurements indicate that measurements of particle number and size distribution do not accurately predict turbidity (Treweek, 1975; Treweek and Morgan, 1978; Kavanaugh and coauthors, 1978).

Because of this, it is suggested that more information needs to be developed on the presence and characteristics of particulates in U.S. water supplies and that particulate measurements be considered in the future as a possible supplement to turbidity standards. At a minimum, more understanding of the particulates in water supplies and emphasis on developing such information is needed.

Radioactivity. A review of the information in the NAS study leads to a conclusion that radioactivity in U.S. water supplies is at low levels and constitutes a very low risk to the health of most consumers. According to the NAS study, levels of radioactivity and exposed populations are such that developmental and teratogenic effects are not measurable. Additional genetic disease cases in the United States attributable to radioactivity were calculated at less than one per year. Similarly, excess bone cancers in the country caused by bone-seeking radionuclides (principally strontium-90, radium-226 and radium-228) in drinking water were estimated by NAS at less than one per year. Therefore, it can be concluded that radioactivity in drinking water is not a major problem and that an increase in regulatory activity in this direction is not necessary unless the situation changes markedly.

Inorganic Chemicals. Inorganic chemicals (or "inorganic solutes" in the NAS study) are unique among water contaminants in that some are essential to health and that for one, hardness, there is apparently an inverse relationship between concentration and detrimental health effects. Although inorganic constituents have been regulated for many years by various standards, not all standards are based on health effects data nor are all health effects adequately understood.

Two major comments are in order. First, the maximum contaminant levels of several inorganic chemicals were compared and found to be inconsistent in the margin of safety provided for different species. This suggests that some of the maximum contaminant levels now included in the drinking water regulations should be adjusted. This might also apply to contaminants other than inorganic chemicals. Second, some inorganics not in the regulations, particularly those responsible for water hardness, appear to be capable of causing health effects substantial enough to merit concern and increased efforts at regulation.

Specific Inorganics and Margins of Safety. Comparisons were made of concentrations allowed by drinking water regulations and NAS judgments on levels at which no adverse health effects have been found. As will be shown in the following discussion, safety factors for all inorganic chemicals are not the same. When it is recognized that these water quality standards were initially set on the basis of detection limits, background limits, technological feasibility, health effects and other factors, these differences are understandable. However, now that more information is available on health effects through the NAS study, a reexamination of the regulations is in order. In addition, future studies would be useful if they could define safety factors for inorganic chemicals where such information is not now available.

The NAS reported concentrations of various inorganic chemicals in drinking water for which maximum no-observed-adverse-health-effect levels could be estimated, based on water intake, toxicological studies, occupational exposures, susceptibility of various groups of people, and total body burden from water, food, air, or other sources. Next, the NAS and this author compared those concentrations with the regulated maximum contaminant levels for these constituents in drinking water (EPA, 1975) in order to calculate the safety factors inherent in the present regulations. Division of the maximum no-observed-adverse-health-effect level by the regulated maximum contaminant level indicates the safety factor for a given contaminant in drinking water. For example, if the maximum no-observed-adverse-health-effect level is 0.10 mg/liter and the regulated interim maximum contaminant level is 0.01 mg/liter, the safety factor for that contaminant in the current drinking water regulations is ten.

Table 2-4 shows the maximum no-observed-adverse-health-effect level reported or estimated from the NAS study, the maximum contaminant level from the National Interim Primary Drinking Water Regulations (EPA, 1975), and the estimated safety factor (reported by NAS or calculated) for the six organic chemicals for which all such information is available. As the table indicates, there is a wide range (from ≤ 1 to 100) in the safety factors allowed in the interim regulations

Table 2-4. Inorganic Chemical Regulatory Safety Factors

Inorganic chemical	(1) Maximum no-observed-adverse-health-effect level[a] (mg/liter)	(2) Maximum contaminant level[b] (mg/liter)	(3) Safety factor[c]
Chromium	5	0.05	100
Lead	≤0.05	0.05	≤1
Silver	0.05	0.05	1
Arsenic	<0.10	0.05	<2
Selenium	0.10 - 0.50	0.01	10 - 50
Nitrate-N	10	10	1

[a]From NAS (1977)

[b]From EPA (1975)

[c]Column (1) divided by column (2)

for these inorganic chemicals. This variation leads to a conclusion that some further work is in order, either to put all health-related inorganic contaminants on a common regulatory basis or to explain why this is not advisable.

Comments on individual inorganic constituents are offered below. Of particular concern in table 2-4 are lead and nitrate, for which the estimated factor is one or, for certain segments of the population such as children and infants, even less than one.

For lead, the NAS study states that the maximum no-observed-adverse-health-effect level cannot be set with assurance at any value greater than 0.025 mg/liter. This would argue for a revised maximum contaminant level of lead at 0.025 mg/liter or less rather than the interim level of 0.05 mg/liter. Based on the evidence contained in the NAS study, I would agree with a recommendation to lower the maximum contaminant level for lead.

For nitrates, the NAS study indicated that available evidence tends to confirm a value near 10 mg/liter nitrate-nitrogen as a maximum no-observed-adverse-health-effect level, but this is also the value of the interim standard, implying a safety factor of one. Therefore, it seems that the nitrate standard should be reexamined for possible lowering in the revised drinking water regulations.

For silver, table 2-4 also indicates that interim regulations only provide a safety factor of one. However, this should not be cause for concern to community water suppliers since the NAS study indicates silver ion has not been detected in water supplies in concentrations greater than half that no-observed-adverse-health-effect level. Nevertheless, a consistent safety factor should be applied.

For arsenic, the NAS study states that the current mandatory drinking water limit of 0.05 mg/liter provides only a meager safety factor (two or less). Therefore, I believe that the arsenic regulation should also be critically reexamined to determine if a change is warranted. Furthermore, explicit allowance should be made in the standards for the various forms of arsenic [As(III), As(V), and organic arsenic] and their relative toxicities.

For selenium, conflicting information in the NAS study gives no basis for modifications to the regulations at this time. On one hand, the presently allowed concentrations just barely provide for the minimum nutritional requirements for selenium, and there is evidence to suggest a protective effect of selenium against certain types of cancers. On the other hand, there is some tenuous evidence cited for carcinogenic effects of selenium. Therefore, I believe that more information needs to be developed on selenium, including forms of selenium in water, before raising or lowering the selenium limit in water.

For chromium, the NAS study suggests that more information is needed on adverse effects of chromium (VI) vs. chromium (III) before any modifications are made to drinking water regulations. I concur.

Guidelines on Safety Factors. Some general guidelines on appropriate safety factors are contained in the NAS study and could be useful to decision makers. Unfortunately, these guidelines are presented in the organics section of the NAS study, but with some exceptions, would appear to apply to inorganic chemicals as well, so long as no evidence of carcinogenicity exists. The general principle underlying the NAS guidelines is that the safety factors should be greater where the quality and quantity of data are more uncertain, inadequate, or equivocal. Also, safety factors were judged applicable only where there is no indication of carcinogenicity. Based on this principle, the NAS adopted the general guidelines in establishing safety factors summarized in table 2-5.

It would be useful if the NAS, in its current study, could determine whether these guidelines are applicable to inorganic chemicals and other contaminants in drinking water or whether they need some modifications. If they are thought to be applicable, the NAS should use them, judge the status of experimental evidence, and suggest safety factors for inorganic chemicals. Then, the EPA could weigh these recommendations on health effects along with costs and

Table 2-5. NAS Guidelines for Establishing Safety Factors in Regulations

Human data	Experimental evidence Animal Data	Carcinogenicity	Suggested safety factors
Valid experimental result from studies on prolonged ingestion by man.	Not applicable	No indication of carcinogenicity.	10
Experimental results of human ingestion not available or scanty (e.g. acute exposure only)	Valid results of long-term feeding studies on experimental animals. In absence of human studies, valid animal studies on one or more species.	No indication of carcinogenicity.	100
No long-term or acute human data.	Scanty results on experimental animals.	No indication of carcinogenicity.	1,000

technological feasibility in considering revised maximum contaminant levels for inorganic contaminants.

Hardness and Related Factors. A second major comment on the NAS study relates to inorganic substances in drinking water, such as mineral salts causing hardness and sodium that apparently could substantially affect the health of users. According to the NAS study, there have been more than fifty studies in nine countries that indicate an inverse relationship between hardness in drinking water and mortality rates for cardiovascular disease. Thus, where the water is hard, cardiovascular disease rates tend to be lower than in areas where water is soft. The credibility of these studies, according to the NAS judgment, depends more on their consistent trends than on individual correlation coefficients or postulated biological mechanisms.

Despite uncertainties in the data and debate on mechanisms, the NAS has concluded that the body of evidence is sufficiently compelling that the possibility must be taken seriously, particularly when the number of preventable deaths in the United States is considered. On the assumption that water factor(s) are causally implicated, the NAS estimates that optimal conditioning of drinking water for hardness could reduce the annual deaths from cardiovascular disease in the United States by as many as 150,000 persons per year. This represents approximately 8 percent of the annual mortality rate in the country. The NAS is currently investigating the issue of hardness and health in more depth in its second study. It is to be hoped that the study will settle the question of hardness and define an optimal hardness range or a minimum hardness limit.

However, even in the absence of more definitive information, the evidence to date is strong enough that I would suggest that regulatory agencies begin to consider hardness more seriously. After deliberation, potential actions may include establishing a minimum or optimal hardness regulation, prohibiting or at least not encouraging municipal water softening, evaluating the advisability of allowing bottled water sales of water with low hardness or advising users of possible risks, and curtailing use of home water softeners or advising users of the associated risk.

An inorganic chemical discussed in the NAS report that is related to health and often indirectly related to hardness is sodium. According to the NAS figures, approximately 15-20 percent of the U.S. population is at risk of developing hypertension. Genetic susceptibility, sodium content of drinking water, dietary intake, and other factors could influence whether an individual develops hypertension. In general, drinking water represents a small percentage of the daily sodium intake. However, among the small segments of the population that are hypertensive and on restricted sodium intake diets, the amount of sodium in drinking water can be a substantial portion of the daily sodium allowance unless persons use low sodium water.

Based on the NAS report, I concur that there is no immediate need to limit sodium in drinking water for the general population. At the same time, there appears to be no reason to encourage treatment of drinking water by sodium ion exchange or other methods that would increase sodium content and, at the same time, reduce hardness.

Organic Chemicals. The issue of organic chemicals in drinking water is a major topic of interest today, particularly in light of EPA's proposed regulations of organic chemical contaminants in drinking water (EPA, 1978), and the NAS should be commended for its thorough evaluation of the problem. In particular, the NAS decision to base its work on a concept of risk rather than to the congressional request for regulating absolute safety in drinking waters appeared to be wise.

The NAS study contains an up-to-date synopsis of current information on the presence and significance of organic chemicals in drinking water. For persons not familiar with the fields of health risk assessment, particularly extrapolation of animal test data to man, the NAS study provides a cogent discussion of guiding principles and rationale for using these methods. The debate on thresholds in carcinogenesis is explained in terms that will be understood by most readers. Finally, the reasons for rejecting the threshold concept in remaining risk assessment descriptions are presented.[2] For a variety of organic chemicals, the NAS study represents a valuable summary of toxic, mutagenic, and other effects.

The NAS study indicates the upper 95 percent confidence estimate of individual lifetime cancer risk per μg/liter of sixteen known or suspected carcinogens. The risk figures assume a continued daily ingestion over a 70-year lifetime of 1 liter of water containing 1 μg/liter of the substance in question. For higher concentrations, it is assumed that the risk can be multiplied correspondingly. A question arises on the advisability of using 1 liter per day rather than the 2 liters per day of water ingestion assumed in the EPA regulations (EPA, 1975). However, on reflection, it would appear that the accuracy of other numbers in the estimates of cancer risks is such that a factor of two is probably not significant in the overall accuracy of risk projections, though using 2 liters per day would at least be internally consistent.

As an example of the risk from one organic compound in water, we can consider chloroform, a trihalomethane. The NAS study indicates that for each μg/liter of chloroform in

---

[2]Ed. Note: That is, the committee rejected the hypothesis that there are dose levels for carcinogens below which no effects will be observed. If such thresholds do exist and can be identified, it clearly makes regulation considerably easier.

drinking water, the upper 95 percent confidence limit for lifetime cancer risk per individual is 0.0000017. If a person drinks water with a chloroform concentration of 100 μg/liter, the additional risk of cancer is 0.0000017, or approximately one in 600,000. Certainly there can be debate about the absolute number, but it can be used to compare the risk of cancer from chloroform in drinking water with other risks.

To put these numbers in perspective, the extra risks to the individual from drinking water containing chloroform can be compared with the overall cancer risks in the United States today. There are in this country approximately 350,000 cancer deaths per year (EPA, 1978), compared with about 2 million total deaths per year (NAS, 1977). Thus, cancer from various causes accounts for approximately 18 percent of the deaths in this country each year. On the individual level, each person faces a risk of about one in six of eventually dying of cancer. When the occurrence of other, nonfatal cancers is considered, the general risk of an individual developing cancer in a lifetime is about 25 percent, or one in four. Compared with this overall cancer risk, the risk of cancer from chloroform in drinking water is very small.

Information in the NAS study can be used to evaluate nationwide risks for organic chemicals in relation to other contaminants in drinking water. With an estimated lifetime cancer risk of 1.7 x 10^{-6} per individual per μg/liter, a total U.S. population of 160 million, and ingestion of 1 liter of water per day, a chloroform concentration in drinking water of 100 μg/liter could be estimated to produce cancer in approximately 27,000 individuals who otherwise would not have developed it. Assuming a 70-year lifetime, this would be approximately 400 additional cancers in the United States per year from chloroform in drinking water. This number is similar to one developed by Tardiff (1977), who considered a variety of models for estimating risk. He estimated that a maximum allowable concentration of chloroform in tap water of 70 μg/liter, might cause a maximum of 252 cancer deaths per year in the United States.

The picture becomes more complicated when all organics found in drinking water are considered together. To date, more than 700 specific organic chemicals have been identified in drinking water supplies in the United States (EPA, 1978). For the majority of these compounds, data are not available on toxicity, carcinogenicity, mutagenicity, teratogenicity, and other health-related effects that would allow for an estimate of risks to the general population at levels found in drinking water. Furthermore, not all organic chemicals in drinking water have yet been identified, nor are synergistic or antagonistic effects understood. Therefore, the total risk from organic chemicals in drinking water cannot be calculated at this time, but that risk may be large.

Relative Risks from Contaminant Groups

Table 2-6 summarizes the information described earlier in this discussion and is based on data in the NAS study. Of course, all estimates on relative risks are highly qualified due to uncertainties in information, extrapolations in data, and other factors. Nevertheless, the information in table 2-6 does allow us to look at relative risks from the contaminant groups in drinking water as a first approximation, particularly for order of magnitude estimates.

The information on relative risks, especially for mortality, indicates the contaminants may be placed in three broad categories:

- Low risk (radionuclides, chloroform, some microorganisms, some inorganic chemicals)
- High risk (low hardness)
- Unknown risk (organic chemicals, viruses, particulates, some inorganic chemicals)

There are two major conclusions that can be drawn from this information. First, we need to know more about the risks associated with certain contaminant groups, with specific contaminants within a group, and with synergistic or antagonistic effects among specific contaminants. Second, on the basis of relative risk, it appears that the effect of low hardness in drinking water is substantial, particularly in comparison with risks from chloroform which may be regulated soon. The risk of 150,000 deaths per year from cardiovascular disease is considerably greater than estimated risk

Table 2-6. Relative Risks from Various Contaminant Groups in Drinking Water (Based on NAS Study Information)

Contaminant group	Approximate annual risk to U.S. population
Microorganisms	10,000 reported illnesses Mortality low to unknown Risks from viruses unknown for latent diseases
Particulates	Unknown
Radionuclides	Less than one additional cancer death
Inorganic chemicals	
General	Uncertain, but low risk likely and some benefits possible
Hardness	150,000 additional deaths from cardiovascular diseases due to low hardness
Lead and nitrates	Risks to infants and children at present regulated levels
Organic chemicals	
Chloroform	200-400 additional cancer deaths
All organics	Unknown

of 200 to 400 cancer deaths from chloroform, yet there is a major effort in the direction of regulating chloroform, but not hardness. I suggest that the EPA should begin to consider a regulation for minimum and/or optimal range of hardness. It is to be hoped that the information on relative risks contained in table 2-6 will be useful in making future decisions on revised drinking water regulations or will at least stimulate preparation of a better synopsis of relative risks.

Future Course of Action

Since the NAS study was requested by Congress in an attempt to consider all known information on the health effects of drinking water contaminants prior to the EPA's decisions on revised drinking water regulations, it is instructive to reflect on a future course of action now that the NAS study is available. Recommended steps in determining future actions are as follows.

- Compare the relative risks of contaminants or contaminant groupings to determine where revised regulations are most needed.
- Establish consistent criteria for establishing safety factors for classes of contaminants.
- Consider uncertainties inherent in the risk estimates.
- Reduce uncertainties where possible by more toxicological or epidemiological studies with priority for studies depending on such factors as the following: (1) relative concentration of a contaminant, (2) size of population exposed, (3) susceptibility of population exposed, (4) number of water systems involved,

(5) relative dose in water compared with total body burden, (6) positive response of chemical in mutagenic tests or *in vitro* studies, (7) similarity of chemical structure to known harmful chemicals, and (8) relative cost and feasibility of removing chemical contaminant or maintaining a beneficial constituent at optimal levels.

- Determine water treatment costs for constituent.
- Estimated benefit-cost ratios using data developed above, with avoidance of a given health risk used in benefit half of the equation.
- Establish priorities on regulation of contaminants.

It will not be easy to accomplish these suggested tasks. However, considering the size of the population at risk and the magnitude of potential adverse health effects, their importance cannot be overestimated. It was evidently the intent of the Safe Drinking Water Act to guarantee the public a supply of safe (or more accurately, low risk) drinking water, so the task ahead is to complete that mandate. With the NAS study and with the planned updates of the study at two-year intervals, the task will be made somewhat easier for decision-makers, particularly when relative risks are considered. In addition, decision makers should bear in mind that for certain factors (such as those causing hardness) there may be a beneficial rather than an adverse impact of increasing levels in water. If it becomes necessary to amend the act in order to regulate these constituents, that should be done.

References

Beard, James D., and Theodore S. Tanaka. 1976. "A Comparison of Particle Counting and Nephelometry in Evaluating Filtration Plant Performance," Proceedings AWWA Water Quality Technology Conference (San Diego, Calif.).

Environmental Protection Agency. 1975. "Interim Primary Drinking Water Regulations," Federal Register vol. 40, no. 248, pp. 59566-59588 (Washington).

______. 1978. "Interim Primary Drinking Water Regulations: Control of Organic Chemical Contaminants in Drinking Water," Federal Register vol. 43, no. 28, pp. 5756-5780.

Kavanaugh, Michael C; Carol H. Tate; Albert R. Trussell; R. Rhodes Trussell; and Gordon P. Treweek. 1978. "Use of Particle Size Measurements for Water Treatment Plant Process Selection and Control," American Chemical Society, Division of Environmental Chemistry; Symposium on Particulates in Water: Characterization Fate, Effects, and Removal (Anaheim, Calif.).

James M. Montgomery, Consulting Engineers, Inc. 1976. "Particle Analysis of Domestic Water Supplies" (Pasadena, Calif.).

National Academy of Sciences. 1977. Drinking Water and Health. A report of the Safe Drinking Water Committee (Gerard A. Rohlich, chairman), Advisory Center on Toxicology, Assembly of Life Sciences, National Research Council (Washington, National Academy of Sciences).

Tardiff, R. G. 1977. "Health Effects of Organics: Risk and Hazard Assessment of Ingested Chloroform," American Water Works Association Journal vol. 69, no. 12, pp. 658-661.

Tate, Carol H., and R. Rhodes Trussell. 1976. "Optimization of Turbidity Removal by Use of Particle Counting for Developing Plant Design Criteria," Proceedings AWWA Water Quality Technology Conference (San Diego, Calif.).

______. 1977. "Developing Drinking Water Standards," American Water Works Association Journal vol. 69, no. 9, pp. 486-498.

Treweek, Gordon P. 1975. "The Flocculation of E. coli with Polyethyleneimine," Ph.D. Thesis, California Institute of Technology (Pasadena, Calif.).

______, and James J. Morgan. 1978. "Prediction of Suspension Turbidities from Aggregate Size Distributions," American Chemical Society, Division of Environmental Chemistry; Symposium on Particulates in Water: Characterization, Fate, Effects, and Removal (Anaheim, Calif.).

U.S. Congress. 1974. Safe Drinking Water Act, PL 93-523.

______. 1977. Safe Drinking Water Act Amendments of 1977, PL 95-190.

Discussion by Harold E. Pearson

The National Academy of Sciences should be complimented in making this thorough and excellent review of the available literature on drinking water and health. When the Summary Report was issued in July 1977, there was some grumbling on the part of those who like to play the numbers game--the regulators, the environmentalists, and yes, some public water suppliers--that the recommendations were not specific enough.

They seemed to forget that the Congress had mandated the NAS study in order to get a dispassionate, objective review of the evidence now available to make a specific judgment concerning contaminant levels that may have an adverse health effect on persons drinking the water. Moreover, in accepting the assignment to complete the study, NAS agreed to assemble and describe the available data, but if there was insufficient evidence, both as to quality and quantity, it would not exercise its judgment to propose a "safe" level.

Because of continuing debate on "how safe is safe," the report states that it is more appropriate to speak of "no-observed-adverse-effect" doses and "acceptable risk" when describing permissible exposure to chemicals. While this may prove as difficult to sell as the metric system in this country, I believe it is a step in the right direction.

Now that the complete report is published and available, the consensus is that the academy has discharged its obligation to the American public well. It has made sound recommendations for future research to provide more information on health effects, epidemiological studies, etc., where there is a paucity of reliable data to make a scientific judgment of public health risks.

The academy's assessment of our lack of adequate scientific information in many areas appears to have been accepted by the Congress. When PL 95-190 was enacted on November 16, 1977 to extend the Safe Drinking Water Act, it contained an amendment charging the NAS to make revisions to its report reflecting new information which has become available since the previous report and to submit these revisions in a report to Congress each two years thereafter.

The Rohlich report, as the first report should rightfully be called, points out that effective water treatment systems in the United States have had a major impact on the reduction of infectious diseases during this century; however, some waterborne infectious diseases still occur.

From the standpoint of gastroenteric disease control the NAS report states that chlorine is the standard disinfectant against which others are compared. While chlorine is not ideal in every respect, much more research is required before substitutes can be recommended to replace it in water treatment. Questions concerning effectiveness of disinfection, toxicity of by-products, and residual in the distribution system must be answered for proposed substitutes, as well as for chlorine.

In the proposed regulations for the control of organics, published in the *Federal Register* on February 9, 1978, the Environmental Protection Agency appears to have sought to resolve this conflict between disinfection efficacy and its objective of reducing trihalomethane production by modifying treatment techniques. Additional microbiological sampling and analytical requirements are contained in the proposed amendments to the interim primary regulations. It would indeed be a tragedy if the federal government were to make a decision that would result in exposing the population to significantly greater risks of infectious disease before adequate scientific estimation of carcinogenicity of chlorination by-products.

As a consequence of the academy's findings, it appears that the decision for setting maximum contaminant levels (MCLs) for organic solutes will have to be made on political and social bases rather than scientific judgment.

I do not challenge the responsibility of the federal government to make decisions regarding the acceptability of public health risks, since the 1974 Safe Drinking Water Act requires EPA to regulate contaminants which "may have any adverse effect on the health of persons." However, since public confidence is at stake, these decisions should rest on scientifically sound risk assessments or the law should be amended to apply such constraints.

Table VI-60 in the report presents a list of twenty-two known or suspected carcinogens found in drinking water. The report states that there is no hard evidence that low-level oral exposure to any of these chemicals produces cancer in humans but risk extrapolations based on animal studies are included in the table. I am concerned with the accuracy of these risk estimates and I feel they should be used with caution for setting maximum contaminant levels. First, assume that the estimates are too conservative. Regulations based upon such risks may force water utilities to try to implement treatment methods which are marginally effective or economically impractical. The cost of such endeavors would obviously be passed on to the consumer. Although the academy was not charged with evaluating the economic impact of averting health effects or organics in drinking water, in reading the report one must keep in mind that a small increase in health protection may not be justified by the cost.

On the other hand, suppose the risk estimates are too liberal, that the true risk is appreciably greater than the estimate. If treatment techniques were adopted which did not eliminate the health risk associated with the carcinogens, then the water consumer would be paying for supposed additional protection that, in reality, he was not receiving. Moreover, water utilities may find themselves in the position of having to comply with these regulations, making large investments which may later prove to be inadequate or worthless. I believe that for the protection of the public's health, and for the best investment of the consumer's water dollar, water quality standards must be based upon reliable estimates of health risks.

Several statements of Dr. Rohlich certainly struck responsive chords in me. One was, "Adverse health effects from trace metals that are not found in excessive concentration in delivered water supplies can be avoided most readily by preventing the discharge of such contaminants into waters in quantities that might increase concentrations to the maximum no-observed-adverse-health-effect level."

I believe this principle is equally valid when considering the control and elimination of many synthetic organic chemicals from our raw drinking water sources. More effective treatment of point-source discharges to prevent the dispersal of organic contaminants and improved containment systems to prevent accidental spills would greatly reduce the problem, but this improvement will not come overnight.

The EPA is currently proposing that public water systems serving more than 75,000 people be required to perform analysis on samples of the raw water intake for more than sixty synthetic organic chemicals in order to get a variance from a granular activated carbon treatment requirement, and 350 to 400 cities will be faced with this monitoring requirement.

This proposed monitoring program will require public water suppliers to submit the analytical results of not less than four samples taken at least 30 days apart for each raw water intake. Cost of identifying and measuring the organics present is estimated to be $1,500 to $2,000 per sample unless the EPA provides the analytical work. For some water systems not vulnerable to contamination by synthetic chemicals, and this applies to those in several regions of the country, this type of monitoring should not be necessary.

Before the enactment of PL 93-523, testimony was presented in support of regulations which might vary according to geographic and other circumstances. Such variation was allowed in the case of the "secondary drinking water regulations," the so-called aesthetic or consumer acceptance standards, but not for the primary health-related regulations. The EPA has not promulgated "secondary regulations," possibly in part because it has not resolved the question of setting levels for nonhealth-related parameters such as total dissolved solids, sulfates, and chlorides, which are usually higher in the arid West than in the eastern and southern parts of the nation.

There is some concern, therefore, about those sections of the academy's study which implicate water hardness as a "water factor" in relation to hypertension and cardiovascular disease. The suggestion is made that optimal conditioning of drinking water could reduce the annual cardiovascular disease mortality rate by as much as 15 percent in the United

States. If this hypothesis could be substantiated and a remedial treatment identified, this would be an important finding. However, I can envision a real controversy in the event that a proposal were made to add calcium to soft waters. But the furor might be even greater if the proposal were to remove sodium. Here, again, is an area where it would be most unfortunate for us to establish a public policy which locks us all into the same pattern nationwide.

Chapter 3

WATER AND EPIDEMIOLOGY

Marvin A. Schneiderman

This paper consists of three parts:

1. Some general remarks on the process of epidemiology, showing some strengths in the method, and some weaknesses;

2. A review of the several epidemiologic studies that have investigated the possible relationship between some contaminants in water and deaths from cancer (I had hoped that the Epidemiology Subcommittee of the NAS/NRC Committee on Safe Drinking Water [the Rohlich Committee] would have issued its final report on trihalomethanes and cancer by this time and that my paper would consist largely of that report. Unfortunately the study is not yet finished, so that although I draw very heavily on the committee's documents, this paper and the opinions expressed are solely my own.[1] No one else should be blamed.);

3. A conclusion based on this review, and influenced by some prominent people in the field whose advice seems to

---

[1]I am particularly indebted to the NAS/NRC Safe Drinking Water Committee professional staff: Dr. Riley Housewright, Peter Pearman, Dr. Robert Golden; to the EPA officers, Dr. Joseph Cotruvo and Dr. Shelley Williamson; and to my colleagues, Dr. Ian Higgins, Dr. George Hutchinson, and Dr. Malcolm Pike for background information, discussion, criticism, comment, and warnings.

me to have been derived from contrasting philosophies. I have contrasted their philosophies and I have argued with most of them.

Ministers usually have a text upon which they base their sermons. A single text is sometimes enough on which to develop several sermons. I'm not that efficient, so I have several texts from which I develop this one sermon. The first two texts are from an article by Christopher Jenks (1978) on education, which appeared in the Washington Post. (1) "If all ideas are equally defensible none is worth bothering with." As you will see, I do bother with some ideas, implying, of course, that I think not all are equally defensible. (2) "There seem to be so many competing interpretations of reality that it is hard to defend one to the exclusion of the others." This is a softening of the first position, and a problem all careful epidemiologists must deal with. A psychiatrist, I once read, wrote "A way of seeing is a way of not seeing." I'm not surprised that the same data lead different people to different conclusions. My conclusion, of course, is the correct one.

The third text is from Hamlet, and I think all schools of public health and all epidemiologists should give it to all their students on little engraved cards: (3) "There are more things in heaven and earth, . . . than are dreamt of in your philosophy." (I have left out a name, to avoid the cult of personality, perhaps.) At any rate, this last quote helps me understand why epidemiologists have such a hard time getting research grants. There is always someone on the study section who can think of something that was not covered in the grant application. This observation also applies to the review of the several studies that come later.

In an address to the American Association for the Advancement of Science (AAAS) Philip Handler (1978), president of the National Academy of Sciences, said, "It is perhaps informative that the examples [of industrial carcinogens] which most readily spring to mind are those . . . in which toxicity to humans was noted before studies in laboratory animals." Handler may have said this to praise the acumen of epidemiologists or to scold laboratory workers. I don't know. The statement is largely, but not literally, correct. One or two of the materials he alluded to had been studied in laboratory animals before their toxicity to humans was seen, and either the work was disregarded, or lost, or otherwise did not see the light of day until too late. Perhaps those examples did not spring to his mind.

However, he does put us on the trail. Epidemiology has been useful and powerful. What more can we learn from and do with epidemiology? And must we learn about potential risks to humans only from epidemiology?

There are several things we can learn from epidemiology and the history of the application of the epidemiologic method. We can learn the etiology of a disease without knowing "the cause." We can learn how to break the chain of Cause → Pathway → Disease, without knowing the basic biological mechanism. We have also learned that the shorter the chain, the easier it is to break. We have learned that some of these chains are linked together in peculiar ways, and the chain may sometimes remain after we have broken what seem to be obvious links. Big chains with big links are easy to see, but small chains with small links require a jeweler's eye. Sometimes the chains are so tangled, so snarled, and so tenuous that you can be relatively certain there was a chain only after you have evidence that you have broken it. That quality of permitting one to take some useful action without knowing all the mechanisms makes epidemiology one of the most powerful of the applied sciences. But it is not all-powerful, and I want to discuss here what epidemiology can and cannot do, and what it might do with limitations.

Epidemiology can most easily find acute effects (in humans)--effects that occur shortly after the causal event. Usually epidemiology can find the linkage between an effect and a presumed cause if the effect is large, if it involves a large portion of the people exposed, if not everyone is exposed, if the cause-effect relationship is simple--that is one effect, one cause; one cause, one effect. There are some logical things one sees in epidemiology. The first one is time sequence. Do something and have an effect occur quickly. Figure 3-1 illustrates an action and a response, and possible trouble, too. The data are Snow's as reported by A. Bradford Hill (1954). Snow was convinced that cholera was transmitted in some way through water (he didn't know about the cholera vibrio). So, in the course of an epidemic, he took the handle off the Broad Street pump, and the epidemic soon ended. The epidemic really did end, and he did take the handle off. Wouldn't it have been nice if he had taken it off September 1 or 2? He didn't of course. It was removed on September 8, as the epidemic was dying down.

Skeptics could easily challenge him by asking, "With this time sequence, how could the water from the Broad Street pump have anything to do with cholera?" If all Snow had to go on was the time sequence of the removal of the handle and end of the epidemic, I suspect that most people would have thought he had not shown much and would have concluded that water had little to do with the transmission of the disease. Hill, in his centenary paper, gives much of the other evidence, including some of the characteristics that Snow said the transmitting agent must possess. Some of this led later laboratory people into the research that identified the cholera vibrio. But so much for time sequences; sometimes the apparent time sequence can be misleading. Think of how much more misleading it might be if we were to wait thirty to forty years after the initiating event

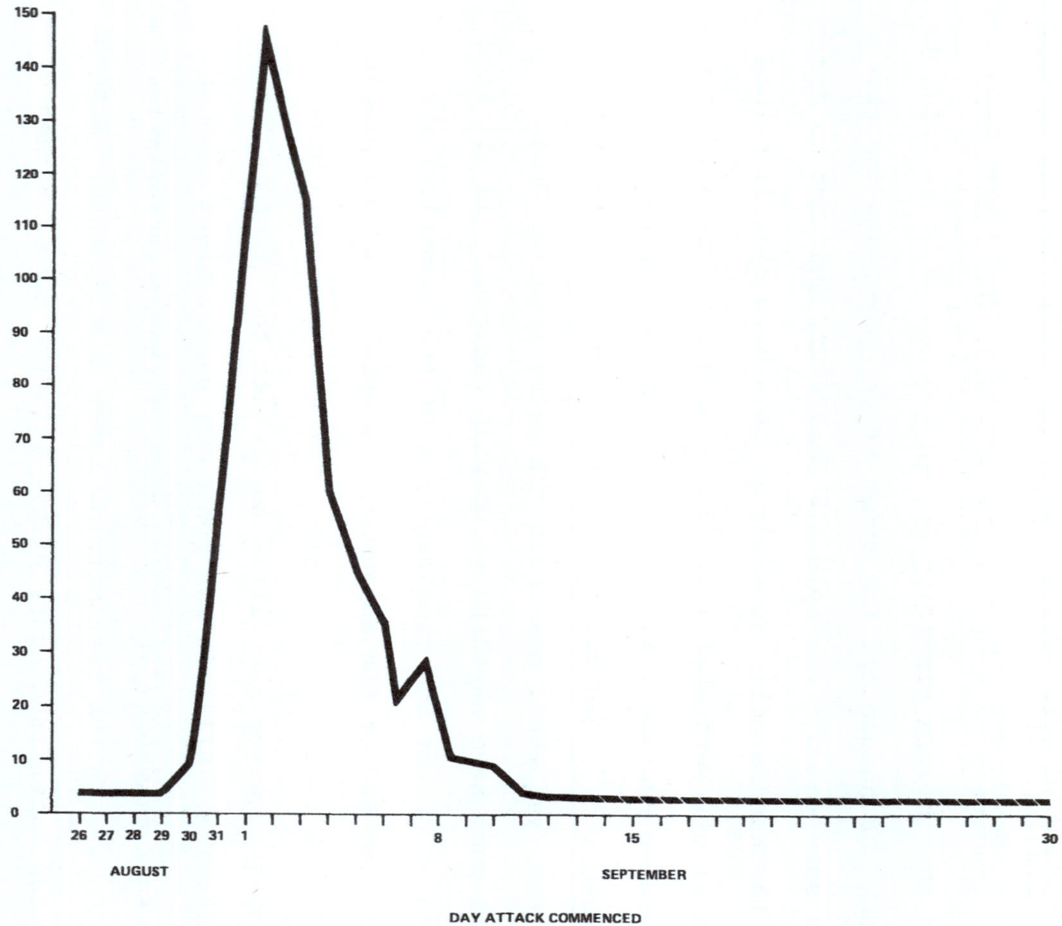

Figure 3-1. Fatal attacks of cholera, in London in 1854.
<u>Source</u>: From Hill (1955).

to see the consequences, and how many other events could intervene.

Cholera is an acute disease. What about chronic diseases? What about diseases that don't follow immediately upon their cause, or turn off immediately when the cause is turned off? Let me list the troubles, and what they mean when we look at chlorine-chloroform and other trihalomethanes and try to associate them with cancer.

1. There is a long latency period for cancer--thirty or more years perhaps. Since people have been exposed to chlorine which has combined with organic materials for a long time, perhaps we should be able to see some different levels of cancer in different places that could be associated with higher and lower levels of exposure.

2. Are we seeing the results in the exposed population? Who are the exposed population? Americans are a mobile people. Are the people who were exposed in Cincinnati thirty years ago still in Cincinnati, so that we might be able to see the consequences, if any?

Some are. Some are not. There has been much migration since World War II off the farms and into the cities and out of the central cities into the suburbs. Did these people who moved have changes in their water supplies? Of course. People who are now of a cancer age who come from the urban centers in the North-Philadelphia, New York, Pittsburgh, Chicago, Detroit, Baltimore--have retired and moved south to Florida, Arizona, or New Mexico. Water dilutes pollutants, and migration dilutes (or disperses) the effects of the pollutants, if any exist.

3. Can we find good comparison populations? Edward Lew (1976), the actuary, has cautioned us that in making comparisons one must try to compare like economic groups who have similar industrial exposures, climates, occupations, habits, and so forth. Can we find such comparison groups whose only difference was whether or not their water was chlorinated?

We have some chance of doing this, but when exposure is ubiquitous, it becomes hard to find effects that can be attributed to exposure alone. Sometimes, if we can find different levels of exposure, we may be able to find gradations in response.

4. Are there other things that could cause or modify the disease we are concerned with? Dr. Lew's admonition about comparing like with like comes out of his knowledge that there are usually confounding factors, and unless these are taken into account, we can mislead ourselves (and others) badly. Cigarette smoking is a confounding factor; it is one of several causes of bladder cancer. So is some industrial employment. Sex is a confounding factor (the fact of it, not the act of it), as is age.

Some corrections for confounding factors can be made by careful choice of comparison populations by stratification, so that like can be compared with like, and by post facto standardization, or adjustment for things about which you can get the data, like age, race, sex. How do you adjust for things for which it is hard (sometimes impossible) to get the data, like cigarette smoking, consumption of specific items, especially alcohol, use of drugs, licit as well as illicit? Did your pharmacist put the name of the drug on the label of your last prescription? Do you know what it is?

Even worse, however, is that perhaps by taking care of some of these confounding factors, by making so-called "clean" comparisons, we could be losing the most important information contained in these data--the effects of interactions. We can overadjust ourselves into triviality. Since, in most cases, cancer appears to be a disease involving several steps, or mutations, or stages, it is possible that one material may act only in an appropriate sequence, following the action of another material (Berenblum, 1964). Perhaps both need to be present for an effect to take place. Thus, if we look for increases in, let us say, bladder cancer, that we might ascribe to a suspect chemical, and look for this increase only in nonsmokers, or only in persons not industrially exposed, we could lose a possible initiator-promoter effect. There are some important implications for animal experimentation in this and I well get to these later.

5. How good are the dose data? If we believe there should be such things as a dose-response effect, can we get good dose data to place against our responses to see if they relate? Generally, no. Looking at the trihalomethanes, measures of their presence in water have been made only over the last several years. Two analytic surveys have been undertaken by EPA--the National Organics Reconnaissance Survey (NORS) and the Region V Organics Survey. These are recent surveys. We really wanted to know what was in the water thirty years or so ago. Even today we're not quite sure what specific compounds, or what concatenation of compounds we want to know about. I wonder if what is being measured today is what people will want to know about tomorrow. I think we will always be caught in the McNamara paradox, in which we will measure what we can measure, whether that's really what we want or not. How stable are these measurements? Do they change with season? With water flow? With the presence of other contaminants in the water?

6. Can we trust the animal models to lead us to the right epidemiology? In addition to the general indication that a material is a carcinogen for a given species, do the animal experiments tell us that it is a carcinogen for other species or for the same sites in that other species? (That other species being humans, of course.) Sometimes; but we must always remember that there are species differences which we have yet to explain. It is safe to presume

that a material which is a carcinogen in an animal is likely, or possibly, a carcinogen for man. It is not safe to presume that the sites of action will necessarily be the same. The one-to-one site specificity often does not exist between rodent species, and there is no reason to assume that it exists between some rodents and man. A concern about differences in metabolism and site specificity makes some people reluctant to extrapolate animal carcinogenicity data to humans (in any way other than by saying yes or no). If one cannot extrapolate, then one cannot estimate risks. If one can't estimate risks, then one can't do risk-benefit analysis. If one can't do risk-benefit analysis, then one has to use some all-or-nothing decision rule, like the Delaney clause.

7. How small an effect can an epidemiologic study discover? Are there limits to the sensitivity of the method? How small an effect you can find depends in large part on how big a study you can do. It also depends on how clean the study is, how many confounding factors are around to mess things up. Everything that dirties up the relationship, everything that permits unwanted or unknown or unmeasured or unmeasurable things to enter into the studies adds "noise." The more noise, the harder it is to hear the signal or to distinguish it from the noise. Most epidemiologic studies have a lot of noise. As a rough rule of thumb, most epidemiologists working on cancer problems have strong self-doubts unless they find a twofold or greater risk.

So far, I've talked about some of the troubles of epidemiologic studies but only some of them, because I really don't know all of them. Epidemiologists are often tempted to sing the old spiritual, or at least one line from it, "Nobody knows the trouble I've seen. . ."

On the other hand, epidemiologic studies have enormous power in doing things that the laboratory usually cannot come near doing. They deal with the complexities of real life, not abstractions from real life. They deal with the interactions as they are. Life may be a tale told by an idiot, full of sound and fury, as Mr. Shakespeare suggests, but the epidemiologists would probably disagree with W.S.'s conclusion that the tale is "signifying nothing." To make the studies signify something is usually not an easy job, but it can be done. Nature contributes enormously by doing the experiments for us without a grant from NIH. (After all, who could be on Nature's peer review committee?) She often conducts large experiments on a usually intractable animal--man--but she can usually do her experiments without paying for animal husbandry, housing, or care and feeding. There is the added problem that Nature preceded Sir Ronald Fisher and never read his "Design of Experiments," and so her experiments are not as neat as we would like. I guess awareness of this once led a Cutter lecturer (on public health at Harvard), Dr. H. M. Sinclair, to remark, "The use of the experimental method has brilliant discoveries to its credit, whereas the method of observation has achieved little, . . ." Subsequent events have shown that Professor Sinclair underestimated the power of observation which, after all, is the essence of epidemiology. With that, let us turn to the specific observational studies about chlorine products in water and cancer.

Figure 3-2 is a map of total cancer mortality by county among white males in the United States for the years 1950-69. I'm sorry I can't bring you a later map, but my colleague Tom Mason, who has done such a lovely job of putting these together, has been stymied by a little economy measure. Half the death certificates of 1972 were not coded in order to save money. We've not gotten the money to do this (about a half million dollars) and the National Center for Health Statistics (NCHA) will have them coded before too long, and we'll be able to make more up-to-date maps. In addition, NCHS will publish similar maps for many more causes of death, which should add to our knowledge and enable us to look, perhaps, at competing causes of death.

This map upset some people. Among the people upset were Robert Harris and his associates Talbot Page and Samuel Epstein (1976). They put together mortality data from Louisiana contrasting rates in areas they had reason to believe got their drinking water from the Mississippi with mortality rates in areas that got their drinking water from other sources. About this time the National Cancer Institute

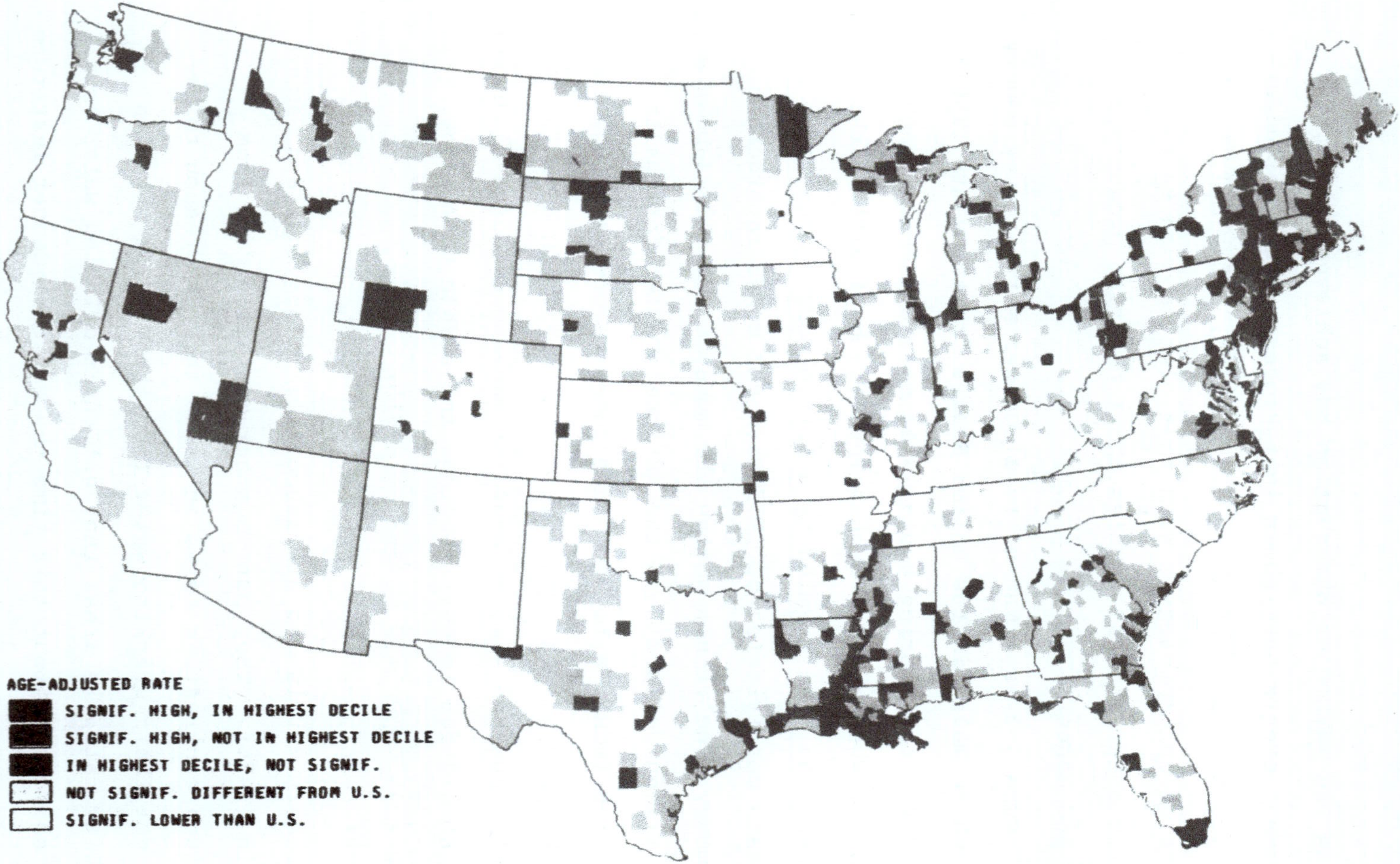

Figure 3-2. Cancer mortality for white males by county, 1950-69, all sites combined.

completed some studies on the carcinogenicity of chloroform in rats and mice, demonstrating an increase in liver and kidney tumors at all doses tested. In keeping with appropriate small-scale animal testing, all these doses were high.

Concurrently, the EPA had been monitoring a large number of chemicals in water supplies in the United States, among them chloroform and other trihalomethanes. Given these several pieces of information, the EPA asked about a dozen different groups of research workers to look into whether there was a demonstrable relation between trihalomethanes and cancer. These workers responded very promptly, and it is their findings, and one or two others, that I discuss here. I have divided the studies into two groups:

1. Studies in which the water contamination-pollution measure is presumptive but not literally measured. Presumptive measures include comparing surface water with groundwater, or known chlorinated supplies with known nonchlorinated supplies. Surface water is more likely to be chlorinated.

2. Studies in which water quality could be given in terms of measured trihalomethane levels.

Specific Studies

Harris and his associates put together three papers (Harris, Page, and Reiches, 1977) based on the published cancer mortality data by county in the United States (Mason and coauthors, 1975). The proportions of drinking water obtained from the Mississippi River and supplied to each parish (county) in Louisiana were related, in a regression analysis, to cancer mortality, controlling for urban-rural status, income, and employment in possibly hazardous industries. Separate analyses were made by site, sex, and race. The regression of mortality on the proportion of water from the river was computed. Significant correlations, all positive, were found for three of four race-sex groups for total cancer in all sites, and for two of the four sex-race groups for genitourinary cancer, and in all four groups for gastrointestinal cancer. Positive, but not statistically significant, associations were shown for prostate cancer and breast cancer.

DeRouen and Diem (1975a, 1977) from Tulane challenged these results, calling attention to a major difference in cancer mortality rates between northern and southern Louisiana (also noted by Buncher, 1975) that could perhaps have arisen independently of the putative cancer-water association. After controlling for this north-south factor, their analysis still showed some cancer excesses associated with the use of river water. DeRouen and Diem, in comparing cancer mortality in southern Louisiana among those receiving river water and those who did not, found positive associations between higher mortality and supply of river water for nine of sixteen sites in white females, with one of these significant at the 5 percent level. For the other three sex-race categories, six statistically significant associations ($p \leq .05$)

were found in the remaining forty sex-race-site categories (DeRouen and Diem 1975b). Positive associations in more than one race-sex group were found for cancer of the bladder, breast, rectum, and large intestine, as well as total cancer at all sites. Replicated negative associations (that is, lower mortality in populations using river water) were reported for cancer of the liver and corpus uteri. Cancer of the lung and cervix uteri was found to be significantly positively associated with water source in one population and significantly negatively associated in another. In Table 6 of their Cold Spring Harbor paper (DeRouen and Diem, 1977), they show that the water variable is positive for all race-sex groups for all sites, for gastrointestinal cancer and for urinary cancer--and six of these twelve comparisons are statistically significant.

A similar approach was used by investigators in the Ohio River area (Buncher, 1975; and Kuzma, Kuzma, and Buncher, 1977) in two papers, the first involving the four-state area--Ohio, West Virginia, Kentucky, and Indiana--and the second confined to Ohio. The latter study makes more extensive comparisons than the former one, including all Ohio counties, classified according to whether surface water or groundwater was the predominant source for the population. Excess cancer mortality in counties using surface water was noted in thirteen of sixteen site-sex categories studied. Four of the thirteen were statistically significant at the 5 percent level, and no comparison showed a significant excess mortality for groundwater sources. Sites with significant excess mortality associated with surface water were the stomach, in both males and females; the bladder in males; and the total of all sites in males. Three additional site-sex groups were positively associated with surface water use: the total of all sites in females; and the liver and breasts in females, but none of these were statistically significant at the .05 level. Data were controlled for age, population of county, percentage of urbanization, median income, and an index of manufacturing activity. The data were not adjusted for ethnic groups, a variable that has been shown to be related to gastrointestinal cancer. Kuzma, Kuzma, and Buncher (1977) say, in part, "This study confirms the hypothesis that cancer mortality rates [in Ohio counties] are higher among [those] whose drinking water comes from surface water . . . than among [those] whose drinking water comes from groundwater. . . . While the study . . . by no means establishes a causative factor, it suggests that . . . substances in the water . . . may relate to increased cancer mortality in communities served by surface water supplies."

Harris, Page, and Reiches (1977) enlarged the Ohio study and reported not only on the same cancers but also on those at different anatomic sites. In general, they confirmed the findings of Kuzma, and coauthors (1977). They found statistically significant associations in five of the seven sites reported previously. They failed to confirm the positive association for liver and breast cancers, and they reported positive associations for esophagus in males (not studied before); pancreas in males; gastrointestinal tract in males (not studied before); and total genitourinary system (not studied before). Harris and his associates did not adjust for ethnic group.

Alavanja, Goldstein, and Susser (1976) carried out at least two case-control studies of persons dying with gastrointestinal or genitourinary cancer in several counties in upstate New York. Their first study (1976) was confined to women in an attempt to decrease the potentially confounding factor of occupational exposures, and the cancer sites were selected on the basis of indications from prior reports, so that this may be thought of as an hypothesis-testing study. Cancer cases were matched one to one with women dying of causes other than cancer. Matching variables were county, age, race, and nativity. Matched-pair analysis procedures, however, were not used, and this is likely to have led to an underestimate of effects. Because most of the subjects in this study came from two large industrial counties, industrial carcinogens may have confounded results, although there is no known evidence of association of industrial carcinogens with water quality within these counties. Water supplies were determined from addresses and were characterized as chlorinated or non-chlorinated, and as surface water or groundwater. A doubling of risk was found to be associated with chlorinated supplies in urban areas. A 50 percent increase in risk was found to be associated with chlorinated groundwater compared with nonchlorinated water in rural areas. No increase (risk ratio 1.01) was seen in rural areas supplied with chlorinated surface water compared with nonchlorinated water. Dr. Alavanja (1978) later extended his work to men.[2]

Krusé (1977) studied cancer of the kidney and liver, and bile ducts in a population in Washington County (Hagerstown), Maryland, that had previously been extensively investigated as an epidemiologic community model. Information as to source of home drinking water had been obtained in an earlier survey. This information was then used in a total population study designed to correlate water source (characterized as chlorinated or nonchlorinated) with the presence or absence of cancers at the specific sites. The approach involved identifying the effects of water source in a multivariate analysis including eight other potentially confounding demographic and sociological variables. An increased incidence of liver cancer with chlorinated water supply, with a risk ratio of 1.5, was found, but this was not statistically significant. Cancer of the kidney showed a

---

[2]The study on males lends support to the conclusions from the study on females, and the two studies together should now be considered as one unified piece of work.

nonsignificant decrease with chlorination, with a risk ratio of 0.96. The number of cancer cases in this study was small (total ninety-one). I find it unfortunate that Professor Krusé confined himself to the two "animal" sites, for other people have found increases in other sites.

A study in Los Angeles County (Mah, Spivey, and Sloss, 1977) correlated cancer mortality rates and incidence rates with a qualitative description of the chloroform content of drinking water. The county was divided into nine subareas, categorized as low, medium, or high with respect to possible chloroform content. Eight cancer sites were studied: esophagus, stomach, colon, rectum, liver, lung, kidney, and bladder. Limited analysis suggested no association of cancer rates with chloroform content of water, and the investigators believed no more detailed analysis was justified in view of (1) the lack of suggestion of relationship, and (2) certain limitations in the data. These limitations included the absence of precise data on chloroform content, lack of knowledge of what sort of water people did drink (there is extensive use of bottled water and water from other nonlocal sources), large immigration into the Los Angeles area in recent decades, possible variations in diagnostic accuracy within the study area, and small numbers of cases in specific sites and local areas within the county.

A further study (Salg, 1976), involving areas served by water from the Ohio River, included all counties within the boundaries of the Ohio River Valley Basin, as determined from water drainage maps. These counties lie within seven states--Illinois, Indiana, Kentucky, Ohio, Pennsylvania, Tennessee, and West Virginia. In an attempt to categorize possible intake of pollutants in the water supply, the exposure variables chosen were (1) percent of county population served by surface water, and (2) percent served by prechlorinated water. The outcome variable was cancer mortality, by county, for 346 counties, specific for nineteen cancer sites. The relationships among these main variables were investigated in a multivariate analysis using, as possible confounding factors, nine variables representing demographic and socioeconomic characteristics by county.

Positive associations (at a 20 percent significance level or better) were found for thirteen of the nineteen site categories in one or more of four sex-race population groups. Negative associations were found for seven site categories in one or more populations. For five categories (esophagus, rectum, breast, Hodgkin's disease, and larynx) positive associations were found in two or more population groups.

The investigators have interpreted their findings to indicate the need for further study of carcinoma of the large intestine, rectum, and bladder. Large intestine cancers were found to be significantly increased in white males.

Bladder cancer was significantly elevated in white males and significantly decreased in nonwhite females.

A study was carried out in census tracts in the Pittsburgh region (Carlson and Andelman, 1977) associating site-, race-, and sex-specific, age-adjusted cancer incidence rates with drinking water, characterized by (1) source of raw water--six categories defined by surface versus groundwater, and by river, Allegheny, Monongahela, or Ohio--(2) forty-two water treatment plants; and (3) a joint source-treatment classification using fourteen categories.

This paper is principally an investigation of statistical methodology for studying data of the complexity involved in all the water quality-cancer studies that various investigators have carried out. Extensive computer tabulations were used to assess the importance of various analytical procedures. The general conclusion was that clearly significant associations are found between water quality and cancer incidence. However, the data were not presented in a way that permits identification of specific cancers as related to an ordering of the intensity of water pollution.

A study of counties in New Jersey (Vasilenko and Magno, 1975) was carried out to determine the relation between water source and the age-adjusted cancer mortality of white males. Water quality was estimated from the ratio of number of households served by public systems and private companies to the number served by individual wells. The presumption was that public sources were more likely to be chlorinated than individual wells. Confounding variables included in the analysis were age (proportion of county population over age sixty-five), race, urban-rural characteristic, poverty, income, education, migration, occupation, and industrialization.

In a multiple regression analysis, the water variable contributed significantly and positively (that is, a high proportion of water from public water systems and private companies was associated with high cancer mortality) to mortality for total cancer, respiratory cancer, and stomach cancer. A (nonsignificant) negative association was found for bladder cancer.

Summary

These ten studies, involving a variety of indirect indicators of water quality, showed a number of statistically significant associations between water quality and measures of cancer frequency, incidence, or mortality. One of the ten studies (Mah, Spivey, and Sloss, 1977), which failed to identify any positive associations, appeared to have limitations greater than those of any of the other studies, most particularly regarding problems of great population movement.

The studies involved several geographically distinct areas in the United States, namely, the state of Louisiana (two studies); the Ohio River Valley Basin (three studies); Washington County (Hagerstown), Maryland; the city of Pittsburgh; the state of New Jersey; upstate New York; and the county of Los Angeles.

Numerous specific cancer sites were identified in positive associations in one or another study, but no single site stood out as being consistently positively associated with water quality. Sites identified as possibly associated with water contaminants in two or more of the broad geographic areas included bladder, stomach, large intestine, rectum, all gastrointestinal sites, and lung.

Known confounding factors of demographic variables, socioeconomic characteristics, and other carcinogenic exposures were taken into account in several of the studies. More direct measurements of water quality were used for the three papers discussed in the following section.

Studies Based on Halomethane Concentrations

Cantor and coauthors (1977) (Cantor is from EPA and is working at NCI) studied age-standardized cancer mortality for 1968-71 of white men and women in all 923 U.S. counties categorized as "urban," on the basis that more than half the county population in 1970 lived in urban areas.

Estimates of county trihalomethane concentrations were made from data obtained in the two surveys conducted by the Environmental Protection Agency (EPA, 1975, 1978). The

National Organics Reconnaissance Survey (NORS) sampled finished water in eighty water treatment plants across the country. A survey conducted by EPA's Region V office covered eighty-three plants in the states of Illinois, Indiana, Michigan, Minnesota, Ohio, and Wisconsin.

Allowance was made in the analysis for nine U.S. regions, median school years completed by inhabitants of county over age twenty-five, foreign stock composition of county, county size, ratio of 1970 to 1950 county population, percent of county that is urban, and percent of the county work force engaged in manufacturing industries. First, these variables were fit by a multiple regression to the mortality rates for each site of cancer with sex-specific mortality rates greater than 1.5 per 100,000 per year. Then, residual mortality rates were correlated with measured trihalomethane levels (logarithms of the concentrations) for all seventy-six counties in which 50 percent or more of the population was served by the sampled water. Individual points in the calculation of the correlation coefficients were weighted by the square root of the sex-specific person-years at risk in the population served by the sampled water supply, estimated by the product of the percent of population served and the population at risk.

Questions have been raised about some of the statistical techniques used in this analysis. My colleague Malcolm Pike believes strongly that residual mortality rates should have been correlated with the <u>residual</u> of measured trihalomethane levels after these levels were fitted to the same explanatory variables as those used to explain the mortality rates (Draper and Smith, 1966). Second, there is reason to believe that the points in the regression fit should have been weighted by sex-specific person-years at risk, not by the square root of this quantity. In both these instances the procedures actually used should tend to decrease statistical significance (that is, increase P values).[3] Third, it also seems reasonable that the regression analysis for estimating sex- and site-specific cancer rates should have been carried out using only the seventy-six counties mentioned above as data points.

Among males, positive correlations (the probability that a correlation this large was due to chance alone was <.2, that is P <.2) were found between nonchloroform trihalomethane (NCTHM) concentration and bladder and brain cancer, and between chloroform ($CHCl_3$) concentration and prostate cancer. A weak negative correlation between NCTHM concentration and pancreas cancer in males was also noted. When increasing the percent of county population served by water supply sampled is considered (in order to reduce possible exposure misclassifications), the correlations with bladder and brain cancer

---

[3]Dr. Cantor has since done a reanalysis of these data, using the weighting system carried out here. He writes "...reanalyses show a significant association between bladder cancer in males and females (separate analyses) and levels of the THM indicator ... reanalysis has ... not changed the direction of the results nor [the] conclusions."

increase. Among women, positive correlations were found between total trihalomethanes (TTHM) and lung cancer, and between NCTHM and bladder cancer. The lung cancer correlation coefficients did not show an increase with dose, but the correlations with bladder cancer did, and reached conventional significance ($P=0.02$) in the areas in which 85 to 100 percent of the population were served by the measured water source. A negative correlation between TTHM and stomach cancer was seen.

The authors also computed separate correlation coefficients for each of three large regions (North, South, and Mountain Pacific) for the fifty-one counties in which 65 percent or more of the population received measured water. Positive correlations between NCTHM and bladder cancer in all regions combined were found for males ($r=0.30$, $P=0.03$) and for each region separately, as well as combined for females (combined, $r=0.33$; $P=0.02$). For men in the North, a highly significant correlation was observed between NCTHM and bladder cancer ($r=0.52$, $P=0.002$).

Hogan and coauthors (1977) carried out a similar study in which they related a different set of cancer mortality statistics to chloroform levels in finished water as determined by the same surveys. These investigators used as independent variables the National Cancer Institute's twenty-year age-adjusted county cancer mortality rates for white men and women (Mason and coauthors, 1975) and county chloroform concentrations in drinking water estimated from the EPA's NORS and Region V surveys (EPA, 1975). Regression equations were fitted to the mortality rates, with chloroform concentrations, size of county, county population density, percent urban population in county, percent nonwhite in county, percent foreign born in county, median number of school years completed by county population over age twenty-five, median family income of county, and percent of work force of county engaged in manufacturing. The data were weighted by the method recommended by Cantor and coauthors (1977).

There were differences in the results based on NORS data in contrast to the results based on Region V data, which raise questions about the adequacy of the water pollution measures. The data are consistent with an increase in cancer rates of the large intestine in both men and women with increased chloroform concentration.

From the Cantor and Hogan studies, the tentative conclusion emerges that bladder cancer appears to be correlated with something in water. It would appear that this "something" is NCTHM (Cantor and coauthors, 1977) and $CHCl_3$ (Hogan and coauthors, 1977). Cantor has confirmed (in a personal communication) that $CHCl_3$ was correlated with bladder cancer mortality in males. This was not reported in his paper because where there was more than one positive correlation, Cantor and coauthors reported only the strongest.

McCabe (1975) showed a positive correlation of age-adjusted total cancer mortality rates with estimated chloroform concentrations in eighty cities. Since no allowance appears to have been made for any of the confounding factors, it seems inappropriate to draw conclusions from this study in order to compare them with the Cantor and Hogan studies.

Even these quantitative studies are limited by the absence of data on exposure in past time periods, and, of course, earlier exposure is directly relevant to cancer that has already been diagnosed. Similarly, all studies are deficient in identifying populations that were stable in the areas where present water quality is studied in these reports.

So, there we are--a mixture of results--none terribly strong; some even negative--going in the opposite direction. What does all this tell us?

Based largely on the quantitative studies (Cantor and Hogan), I come to the conclusion that higher concentrations of trihalomethanes in drinking water are most likely to be associated with cancer at at least two sites--bladder and large intestine. These don't correspond to the "animal sites." The negative relationship reported for stomach cancer in women bothers me, but perhaps more than it should, since removing one county from the data for each of two dose groups cuts the negative correlation coefficient in half. Is it because stomach cancer is most rapidly decreasing in women in the United States that these studies have given us the peculiar result? Or, have the epidemiologic studies really discovered _why_ stomach cancer is decreasing so rapidly in the United States? If the laboratory workers had a good animal model for stomach cancer, we might be able to test this. Perhaps it might be worthwhile looking to see where stomach cancer has decreased most rapidly (rather than examining just absolute levels) and if this is correlated with trihalomethane levels. Or we could look at migrant populations--populations who migrated from high stomach cancer areas to low stomach cancer areas (Japanese, Finns, perhaps rural Blacks in the United States, migrating to cities) to see if the changes in mortality are correlated with changes in water supply.

For bladder cancer, at least, the effects of certain confounding factors, such as cigarette smoking, have not been worked out. Perhaps the large National Cancer Institute-Food and Drug Administration (NCI-FDA) bladder cancer study will help sort these things out.

One of these studies--Hogan's--allows us to give estimates of what changes in the chloroform levels in water might do to cancer rates. Crudely, his data indicate that a decrease of 100 μg/L of chloroform could lead to a decrease in cancer mortality rates as follows:

Location of cancer	Men (%)	Women (%)
Bladder	1.3 - 7.5	3.3 - 10
Large intestine	4 - 8.5	3 - 7.5

These decreases would probably be too small to be seen in the usual epidemiologic study, given the possible confounding effects of cigarette smoking, industrial exposure, and dietary changes.

Given these data plus the animal data, what advice could we give EPA, or other scientists, or other regulators?

Conclusion

The administrator of EPA, Douglas Costle (1977), said in an interview, "I am concerned . . . about the increasing number of man-made chemicals that we're finding in drinking water. Some of them, like chloroform, for example . . . have been shown to cause cancer in animal tests. We need to better control the introduction of . . . organics into our drinking water."

Edward Lew (1976), the famous actuary whom I quoted earlier, said, "Programs for controlling contamination of the environment come into conflict with the desire of most Americans to maintain high standards of living. A realistic balance must be struck between a minimal degree of biological damage, that is an acceptable degree of pollution, and an acceptable reduction in living standards." Dr. Lew's remarks bother me, because it is not obvious to me that reducing pollution must be done at the expense of living standards. In fact, I think of less pollution as an indicator of higher living standards--even if economists do not.

Philip Handler (1978), in the speech to the AAAS, said, "For only a handful of compounds is there a convincing consistent body of laboratory and epidemiologic data." Talking about the work going on, he added, "I wish that those engaged in such studies would behave as do scientists generally and refrain from publication until they have completed a sufficient series of coherent studies to enable rational discussion, rather than announcing each experiment in turn, generating public alarm that can neither be justified nor assuaged." Now that's political, rather than scientific advice, and I think it is appropriate for Dr. Handler, in his official position, to give political advice.

I don't know whether Professor Handler would scold us here today for being premature in our review of these thirteen studies or perhaps for generating unjustified alarm. I do wonder how we would get each other to do the things we need to do, if we don't call to each other's attention what we have done. Joyce McCann doesn't do epidemiology, but some of her results place an imperative on some of us who do to get on with the job. How do the findings of the epidemiologists, mostly about interactions such as smoking and radon exposure, tobacco and alcohol, smoking and asbestos, urban air pollution and smoking, and lung cancer, get back to the laboratory so that the laboratory workers can start designing more meaningful screens of combinations of exposures? When have enough experiments been done so they can be talked about rationally? Should on-going experiments influence what new experiments are done? How should experiments influence epidemiology? Should this be a two-way street?

In this whole area I think I prefer Michael Shimkin's (1977) political advice to Handler's. Shimkin said, "Public policy often demands decision and action before scientific information is complete." Gus Speath (1978) of CEQ comes down on the Shimkin side:

> Of all the chemicals that have been released into the human environment, we _know_ that a small number cause cancer in humans. We know this for the best and most tragic of reasons: they _have_ caused cancer in people . . . But we cannot _wait_ patiently for other carcinogens to make themselves known in this manner. Having learned all we can from our technicians we are forced to realize that . . . social value judgments frequently exceed the decision making prerogatives of any profession or discipline. It is both the glory and the burden of democracy that lay citizens must make the final choice.

In another section of his speech, Handler calls for quantification of risks. "The human consequences can appear only after so long that, if they are to be forestalled, they must be predicted from theoretical considerations or laboratory observations. . . ."

Shimkin says, "I am also willing to listen, but remain unimpressed with mathematical games on rodents that extrapolate to one-billionth of the dose as a measure of safety for man." I do not know if that is disagreement (I think it is, but who knows with colorful writers who are scientists-politicians), but I think I'm closer to Handler on this one. I also think it shows Handler to be not quite so rigid as Dan Greenberg makes him out to be.

Greenberg (1978), in The Washington Post, scolded Handler for his speech saying, "Handler's argument proceeds from the premise that we should know everything before we do anything." I don't know if Handler proceeds from that premise or what premise he proceeds from. I don't even know if he is to be chided, but if he were, I would do it for his not chiding the laboratory people about their slowness in developing experiments that reflect the real and complicated world that the epidemiologists have uncovered. No matter how long Dr. Handler waits, more laboratory experiments of the kind we are now doing will not enable us to discuss this issue more rationally.

Perhaps Dr. Lijinsky, who is the lead speaker at the next session of this conference, will tell me that the laboratory people have already learned how to do more realistic experiments. I hope he will.

At any rate, I find some evidence that trihalomethanes, perhaps chloroform, perhaps some of the others, are likely to adversely affect the health of people who drink water with that stuff in it. And derived from a basic prejudice that clean water is likely to be better than dirty water, I would recommend (as a scientist, politician, and drinker of water) that EPA get on with the process of improving our standard of living and our chances of living, by improving our standards of water.

References

Alavanja, M. 1978. "G.I. and U.T. Cancer Mortality and Drinking Water Chlorination." Unpublished.

Alavanja, M.; I. Goldstein; and M. Susser. 1976. "Report of Case Control Study of Cancer Deaths in Four Selected New York Counties in Relation to Drinking Water Chlorination." Report of EPA Contract 76-224.

Berenblum, I. 1964. "The Two-Stage Mechanism of Carcinogenesis as an Analytical Tool," in P. Emmelott and O. Muhlbock, eds., Cellular Control Mechanism and Cancer (Amsterdam, Elsevier pp. 259-267).

Buncher, C. R. 1975. Cincinnati Drinking Water--An Epidemiologic Study of Cancer Rates (Cincinnati, Ohio, Division of Epidemiology and Biostatistics, University of Cincinnati Medical Center).

Cantor, K. P.; R. Hoover; T. J. Mason; and L. J. McCabe. 1977. "Association of Halomethanes in Drinking Water with Cancer Mortality" (Bethesda, Md., Environmental Epidemiology Branch, National Cancer Institutes).

Carlson, W. S., and J. B. Andelman. 1977. Environmental Influences on Cancer Morbidity in the Pittsburgh Region, Final Report prepared for EPA Health Effect Research Laboratory, Cincinnati, Ohio (Pittsburgh, Pa., Environmental Health Section, University of Pittsburgh).

Costle, D. M. 1977. Interview with U.S. News and World Report, December 19, pp. 31-32.

DeRouen, T. A., and J. E. Diem. 1975a. "The New Orleans Drinking Water Controversy. A Statistical Perspective," American Journal of Public Health vol. 65, pp. 1060-1062.

______. 1975b. Unpublished manuscript.

______. 1977. "Relationships Between Cancer Mortality in Louisiana Drinking-Water Source and Other Possible Causative Agents," in H. H. Hiatt, J. D. Watson, and J. A. Winsten, eds., Origins of Human Cancer. Book A, Incidence of Cancer in Humans (Cold Spring Harbor, N.Y., Cold Spring Harbor Laboratory).

Draper, N. R., and H. Smith. 1966. Applied Regression Analysis (New York, Wiley).

Environmental Protection Agency. 1975. Preliminary Assessment of Suspected Carcinogens in Drinking Water: An Interim Report to Congress (EPA, June).

______. 1978. The National Organic Reconnaissance Survey (EPA, February).

Greenberg, D. 1978. Washington Post, February 27.

Handler, Philip. 1978. "Pangs of Science." Lecture given at Annual Meeting, American Association for the Advancement of Science, Washington, D.C.

Harris, R. H.; T. Page; and N. A. Reiches. 1976. Drinking Water Contamination and its Relations to Human Cancers (Abstract of papers presented at the meeting on Origins of Human Cancer, Cold Spring Harbor, N.Y., sponsored by the Environmental Defense Fund and Resources for the Future, both of Washington, D.C., and the Department of Preventive Medicine, Ohio State University, Columbus) p. 86.

Hill, A. B. 1955. "Snow, An Appreciation," Proceedings of the Royal Society of Medicine vol. 48, pp. 1008-1016.

Hogan, M. D.; P. Y. Chi; T. J. Mitchell; and D. G. Hoel. 1977. "Association Between Chloroform Levels in Finished Drinking Water Supplies and Various Site-Specific Cancer Mortality Rates." Unpublished.

Jenks, C. 1978. Washington Post, February 19.

Krusé, C. W. 1977. "Chlorination of Public Water Supplies and Cancer." Preliminary Report for Washington County, Maryland. Submitted to EPA.

Kuzma, R. J.; C. M. Kuzma; and C. R. Buncher. 1977. Ohio Drinking Water Source and Cancer Rates (Cincinnati, Ohio, Department of Environmental Health, University of Cincinnati).

Lew, Edward. 1976. Paper in the Seventh Water Quality Symposium, p. 176.

Mah, R. A.; G. H. Spivey; and E. Sloss. 1977. "Cancer and Chlorinated Drinking Water" (Los Angeles, University of California).

Mason, T. J.; F. W. McKay; R. Hoover; W. J. Blot; and J. F. Fraumeni, Jr. 1975. Atlas of Cancer Mortality for U.S. Counties: 1950-1969, U.S. Department of Health, Education and Welfare Pub. No. (NIH) 75-780 (Bethesda, Md., National Institutes of Health).

McCabe, L. J. 1975. "Association Between Halogenated Methanes in Drinking Water and Mortality (NORS Data)" (Washington, D.C., Water Quality Division, EPA).

Page, T.; R. H. Harris; and S. S. Epstein. 1976. "Drinking Water and Cancer Mortality in Louisiana, " Science vol. 193, p. 55.

Salg, J. 1976. Cancer Mortality Rates and Drinking Water in 346 Counties of the Ohio River Valley Basin. Final Report #PO-5-03-4528-J. Submitted to the EPA.

Shimkin, M. 1977. "Ethics for Oncologists," Medical and Pediatric Oncology vol. 3, pp. 121-122.

Speth, G. 1978. "Waiting for a Philosopher King," Sierra (February-March).

Vasilenko, P., and L. Magno. 1975. "Factors Relating to the Incidence of Cancer Mortality of New Jersey." Internal Princeton University Report. Public Affairs 510.

Discussion by Michael Alavanja

To an audience that may not be familiar with the history of epidemiology, Dr. Schneiderman's opening remarks may seem to discredit the epidemiological method. Dr. Schneiderman seems to imply that Dr. John Snow's fame, stemming from his work on the epidemiology of cholera, was ill founded because his attempt at controlling the cholera outbreak, by the removal of the handle from the Broad Street pump, occurred after the epidemic had already subsided. However, by concentrating on this one issue, the audience may miss the crucial historic lesson: That is, Snow was able to conclusively establish the mode of transmission of cholera and predict that the agent of this disease was alive, heat labile, and microscopic in size some twenty years prior to the formulation of the germ theory of disease.

Snow's work is not the exception to the historic rule. In the early twentieth century, Dr. Joseph Goldberg of the U.S. Public Health Service was able to demonstrate during an epidemiological investigation of pellagra that it was a deficiency disease. Goldberg made his findings during a period when the medical profession and laboratory scientists were convinced that the cause of pellagra was an infectious microorganism that simply defied attempts at isolation. Moreover, Goldberg's findings were made twenty-two years prior to the discovery that nicotinic acid was the pellagra preventative factor in animal studies.

Lack of time prevents me from describing dozens of additional historic examples from the nineteenth and twentieth centuries that would document this point. Epidemiological investigations have frequently pointed to the cause or causes of a disease prior to the formulation of an exact biochemical or pathological mechanism in animal experiments.

Focusing on the topic of drinking water and health, I believe that it would be dangerous for us to use animal studies exclusively for the assessment of risk, for two reasons.

First, studies by Sontheimer and Kuhn indicate that the trihalomethanes represent only a small portion of the total halogenated products of water chlorination. The bulk of the organic matter in water is not amenable to detection by methods commonly available today. Moreover, by using animal studies alone it is impossible to assess the interaction of all these chemicals. In the National Academy of Sciences study cited here today the risk estimates are made on only one by-product of chlorination, namely, chloroform. These estimates do not consider the presence of other substances which might be present prior to or introduced concurrently with chloroform that would contribute to a higher level of risk.

Second, extrapolating human risk for a single compound from animal studies has sometimes proved to be grossly erroneous. Arsenic and benzene, two known human carcinogens, have not been convincingly demonstrated as carcinogenic agents in animal studies. Also, humans have been shown to be sixty times more sensitive than mice and seven hundred times more sensitive than hamsters to the teratogenic effects of thalidomide.

To date, the epidemiological evidence relating trihalomethane concentrations, or other drinking water quality factors, to cancer mortality is not conclusive. However, given the diversity of the study populations, measurements of water quality, and statistical methods used in the nine completed investigations, the similarity of results is noteworthy and suggests a health risk larger than that which can be extrapolated from the animal-chloroform data.

What can be done to resolve this inconsistency? It is imperative to undertake a second phase of more sophisticated analytical epidemiological studies. Several of these studies are now being conducted by Dr. Gottlieb at Tulane University, Dr. Brenneman at the University of Illinois at Chicago, Dr. Bahn at the University of Pennsylvania, and by my study group at the City University of New York. With the completion of these studies, epidemiologists will be able to provide reliable estimates of the health risk of the by-products of chlorination in drinking water.

Discussion by Nancy A. Reiches

Like many other sciences, epidemiology is largely comprised of detective work--finding suspects (in this case, chemicals in drinking water) and somehow relating them to the crime (in this case, cancer). Of course, along the way, there appear to be various leads or clues, various suspicions (or hypotheses), and occasionally (when we are very lucky), some sound conclusions. But in reality, many of the leads are false, and many of the conclusions tenuous--since in epidemiology, the suspect is rarely capable of confessing, and much of the evidence is circumstantial.

Despite the risks inherent in this sort of detective work, it is still a reasonable business. Schneiderman has alluded to many of these risks, but I sense that he and I share the impression that epidemiology can be an exceptionally powerful tool, especially in the context of public decision making.

In these comments, I am first going to contribute a bit to the ambiguity in the epidemiologic findings regarding drinking water and cancer. Second, I shall discuss what I believe to be the source of some of these uncertainties in our findings. And finally, I would like to reflect briefly on a few of the policy implications of this body of epidemiologic research.

Recently, I completed an analysis using data from the NORS and the Region V Organics Survey, the same chemical data used by Cantor and coauthors (1977) and by Hogan and coauthors (1977). Like many of the other studies to which Schneiderman referred, my analysis attempted to control for the effects of certain potentially confounding factors, such as industrialization, urbanization, education, and income. When this study was undertaken, it appeared to have several characteristics that would strengthen previous findings. For example, many of the earlier studies, including my own (Reiches, 1977), utilized the twenty-year average mortality rates compiled by Mason (1975). The second study (Harris, Page, and Reiches, 1977) employed mortality data for the period 1966-74. The assumption was that more recent cancer data would better reflect the relationship to drinking water, since many of the suspected carcinogens may not have been in the water long enough to reflect mortality as early as 1950. Clearly, then, it seemed advantageous to rely on the most recent cancer data available.

A second presumed advantage at the onset was that a careful survey of water treatment facilities was conducted to determine certain characteristics of the water treatment process and to assess accurately the size and location of the population served by each water system. Since water districts rarely follow county boundaries, it was important to gain precise estimates of the size of the county population that actually received its water from the source analyzed by the EPA.

These are just two examples of the ways in which an attempt was made to introduce as much precision as possible into the study design. Investigation of the data consisted largely of multiple linear regression analyses for nine site-specific cancers of the gastrointestinal and urinary tract organs. Stepwise regression equations were constructed in which the control variables (industrialization, urbanization, education, income, and percentage of the population served by the measured water source) were entered first, followed by the chemical data.

The results of this analysis were most surprising. Few regression coefficients reached a satisfactory level of statistical significance. In fact, in the entire set of tests, no more coefficients were positive and significant than would be expected by chance alone. Many were negative. Furthermore, the few coefficients that were significant, together with those that had positive signs (but levels of significance less than 0.95), formed no cohesive pattern. That is, even stretching the usual rules of scientific and statistical judgment could not convince me that this investigation yielded confirmatory evidence of the drinking water hypothesis. At best, what I saw were very small trends that in and of themselves did not constitute a confirmation.

I was then confronted with the question, What has this analysis contributed to our knowledge of drinking water and cancer? Have these findings done any more than add to the growing confusion and ambiguity? While the results certainly are ambiguous, it is my feeling that we have learned something important--both substantively and methodologically.

In terms of content, it is significant to note that the suggestions of a drinking water effect in this study are consistent with findings of several other studies, namely with respect to the large intestine and bladder.

What is as intriguing as the entire set of ambiguous findings, in my work and that of others, is the search for the sources of the uncertainty. Schneiderman has discussed the problem of variables unspecified in the model, the difficulty of testing interactive effects, and the problem of discovering small effects. I concur thoroughly with his recognition and analysis of these problems. To be more specific for a moment, I would like to select one variable and discuss the problems it presents to an epidemiologic investigation.

The unknowns that surround the definition of a drinking water exposure variable comprise a formidable list. Each of the items noted contributes significantly to our inability to assess with great accuracy who is exposed to what; for example:

1. Existing analytic chemical techniques can only identify approximately 5 percent of the organic contaminants in drinking water.

2. Of the identified fraction, only about 10 percent has been adequately tested for carcinogenicity/ mutagenicity.

3. Of the trihalomethanes, the brominated species lack available carcinogenicity data, although they may be as potent as chloroform.

4. Levels of these compounds in drinking water vary with season, thereby altering exposure over time.

5. Exposure to trihalomethanes of a population served by a single water treatment plant is dependent on the travel time of the water from plant to tap (exposure goes up with travel time).

6. Changes in exposure over a fifteen- to twenty-year period may be a function of changing or increasing precursors (from industry, and so forth) leading to the production of chloroform.

7. Knowledge of the interaction of contaminants is almost nonexistent.

8. Migration within the United States makes it difficult to determine accurately who was exposed to what, and for how long.

9. There may be considerable individual variation in water consumption.

10. In many communities there is more than one water source, and persons may live in one water district yet work in another.

I am certain that other unknowns could be added to this list; but even these point clearly to the difficulties we have encountered. Presumably, our best solutions so far are the ones expressed in the various studies reported.

Compounded by the other known (and unknown) measurement and analysis problems, the magnitude of the difficulty of pursuing this type of research becomes readily apparent. This, however, is <u>not</u> an argument for abandoning work in this area. Rather, it is offered as an aid for interpreting our findings. It is an "in spite of" argument. In spite of the problems, and in spite of our inability to make Nature behave as if she were in a laboratory, I believe we have made significant progress and that we epidemiologists are in an advantageous position to advise policy makers and regulators.

If we summarize just a few of the things of which we are fairly certain, the weight of the evidence is in favor of pursuing a strong drinking water policy, even in light of the obvious ambiguities.

We can divide the evidence regarding drinking water and cancer into three categories: chemical, toxicologic, and epidemiologic. The chemical evidence is rather clear. Even if all the compounds are not identified and quantified, there is little doubt that many of the organics of interest do indeed exist in measurable concentration in our nation's waters. Given reasonable evidence that large populations are actually exposed to these substances, on what basis can we assess their health effects? One predictive model, of course, comes from the laboratory. While I am sure that Doctors Lijinsky and McCann will address this issue in more detail, let me comment that substantive evidence exists which implicates chloroform and other hydrocarbons in the etiology of cancer. Therefore, given that the compounds exist in water and given that we have adequate reason to believe they are related to cancer in animal populations, do we have sufficient knowledge of their effects on human populations to argue for their regulation? To this question, I must say yes, in spite of the methodologic and other difficulties we have encountered. My reasons for this position derive not only from a strict interpretation of epidemiologic findings, but also from a general consideration of other epidemiologic clues to the etiology of cancer.

One possible explanation for the weak effects that have been measured is that it may simply be too early to be noticing stronger effects. Most of the chemical contaminants in drinking water, including those that may serve as the precursors to chloroform, have been introduced as a result of rapid growth in the synthetic chemical industry since the period 1950-55. Much of the growth in chemical production has occurred only in the last ten to fifteen years. If these are some of the agents responsible for cancer, it is feasible to believe that we have not yet entered the time period in which their effects would be most visible. And considering the presumably long latency period in the development of neoplastic disease, it should not be surprising that the greatest effects are still hidden from us.

It is largely on the basis of this last argument that I would suggest the need for instituting regulatory action at this date. Planning and policy making is an activity with an obvious future dimension, and we should not abrogate our responsibility to act rationally for the future as well as well as the present. Epidemiology is well suited to operationalizing these values, since we like to think of our models, imperfect as they may be, as useful prospective or predictive tools, or both. If we have reason to believe that risks from chemicals in drinking water may increase in the future, then that constitutes sufficient evidence for regulating these substances in public water supplies.

Our policy in these areas must be judicious, however, and should be accompanied by intense scientific investigation to illuminate further the potential risks. Large investments in epidemiologic research will be required to remedy the problems to which Schneiderman and I have alluded. More refined and precise investigations should be planned in carefully selected communities; case-control studies and prospective data-gathering mechanisms should be instituted in a systematic and organized fashion.

Although epidemiologists still lack some of the clues, or lack the methods with which to tie up all the loose ends, it is becoming a good bet to believe that regulation of chemicals in drinking water would improve health. The real beauty of epidemiology lies in its ability to disrupt the causal path, even when the entire path is not understood. That is, short of explicating all the causal mechanisms involved, epidemiologic findings permit us to intervene in the disease process and affect an outcome, in spite of the ambiguities.

References

Cantor, K. P.; R. Hoover; T. J. Mason; and L. J. McCabe. 1977. "Association of Halomethanes in Drinking Water with Cancer Mortality" (Bethesda, Md., Environmental Epidemiology Branch, National Cancer Institutes).

Harris, R. H.; T. Page; and N. A. Reiches. 1977. "Drinking Water Contamination and Human Cancer," in H. H. Hiatt, J. D. Watson, and J. A. Winsten, eds., Origins of Human Cancer (Cold Spring Harbor, N.Y., Cold Spring Harbor Laboratory) pp. 309-330.

Hogan, M. D.; P. Y. Chi; T. J. Mitchell; and D. G. Hoel. 1977. "Association Between Chloroform Levels in Finished Drinking Water Supplies and Various Site-Specific Cancer Mortality Rates." Unpublished.

Mason, T. J.; F. W. McKay; R. Hoover; W. J. Blot; and J. F. Fraumeni, Jr. 1975. Atlas of Cancer Mortality for U.S. Counties: 1950-1969, U.S. Department of Health, Education and Welfare Pub. No. (NIH) 75-780 (Bethesda, Md., National Institutes of Health).

Reiches, N. A. 1977. "The Relationship of Halogenated Organic Chemicals in Drinking Water to Human Cancer" (Ph.D. thesis, Department of Preventive Medicine, Ohio State University).

Chapter 4

TOXICOLOGY RELEVANT TO DRINKING WATER POLICY

William Lijinsky*

The problems of assessing the toxicology of drinking water are made more difficult than usual by the complexity of the mixture of chemicals found in water, by their presence at very low concentrations, and by the absence of good toxicity tests for most of them, even those industrial chemicals which are so widely used that they appear in water because they are discharged as effluents from factories. The possible adverse health effects of drinking contaminated water are aggravated by the chronic nature of the exposure. Because of this lifetime exposure, the presence of chronic toxicants, including carcinogens, even at very low levels, becomes more important than intermittent exposures to the same type of compound, even if they are at higher concentrations. A further complication is that water is not only consumed as such, but is used to prepare many other foods, both domestic and manufactured, including commercial bottled beverages other than wine. Since the exposure of many people to water, perhaps a majority, is in the form of commercial beverages, it

---

*Chemical Carcinogenesis Program, Frederick Cancer Research Center.

becomes quite difficult to conduct meaningful epidemiological studies on the relation of water to human chronic disease. This difficulty is compounded by the shifting populations and changes in the source and quality of water over the long periods of time involved.

Therefore, we must rely on toxicological tests to tell us of the possible hazards in water, as in the other parts of the environment. Once a hazard is indicated, particularly a carcinogenic one, steps should be taken to reduce public exposure to it. Any other response is contrary to the philosophy of protection of the public health.

There have been two approaches to the problem of toxic contaminants in the water supply: chemical analysis and biological assessment of toxicity. The chemical approach enables us to identify foreign substances, many of which have been tested biologically and have known toxic effects, or, in most cases, are free of chronic toxic effects. The chronic effects are of greatest concern, since they include the risk of cancer, while acute toxic effects are displayed by almost all substances if the dose is sufficiently high. The major concern at present is with carcinogenesis, since most human cancer is of unexplained origin, although it is believed that most cancers can be attributed to chemical causes, both endogenous and exogenous. Foreign chemicals in the environment are major suspects, although these are as much chemicals introduced into food by cooking and processing, or as by-products of human habits, such as smoking and drinking, as they are products of the chemical industry. However, not all of these potential sources of carcinogens can be examined and assessed in the same way. It might, for example, be easier to control human exposure to chemicals from factory effluents than to control their exposure to carcinogens in tobacco smoke, the latter being a complex sociological problem not easily subject to legal strictures.

Since water is derived from various sources, including streams, rivers, and lakes, it can contain materials that are dumped into such water. Rivers and lakes, for example, are frequently the repositories for factory effluents, which if they are not carefully controlled, might well contain toxic or carcinogenic substances. Examples are the reports of halogenated hydrocarbon pesticides, carbon tetrachloride, and nitrosamines, to name but a few. In addition, chlorination of potable water leads to the formation of several chlorinated hydrocarbons, most notably chloroform, through interaction of the halogen with a variety of simple organic compounds. Among the compounds which have been found as contaminants in water are herbicides and insecticides, probably from agricultural runoff or industrial operations (atrazine, DDT, aldrin, dieldrin, malathion, methomyl, carbaryl, thiram, and captan, as a few examples), benzene, PCBs, chloroform, carbon tetrachloride, trichloroethylene, vinyl chloride, methyl chloride, and styrene, together with many others. A number of these might be carcinogenic; in fact, many of them are. Since there is such a large number of contaminants in drinking water, how do we test their carcinogenicity or other chronic toxic effects? The best way to test them, although it is far from ideal, is by chronic administration of maximally tolerated doses to experimental animals, with examination of the induction of tumors and other lesions during the course of their lifetime. However, the magnitude of the effort needed to do this with the many hundreds of contaminants in water would be enormous and probably exceeds the resources available for bioassay of all environmental chemicals at present. However, a start must be made and many of these compounds are already being tested in animal bioassay systems. Whatever the form of the bioassay, the most important consequence is risk assessment, or extrapolation of the test results to human cancer risk from exposure to the chemical or chemicals. This aspect of the problem is being examined intensely.

Because all of the potentially hazardous environmental chemicals cannot be tested in animals, a variety of short-term tests have been developed for screening chemicals for potential carcinogenicity. Among these are several bacterial mutagenesis tests, most notably that developed by Ames (McCann, Tomasaki, and Ames, 1975), and several *in vitro* cell transformation tests originally developed by DiPaolo and others (DiPaolo and coauthors, 1973; Pienta, Poiley, and Lebherz, 1977). While results of the Ames test can be correlated qualitatively with the known carcinogenicity in animals and man of many groups of compounds, there are some to which this relationship does not apply (Meselson and Russell, 1977; Rao and coauthors, in press). These include nitrosamines, halogenated hydrocarbons, and hydrazines, to which classes many suspected and known water contaminants belong. Therefore, the chances of failing to detect a carcinogenic contaminant of water when assayed in a bacterial mutagenesis test are fairly high. Furthermore, there is no quantitative correlation between carcinogenic potency in animals and mutagenic potency in bacteria, even among those groups of carcinogens which are bacterial mutagens. Of course, a quantitative correlation between these two biological effects would not be expected, since the mechanisms of carcinogenesis and mutagenesis are not the same and are probably not even similar. The process of carcinogenesis is very complex, and is a cumulative effect, and after many decades of study we still do not know the mechanism of action for a single carcinogen, although progress is being made. On the other hand, the mechanism of mutagenesis is more clearly understood in the case of several mutagens, although even here the mechanisms are not fully elucidated.

The establishment of mammalian cell transformation systems as predictive of carcinogenic properties is far less advanced than that of bacterial mutagenesis tests, mainly because of the relative difficulty of standardizing the procedures and the materials used. Dr. Pienta (1977) of the Frederick Cancer Research Center has made considerable progress in establishing a reproducible system using the hamster embryo cell transforming system *in vitro*, as first suggested by DiPaolo. Pienta's system has few, if any, false positives, although many carcinogens are negative in the assay. However, several carcinogens which are negative in the Ames assay are positive in the cell transformation assay. Therefore, the two types of test, and others that might be developed, might complement each other in providing a qualitative screening test for potential carcinogens. However, the problem of a large number of false positives might arise, that is, noncarcinogens that are positive in the collective screening tests, which might make regulatory solutions more difficult. In particular, the lack of quantitative relationships between the results in any of the short-term tests and carcinogenicity in man or animals leads to confusion. The announcements of mutagens as carcinogens might become so frequent that a backlash of skepticism from the public and from the regulatory agencies could frustrate all actions to protect the public from exposure to carcinogens.

One valuable use of short-term tests could be an assessment of the toxicity of water contaminants as a group. That is, an extract or concentrate of a water sample could be prepared by solvent partition or by adsorption and the concentrate could be tested in a battery of short-term tests.

The results could be compared with those of similar preparations from other water samples. This has been done for some water extracts, using the bacterial mutagenicity test (Tardiff, Carlson, and Simmon, 1976), with results that showed considerable differences between one sample and another.

However, whether this approach would give a good assessment of the carcinogenic risk in drinking water is not certain. Because the quantitative correlation between activity in the short-term tests and carcinogenicity is so poor, even with single compounds of known structure, the chances of being in error when dealing with mixtures of unknown composition are magnified enormously and could be in error, one way or the other, by several orders of magnitude. The purpose of such tests is to determine whether some samples offer less risk than others, and if the assessment could be so much in error, the exercise could lead to the worst action, instead of to the best action.

The assessment of carcinogenic risk is the crux of the problem of exposure to environmental contaminants. Because carcinogenesis is a response to cumulative risks, rather than to single "insults," it is very difficult to establish safe levels of exposure for any carcinogen. On the other hand, it is not possible to avoid exposure to even one molecule of carcinogen in drinking water. So the decision must be made between the two extremes of no exposure and unrestricted exposure. We have not identified all of the contaminants in drinking water, let alone established the carcinogenicity or lack of it in the compounds identified. Carcinogenicity can best be established by assays in animals using chronic administration, preferably by the oral route, since that is the principal means of human exposure to water. The tumor response in properly conducted tests, or the lack of tumor response, can give a better measure of the risk in exposure to the material than can be found by any other means we know at present. This is true in spite of not knowing the relative susceptibilities of man and the experimental animal to the substance or whether the metabolism in man and the animal is the same or even similar.

We do have some background which leads me to believe that rodents and man are similar in their response to many carcinogens, such as aromatic amines, polynuclear hydrocarbons, vinyl chloride, and diethylstilbestrol (Lijinsky, 1977), to give a few examples. We also know that individual animals and individual people vary considerably in their susceptibility to carcinogens. However, we also know that small doses applied continuously over a long period are most effective in inducing tumors. In any particular case, man might be more or less susceptible to a carcinogenic agent and, in protecting the public from the risks of such exposures, we must assume that man is at least as susceptible. There is a probability that several carcinogens act additively, and perhaps synergistically, in increasing the risk of tumor induction. For this reason, all exposure to a carcinogen, even if to very small amounts, increases the risk. Our objective must be to decrease the risk as much as possible, even though it is not feasible to eliminate entirely exposure to all carcinogens.

In designing the animal tests, it has to be borne in mind that a small group of relatively short-lived experimental animals has to represent large human populations, perhaps several millions, exposed to the same agents. Because the exposure (to carcinogens, such as those present in the water) of a rat or mouse is so much shorter than that of people, we have to increase the dose given to the animal considerably above that to which humans are exposed. There is considerable evidence that the probability of developing cancer following exposure to a carcinogen is a function of the dose administered rather than the life span of the animal. Similarly, to induce a statistically significant incidence of tumors in a small group of experimental animals, we must increase the dose administered considerably above that of human exposure, if we are to represent the risk to a large human population, bearing in mind the large variation in susceptibility of individuals, and wishing to protect the most susceptible members of the population.

In practice, if we want to assess a risk of one person in a thousand developing cancer from exposure to a given agent (if that were the only agent to which the thousand were exposed), we would have to treat a group of 5,000 animals with 35 times the dose to which people are exposed, to compensate for the difference between the two-year exposure of the rat or mouse and the possibly seventy-year exposure of people. We would then expect to see tumors induced in 5 of the 5,000 animals. Since it is not feasible to use 5,000 animals with 5,000 untreated controls for comparison to test each agent (which would be enormously costly), we use 50 animals of both sexes and increase the dose greatly. However, we use a maximal dose which is lower than that which produces toxic effects, so that the experiment is not distorted by other effects which might, at the worst, lead to early death of the animals without tumors. We also use two species of test animals in case one might be resistant to the chemical. If a substance in a maximally tolerated dose over a lifetime does not induce a significant incidence of tumors in our 100 experimental animals, we say that the substance is free from carcinogenic risk or "safe," although all we have shown is that the risk is such that fewer than, say, one in a thousand people will develop tumors as a result of lifetime exposure to it. On the other hand, if a significant incidence of tumors is observed, then the risk of cancer resulting from exposure to the substance is substantially greater. In these circumstances, exposure to the substance should be minimized or eliminated. Fortunately, most substances tested have not been carcinogenic in such tests, from which we can take some comfort.

Returning to the case of drinking water, it might be possible to carry out such tests with extracts or concentrates of a few representative water samples, thereby establishing which of them presents a greater cancer risk than others. If this were combined with chemical analysis of the extracts, we might be able to assess the relationship of chemical composition to carcinogenicity of the samples, and take steps to eliminate those agents which are responsible for the carcinogenic risk. This is the approach of the EPA in making regulations to reduce the concentration in water of such known carcinogens as chloroform and other trihalomethanes to a maximal contaminant level, which is easy to monitor. Similarly, sources of contamination with such carcinogens as vinyl chloride must be found and pollution of water with them stopped. The problem of the ubiquitous suspect carcinogens such as DDT, PCBs, and chlorinated pesticides should be approached in the same way, by reducing the water content to the minimum attainable level. Dealing with contaminants of unknown carcinogenic risk is much more difficult, until all have been subjected to bioassays in animals. It is to be hoped that a comprehensive attempt to reduce all contaminants to the minimum level, by the use of activated charcoal or some other adsorbent, will reduce the risk of exposure to those that are carcinogenic. If this could be the approach to all environmental carcinogens or possible carcinogens, we might look forward to reducing significantly the appallingly high incidence of cancer in our population, which carries off one thousand Americans a day.

References

DiPaolo, J. A.; R. L. Nelson; P. J. Donovan; and C. H. Evans. 1973. Archives of Pathology vol. 95, p. 380.

Lijinsky, W. 1977. Chemical & Engineering News vol. 55 (June 27) p. 27.

McCann, J.; B. Choi; E. Yamasaki; and B. N. Ames. 1975. Proceedings of the National Academy of Sciences (U.S.) vol. 72, p. 5135.

Meselson, M., and K. Russell. 1977. In H. H. Hiatt, J. D. Watson, and J. A. Winsten, eds., Origins of Human Cancer (Cold Spring Harbor, N.Y., Cold Spring Harbor Laboratory) p. 1473.

Pienta, R. J.; J. A. Poiley, and W. B. Lebherz. 1977. International Journal of Cancer vol. 19, p. 642.

Rao, T. K.; J. A. Young; W. Lijinsky; and J. L. Epler. (in press) Mutation Research.

Tardiff, R. G.; G. P. Carlson; and V. Simmon. 1976. "Halogenated Organics in Tap Water: A Toxicological Evaluation," in Proceedings of a Conference on Environmental Impacts of Water Chlorination (Sponsored by Oak Ridge National Laboratory).

Chapter 5

EUROPEAN EXPERIENCE WITH PROBLEMS OF DRINKING WATER QUALITY

Heinrich Sontheimer

Introductory Remarks

Problems of drinking water quality, and consequently drinking water policy, are of great importance in Europe, especially in connection with the heavily polluted Rhine river, which supplies raw water for about 10 to 15 million people in Switzerland, Germany, and the Netherlands. The extent of pollution is demonstrated in figure 5-1, which shows the results of gas chromatographic and mass spectroscopic measurements of substances removed by means of an activated carbon filter which treats Rhine water after ground passage through the river banks.

Figure 5-1 gives an indication of the number of different organic compounds removed by adsorption. It is evident that the carbon filters removed many substances which might be dangerous to human health (Park and Sontheimer, 1974). The figure also demonstrates why so many waterworks along the Rhine have carried out extensive studies to develop more sophisticated and advanced treatment steps for obtaining a high-quality drinking water. In most cases these processes

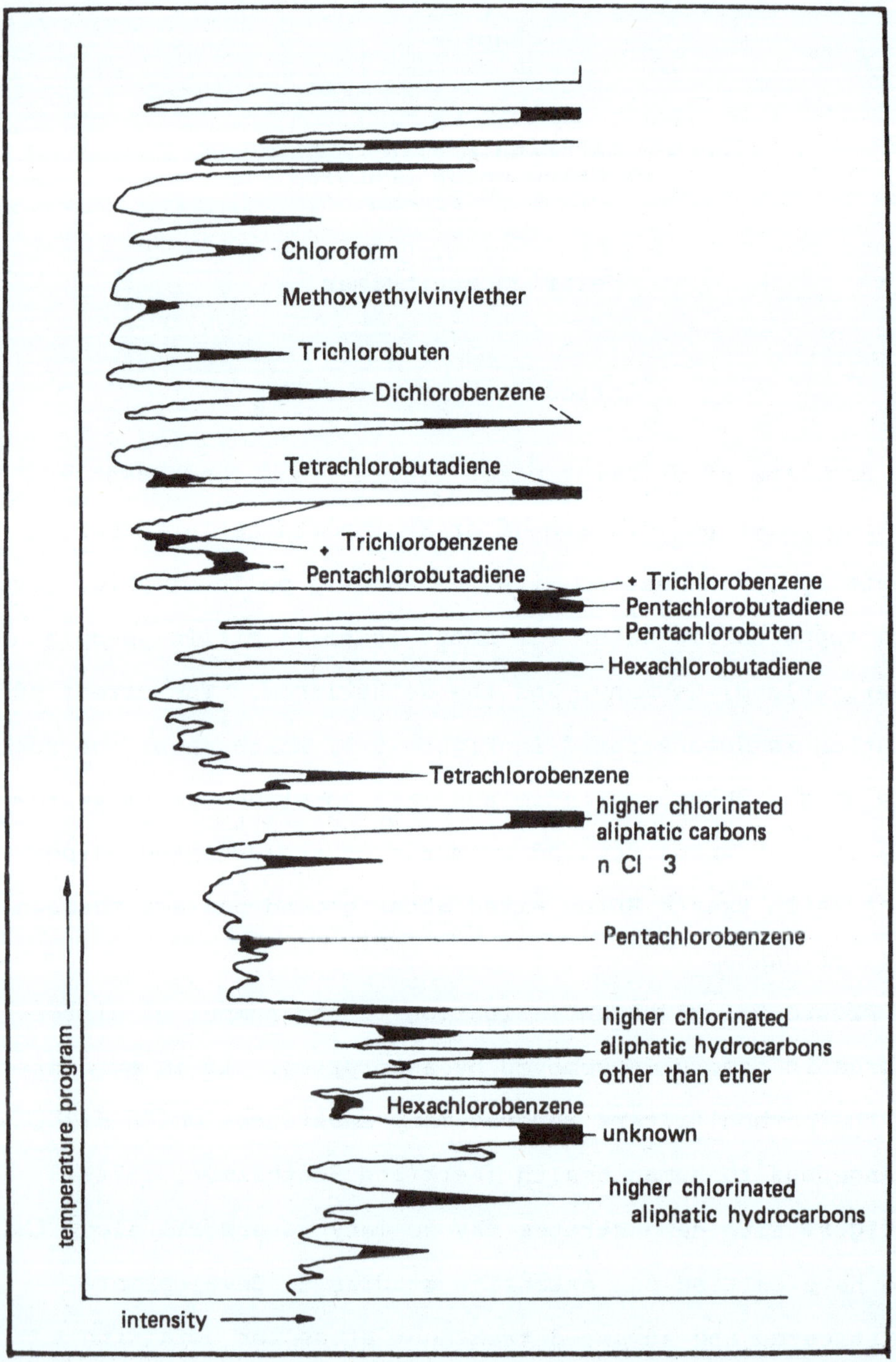

Figure 5-1. Gas chromatogram of extracts from an activated carbon filter.

include ozonation and carbon filtration, and are known, for instance, as the Düsseldorf or the Mülheim process.

However, this is a fairly new development, having started ten to twenty years ago. Until then, the treatment of river water included only a ground passage step, such as river bank filtration or groundwater infiltration through slow sand filters. Most of the river water used thirty years ago did not need any further treatment.

The general opinion in Europe at that time, and especially in Germany, was that on oxygen-containing groundwater, which could be distributed without any pretreatment and which would not even need chlorination, would be the best possible drinking water and should be used in preference to any other type. To this day we have many waterworks in Germany, for example, at Munich and at Karlsruhe, which do not need any chlorine at all for drinking water treatment. Water should be naturally pure, it should not need any chemical treatment or if it is at all necessary, only small amounts of chlorine should be used for disinfection. Treatment processes should be natural ones, such as slow sand filtration and ground passage.

This general idea of natural and pure drinking water was a decisive factor for water technology and water policy development, particularly in Germany.

German Regulations on the Quality of Drinking Water

Up until a few years ago, there were no strict regulations in the Federal Republic of Germany on the quality of drinking water, but merely quidelines such as a rather general but fitting description of natural, pure water, which states that drinking water quality should be similar to water that is found in nature, without anthropogenic impurities, taken from a safe depth after a long retention time, and equivalent to oxygen-containing groundwater.

It may seem that these requirements are fairly vague and noncommittal, but when examined closely, they offer very concrete guidance on the quality of water. For instance, good quality drinking water should be in equilibrium with the most important ground minerals, and it should be polluted as little as possible. Looking at it from this angle, it seems logical that one of the first regulations on drinking water quality in the Federal Republic was concerned with the amount of foreign substances added during the treatment process. The most important regulations are listed in table 5-1.

Care has always been taken to keep the amount of chlorine added to the water as low as possible. Special permission had to be obtained to go beyond the permitted limit of 0.3 mg/liter up to 0.6 mg/liter, measured after a reaction time of between 15 and 30 minutes. If, on the other hand,

Table 5-1. German Regulations for Chemicals Used in Drinking Water Treatment

Free chlorine	
After 15-30 min.	max.0.3 ppm Cl_2
Reaction time	max.0.6 ppm Cl_2 (special allowance)
	min.0.1 ppm Cl_2 (if disinfection is necessary)
Other chlorine products	
Chloramines	max.0.6 ppm NH_4^+ (addition)
Chlorine dioxide	max.0.1 ppm ClO_2^- (recommendation)
Other added chemicals	
Phosphates	max.5 ppm P_2O_5
Flocculants	max.0.2 ppm Al or Fe (after treatment)

chlorine had to be added for disinfecting purposes, a minimum of 0.1 mg/liter of chlorine was prescribed. As can be seen from the table, there are similar regulations for chloramines and chlorine dioxide, even for phosphates, and also for the remaining flocculants when they do not precipitate completely.

In 1975, standards were set up for certain quality criteria on drinking water, and regulations were put into effect on the number of tests to be carried out, on the methods to be used, etc. The present standard values are shown in table 5-2. One can see that the data for the disinfection effects differ, depending on whether the water requires chlorination or not. The table also shows the maximum concentrations of a number of heavy metals and other inorganics, and similar values for polycyclic aromatic hydrocarbons. Before going into further detail, it should be mentioned that the European Economic Community is at present drawing up standards resembling those shown here, but containing some further criteria. However, complete agreement on the standard values has not been reached yet.

The data for polycyclic aromatics shown in the table give rise to certain basic reflections on the importance and usefulness of standards for drinking water. There seems to be no doubt that these compounds may cause cancer and that their concentrations should be limited. But here we have the problem, first that only six single substances of these compounds can be easily analyzed, and of these the

Table 5-2. German Standards for Drinking Water

	with disinfection	without disinfection
Total bacteria counts	20/ml	100/ml
Escherichia coli	0/100ml	0/100ml
Coliforms	0/100ml	0/100ml
Free chlorine	≥ 0.1mg/liter	-

Arsenic	0.04	mg/liter
Lead	0.04	mg/liter
Cadmium	0.006	mg/liter
Chromium	0.005	mg/liter
Mercury	0.004	mg/liter
Selenium	0.008	mg/liter
Zinc	2.0	mg/liter
Sulfate	240.0	mg/liter
Nitrate	90.0	mg/liter
Fluoride	1.5	mg/liter
Polycyclic aromatic hydrocarbons	0.00025	mg/liter

noncarcinogenic fluorine anthrene always has the highest concentration (Kölle and Sontheimer, 1968). On the other hand, measurements of such substances contained in the river Rhine were made by our institute and the results are shown in figure 5-2.

The data clearly indicate that the limit values for polycyclic aromatic hydrocarbons are only reached when solid substances are included in the measurements, that is, when the polycyclic aromatics adsorbed onto the solids are included. Since the removal of suspended solids is a well-known problem, it would seem more practical to measure the turbidity instead of analyzing the polycyclic aromatics. If the turbidity does not exceed certain values, one can be sure that the limit values for polycyclic aromatics are not exceeded either.

Experience with Surface Water Treatment in Europe

The generally accepted opinion that drinking water should be of the quality of a natural, oxygen-containing groundwater has had considerable influence on the use of surface waters for drinking water treatment, which is unavoidable in densely populated areas and for large cities (Sontheimer, 1977). Therefore, river bank filtration has become a significant treatment step whenever the quality of a certain river water permitted it. As a consequence, numerous waterworks were erected along the Rhine at the beginning of the century and

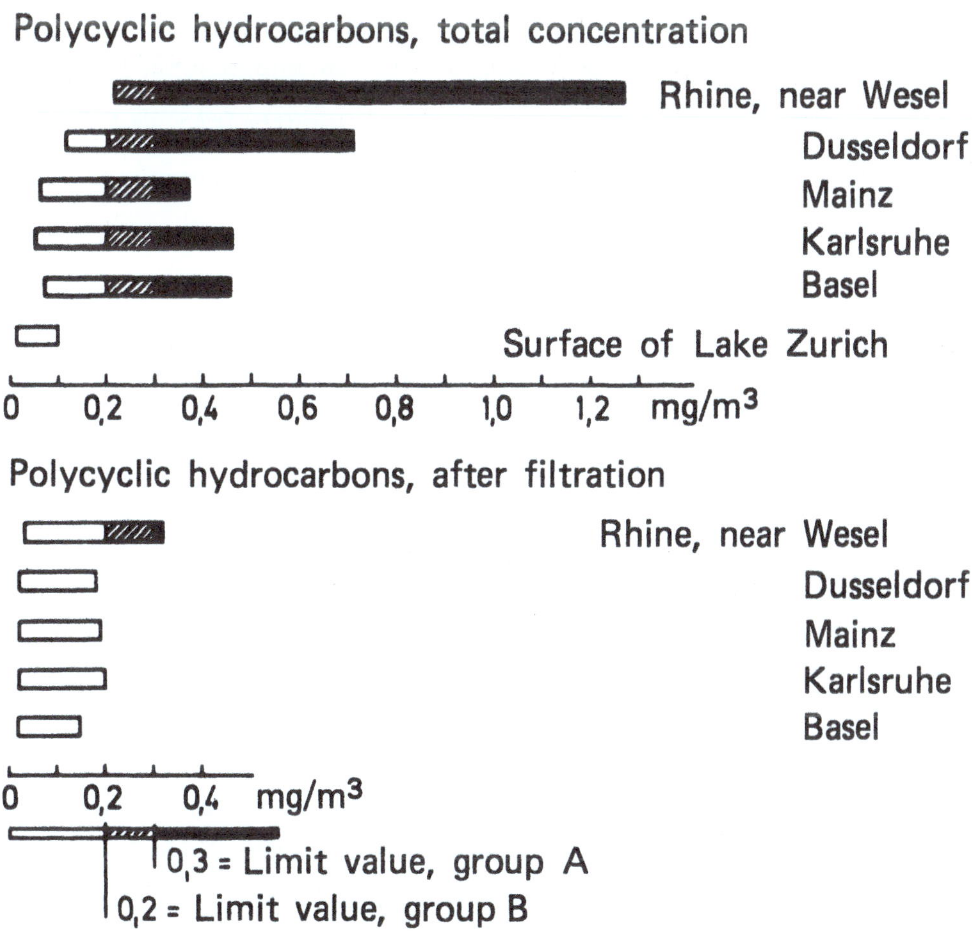

Figure 5-2. Band variations in concentrations of polycyclical hydrocarbons in Lake Zurich and in various sections of the Rhine.

wells were constructed at a distance of about 50 m from the river which, after sufficient lowering of the water level, yielded bank-filtered river water.

Even today quite good results are obtained with this method, as is evident from table 5-3 which shows some results of bank filtration in the Duisburg area (Sontheimer and Nissing, 1977). One can see that it is possible to remove up to 75 percent of all dissolved organics by bank filtration, and even up to 40 percent of the organochloro compounds. The removal of organic sulfur compounds is less satisfactory, on the other hand. Good results were also obtained in removing a number of heavy metals, the degree of effectiveness depending on the actual concentration in the river. Taking into account that the noted concentrations in the river are fairly low and almost never reach the permitted limits for drinking water, it can be deduced that bank filtration is an additional safety factor in the event of a sudden increase of heavy metal concentrations and that it is an excellent aid to improve the quality of the heavily polluted water of the Rhine. A number of process combinations have been developed during the past ten or twenty years for the additional removal of interfering substances. They are listed in figure 5-3.

Bank filtration as the only treatment will be sufficient when the river water is only slightly polluted. In the case of possible contaminants, on the other hand, either an activated carbon filter or a combination of ozonation and activated

Table 5-3. Removal Efficiency of Riverbank Filtration in Duisburg (Values for 1975-76)

		Mean concentration in the river Rhine (dissolved compounds)	Removal efficiency in percent
Dissolved organic carbon	(DOC)	7.0mg/liter	60 - 75
Adsorbable organics	(DOC)	4.5mg/liter	30 - 50
Organic chlorine	(DOCl)	0.12mg/liter	30 - 40
Organic sulfur	(DOS)	0.11mg/liter	20 - 30
Chromium	(Cr)	6 - 12 μg/liter	90 - 95
Copper	(Cu)	15 - 35 μg/liter	70 - 80
Zinc	(Zn)	90 - 250 μg/liter	40 - 60
Nickel	(Ni)	10 - 15 μg/liter	40 - 60
Lead	(Pb)	9 - 20 μg/liter	40 - 60
Arsenic	(As)	4 - 8 μg/liter	30 - 60
Mercury	(Hg)	0.1 - 0.6 μg/liter	30 - 60
Cadmium	(Cd)	0.5 - 2 μg/liter	20 - 50
Selenium	(Se)	0.5 - 2 μg/liter	0 - 20

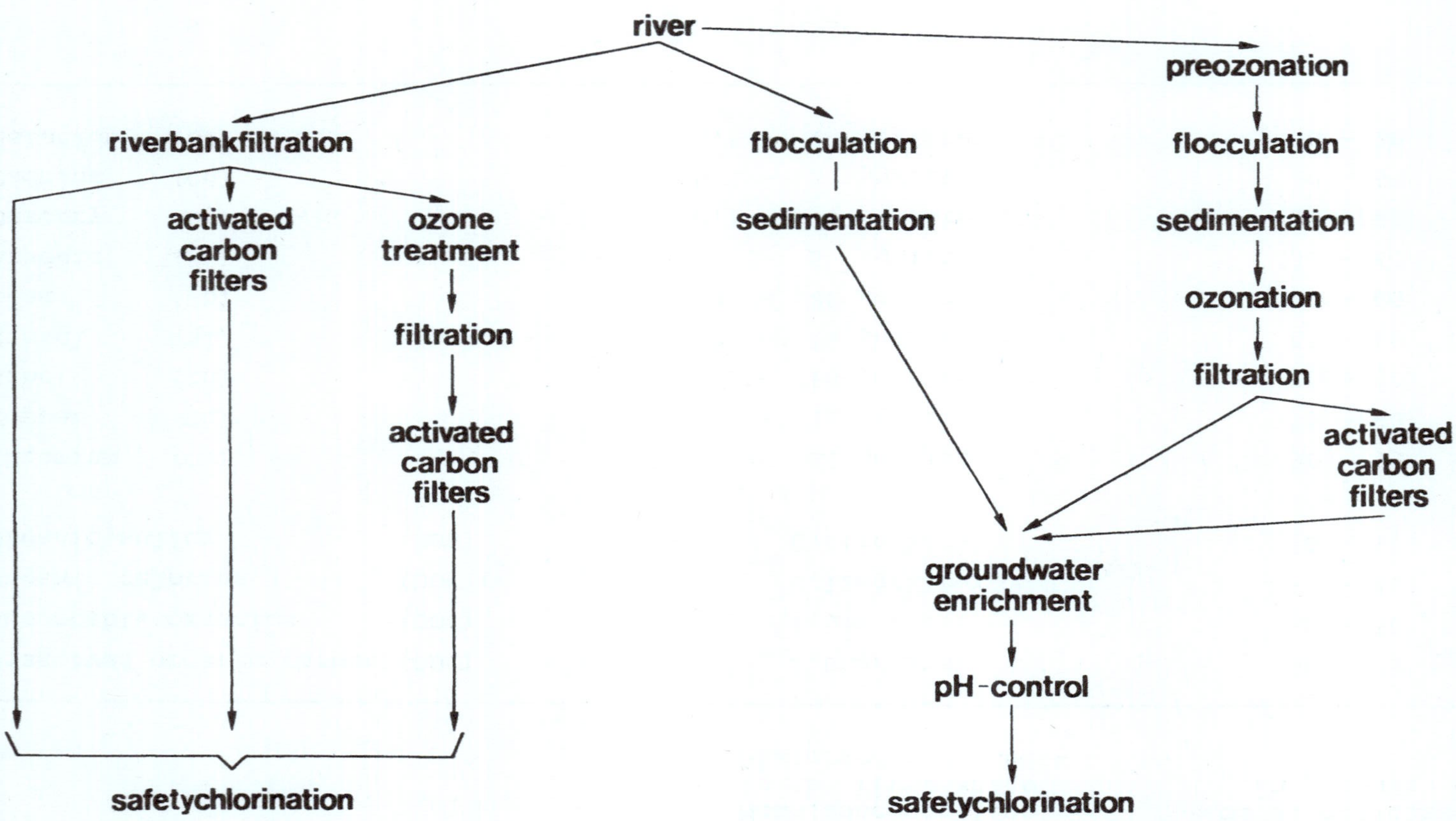

Figure 5-3. Process scheme development for river water treatment in Europe (with underground passage).

carbon filtration will be necessary (Poggenburg and coauthors, 1973). The latter method has proved very successful along the lower Rhine for more than twenty years, and a very good quality of drinking water is still being obtained.

If bank filtration is not possible, treatment generally consists of groundwater enrichment by slow sand filters, usually after flocculation, sedimentation, and filtration. To increase the effectiveness still further, the Mülheim process may be applied, whereby preozonation replaces breakpoint chlorination and ozone is used for the chief treatment after flocculation and sedimentation. This type of pretreatment improves the biological activity within the double-layer filters, in the activated carbon filters, and naturally also within the ground, and leads to a considerable improvement in the quality of the drinking water. Further details of this process will be given later in this paper.

The change at the Mülheim plant from the classical surface treatment using breakpoint chlorination and subsequent flocculation and filtration was made because investigations had shown that during chlorination large amounts of trihalomethane compounds and even larger amounts of other organochloro compounds were formed (Sander, Kühn, and Sontheimer, 1977). This is illustrated in figure 5-4, where pure humic acid solutions have been treated with chlorine. The amount of chloroform produced during reaction is shown, along with values for the total amount of organochloro compounds. It

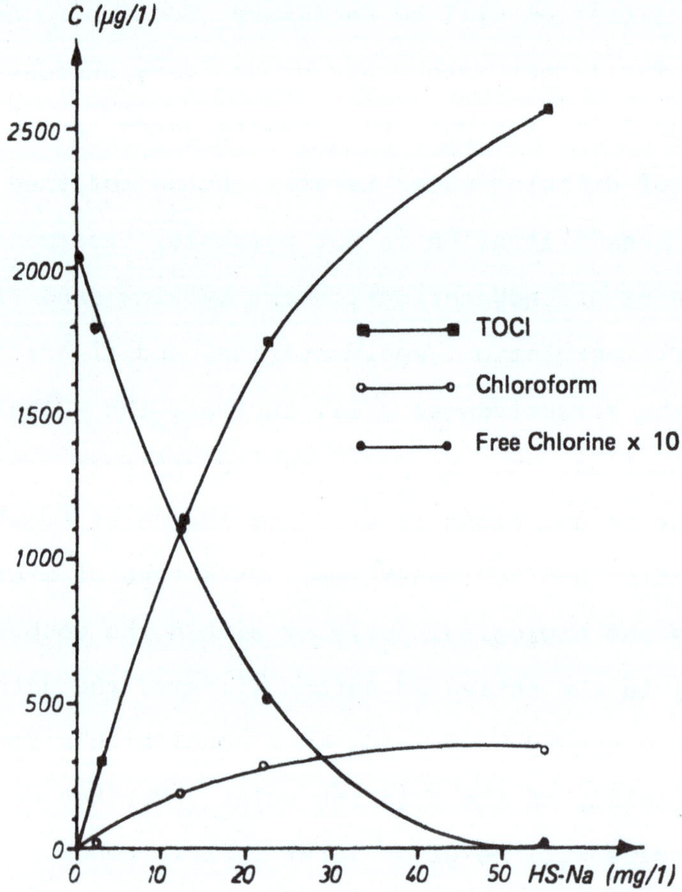

Figure 5-4. Formation of chloroform, change in total organic chlorine, and final chlorine content as a function of humic acid over a 23-hour reaction time.

is easy to see that the amount of organochloro compounds which are too complicated to be identified is usually ten times as large as the amount of haloforms. It is extremely difficult to remove these substances by the normal treatment methods, as can be seen from the data shown in table 5-4.

The data shown here were taken from a pilot plant at Stuttgart for the treatment of water from the river Neckar. Values are shown for the river water before treatment, after breakpoint chlorination, sedimentation, filtration, and after carbon filters using two different types of activated carbon. The data were obtained for a period of 1 month after the pilot plant had been in operation for 6 weeks. It appears that when using the particularly suitable activated carbon type F 300, removal of organics was still satisfactory after the carbon filters had been in use for 6 weeks. Unfortunately this was not the case for the removal of haloforms, where only about 50 percent were removed with the same filter. Furthermore, the concentrations of chloroform and organic bromides found after 2½ months in the outlet of the activated carbon filter plant, using the F 300 carbon type, were higher than in the inlet. This fact was also confirmed by measurements carried out at Cincinnati, Ohio.

The behavior of the other organocholoro compounds is quite different. The degree of removal achieved lies around 60 percent and differs according to the composition of compounds, which again influences the degree of adsorption.

Table 5-4. Effects of Classical Treatment for a River Water (Neckar, Stuttgart) with Breakpoint Chlorination and Activated Carbon Filters

		River water	After breakpoint chlorination, flocculation, sedimentation and filtration	After GAC Carbon LSS	After GAC Carbon F 300
Dissolved organic carbon (DOC)	mg/liter	5.0	4.1	3.1	1.6
UV absorbance at 254 nm	mg/liter	10.9	8.6	5.0	3.6
Sum of haloforms	μg/liter	0.4	50	16	25
Total organic chlorine (TOCl)	μg/liter	33	524	364	296

The effectiveness of the carbon filters also depends on the quality of the raw water and thus on the fluctuating concentrations of organochloro compounds in the inlet (Kölle, Sontheimer, and Stieglitz, 1975). It was partly for this reason that in the Federal Republic and in the whole of Europe the classical treatment process, illustrated on the left-hand side of figure 5-5 was replaced by improved processes. The first aim was to remove all surplus chlorine coming from breakpoint chlorination by means of the additional use of activated carbon filters and to take out the organochloro compounds. However, as the Stuttgart data show, there are considerable limits to this scheme. Quite a different course was adopted in France, where the chief aim was to remove any type of virus by additional ozonation, and to improve the color and taste of the water. This can be done, but it requires higher dosages of chlorine as a safeguard against the regrowth of bacteria.

Based on the results obtained at various plants, a different treatment procedure was first tried out at the plant at Langenau near the river Danube, whereby ozonation followed immediately after flocculation and sedimentation instead of coming at the end of the treatment process. Apart from a safe removal of organics, it was intended to improve the removal of turbidity, and this was further enhanced by adding a secondary flocculant dosage after ozonation. The disadvantage of this method is that, owing to breakpoint chlorination, high concentrations of organochloro compounds

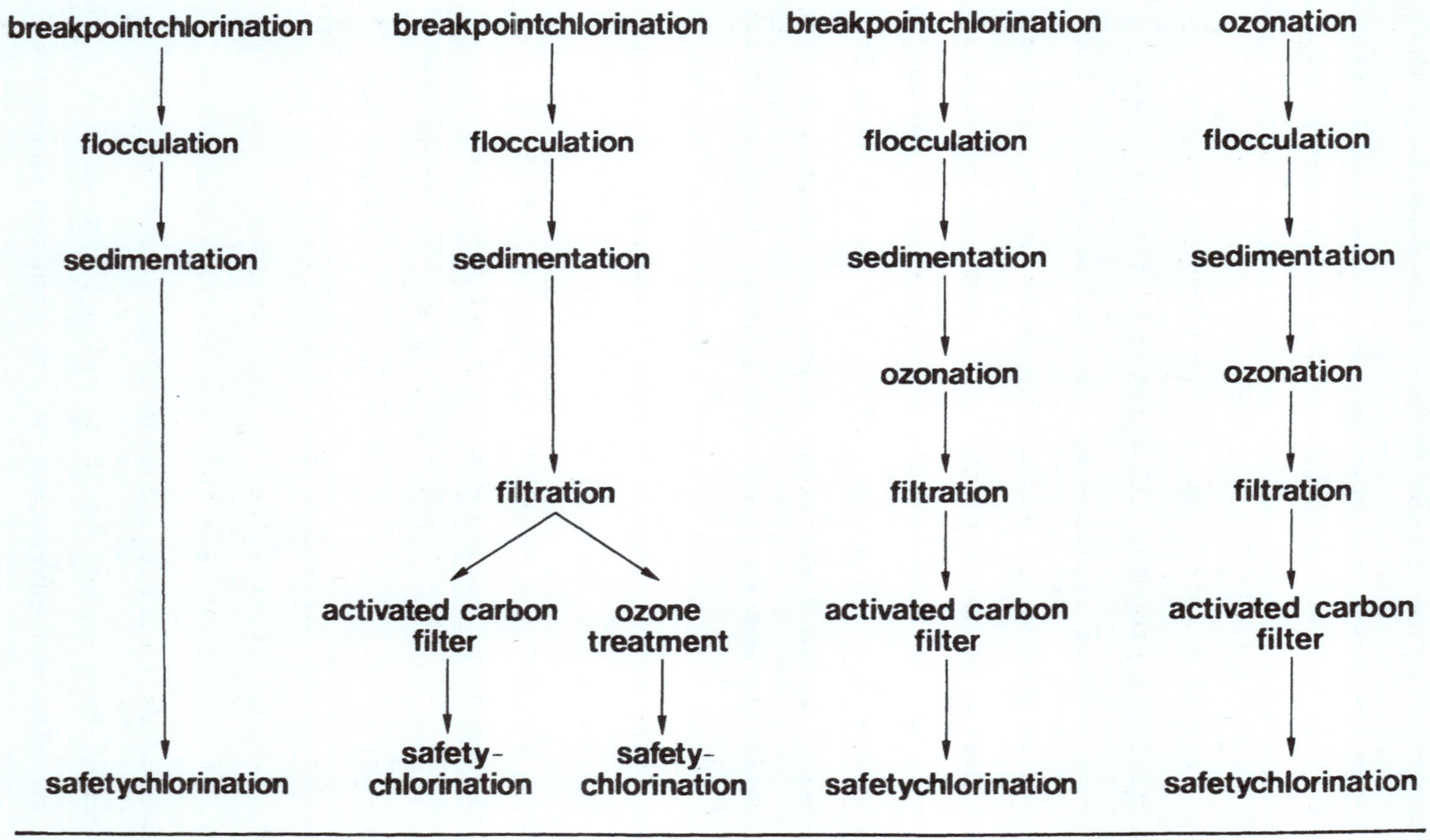

Figure 5-5. Process scheme development for river water treatment in Europe (without underground passage).

still arise, which reduce the time between two regeneration steps of the activated carbon in the filters. This drawback can be avoided, however, by making use of the latest development in surface water treatment, put into practice for the first time with great success in a large-scale plant at Mülheim (Sontheimer and coauthors, 1978). This new method includes preozonation, with the result that flocculation and sedimentation are greatly improved. The principal stage of the ozone treatment follows after the removal of all precipitable substances and prior to filtration. This type of treatment increases the effectiveness of the filter plant and of the carbon filters. Ammonium is completely oxidized to nitrate, and there is extensive removal of the dissolved organics. At the same time this scheme has the advantage that the activated carbon filters retain a certain amount of activity, which means that an unexpected invasion of contaminants would be kept back by adsorption in the carbon filters. The combination of biological oxidation and adsorption in the filters can be controlled by measuring the biological effectiveness, shown by the amount of inorganic CO_2 formed within the filters, which can easily be established by measuring the pH value. These general remarks show that there have been substantial developments in treatment methods of surface waters in Europe during the past ten years, ensuring the supply of drinking water of a good quality.

Examples of Modern Drinking Water Treatment in Europe

The importance of the general aspects discussed so far in this paper will be made more explicit by looking at a few examples of typical European waterworks and their treatment processes. All these works use a different type of raw water and follow a different process scheme, but they rely on the same general conception regarding water quality and water treatment.

The first example comes from the waterworks at the river Ruhr in Western Germany. Here, for almost 100 years slow sand filters, in combination with groundwater infiltration, have been the only but very effective treatment. With increasing water pollution, many attempts were made to obtain a better water quality through some sort of pretreatment before infiltration. At their works at Witten, Gelsenwasser AG, which is Germany's largest water company, have therefore introduced preflocculation for this purpose, followed by sedimentation or flotation. This helped to keep the concentrations of organics low before and after infiltration, which can be seen from figure 5-6. The summary frequency distributions are shown for the entire organic concentrations, and it can be seen that the effect of pretreatment is noticeable also after ground passage. But this is not the only effect produced by preflocculation, as will be explained in figure 5-7. The data presented show that there is a far

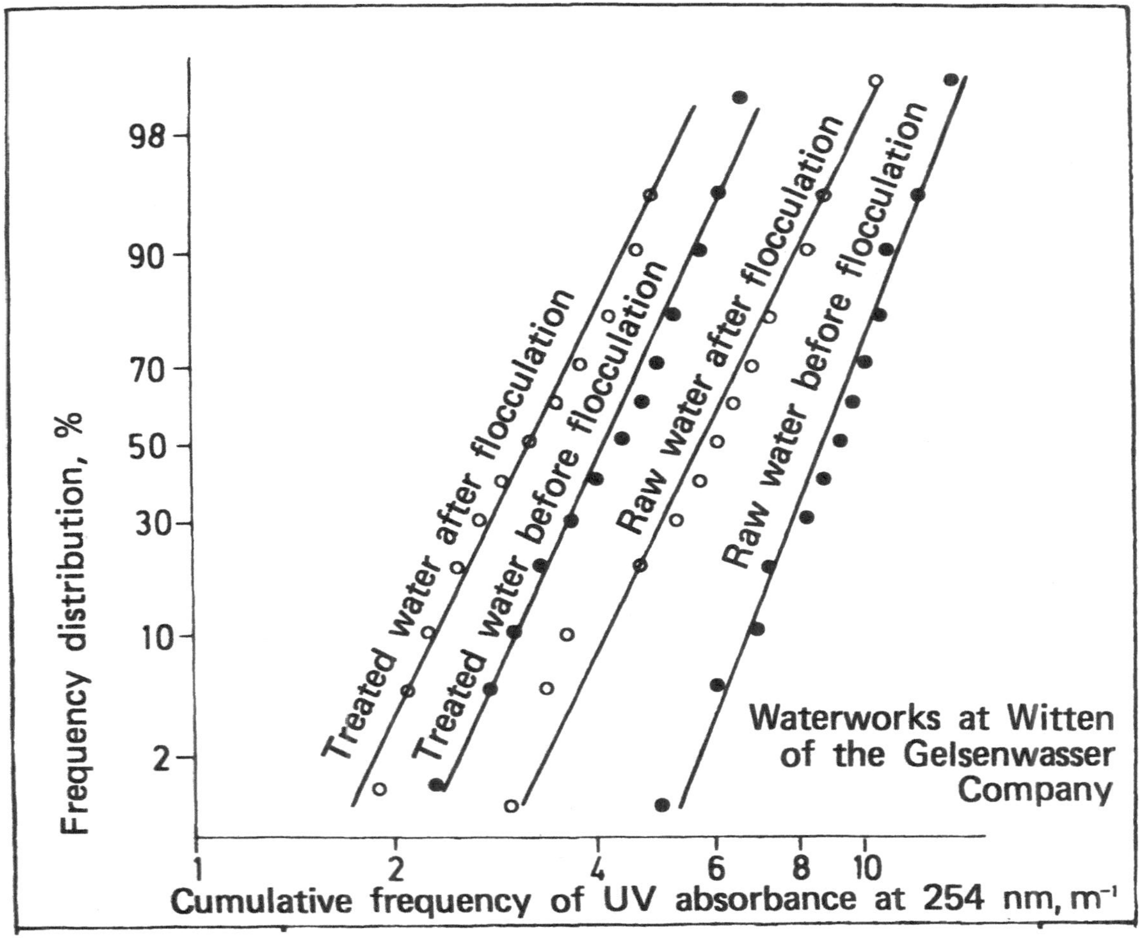

Figure 5-6. Improvement in groundwater quality by means of infiltration of water after flocculation.

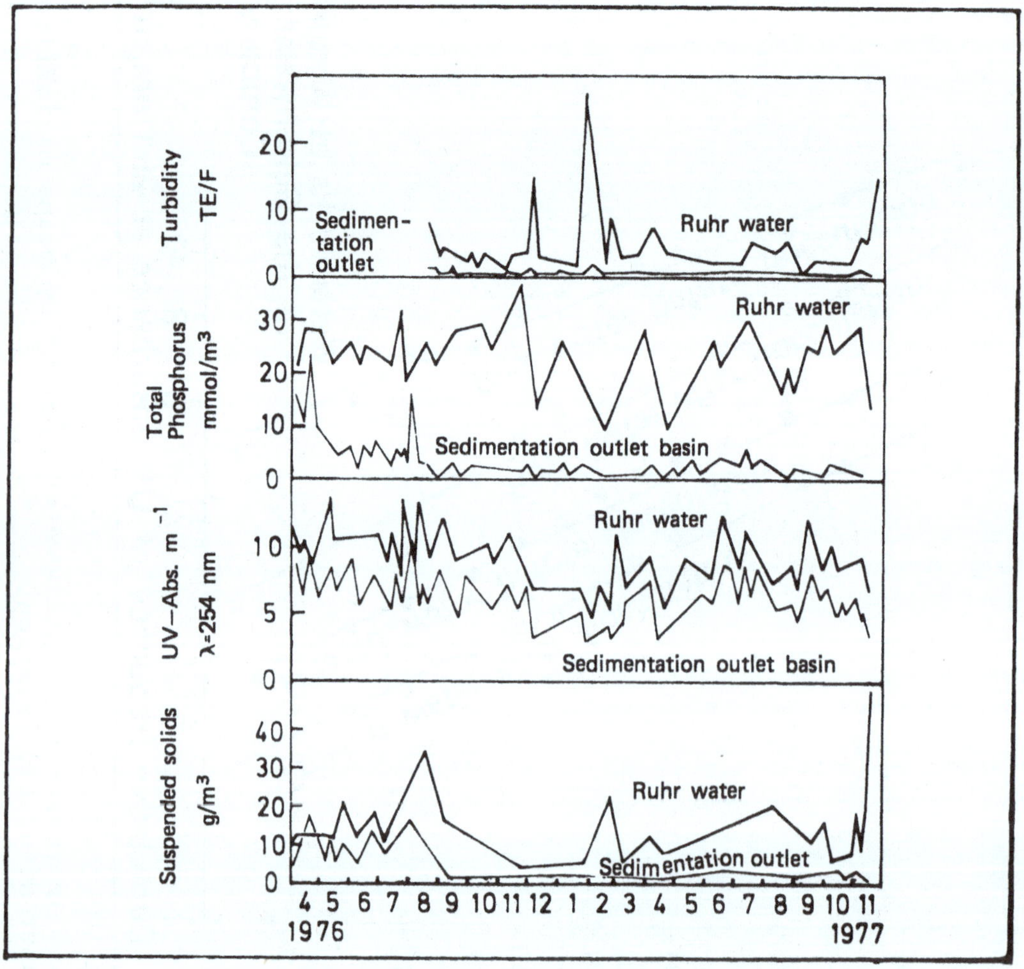

Figure 5-7. Results of preflocculation of water from the Ruhr River in the Gelsenwasser Corporation's Witten waterworks.

more pronounced effect regarding turbidity and the removal of phosphorus. As a result, it is possible to reduce the growth of algae in the infiltration basins, and thus avoid secondary pollution through the by-products of the algae.

Since large infiltration areas are needed for such a treatment process, and since these were not always available in the densely populated regions of the lower part of the river, the Mülheim waterworks (Rheinisch-Westfälische Wasserwerke: RWW) some ten or twenty years ago decided to use the classical pretreatment method, consisting of breakpoint chlorination, flocculation, sedimentation, and filtration prior to ground infiltration. In order to obtain a good biological activity within the ground, dechlorination, using activated carbon filters, was also introduced. This type of pretreatment resulted in a high infiltration rate into the ground, but there were certain problems with the water quality. An example of these is given in table 5-5, showing data on the organochloro compounds for such a process scheme. It can be seen from these data that, apart from chloroform and some bromo compounds, a high concentration of other organochloro compounds resulted from breakpoint chlorination. A certain fraction of these organics is degraded by ground passage, but the remaining concentrations are still fairly high, and there is no reduction of chloroform.

These and other observations have been the starting point for a large research program, with the goal of finding

Table 5-5. Organic Chlorine Compounds for Breakpoint Chlorination Treatment

	DOCl ppb	DOClN ppb	Sum of haloforms ppb	$ChCl_3$ ppb
Raw water (Ruhr)	17	5	5	1
After flocc. & sedimentation	-	-	15	6
After filtration	203	30	23	7
After activated carbon filters	151	17	21	7
After ground passage and safety chlorination	92	18	23	9

a new process scheme that does not require breakpoint chlorination. The newly developed process can be seen in figure 5-8. The treatment starts with preozonation, combined with the addition of the flocculant at high turbulence. This is followed by a sludge blanket clarifier and the main ozone treatment step. After this we have filtration, followed by biological activated carbon filtration and groundwater passage. The results achieved so far are very promising (Sontheimer and coauthors, 1978). Some of these can be seen in table 5-6. The data before and after the process change have been summarized in this table, showing a much better efficiency after using ozonation instead of chlorination. The effects can be explained by remembering that ozonation changes nonbiodegradable organics in such a way that they become biodegradable. It is important to note that the discussed process change not only resulted in a better drinking water quality with very low organic chlorine concentrations, but that the process was easier to control, while treatment costs remained the same or were even lower. The improvement in water quality after the overall treatment is demonstrated by the fact that it was possible to reduce the chlorine required for safe chlorination from amounts between 0.3 and 0.5 ppm to amounts between 0.1 and 0.2 ppm, without any problems with disinfection or any increase in the total bacterial count in the distribution system.

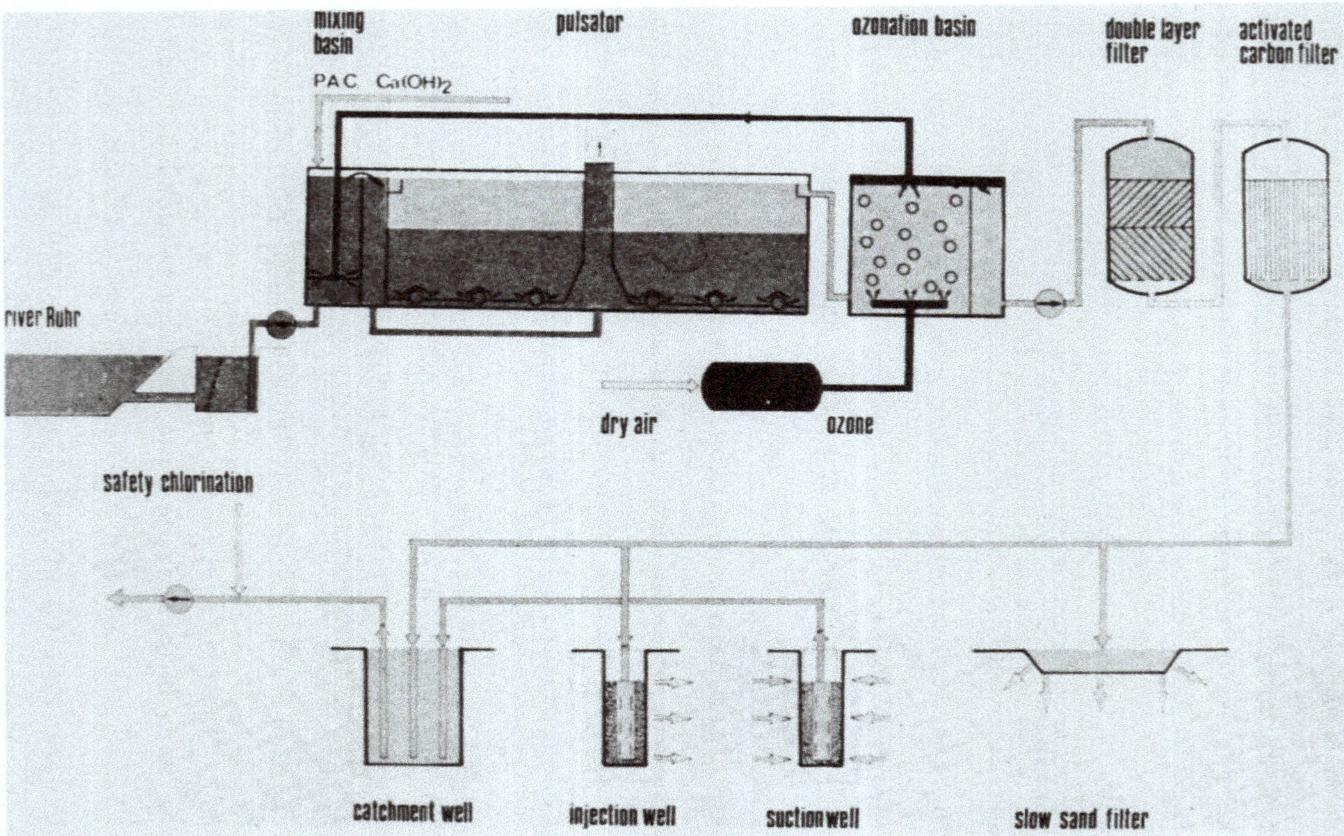

Figure 5-8. Flow scheme - Dohne-Mülheim.

Table 5-6. Analytical Data of the Dohne Plant in Mülheim

	DOC mg/liter 1975	DOC mg/liter 1977	NH_4^+ mg/liter 1977	Total bacteria counts ml^{-1} 1977	E. coli $100ml^{-1}$ 1977
River water	3.9	4.0	1.1	15,000	1,600
After flocculation	3.2	3.1	0.8	2,300	6
After filtration	3.2	2.8	0.2	6,000	0
After activated carbon filters	3.0	1.5	0.0	3,700	0
After ground passage	1.8	0.9	0.0	25	0

If we switch from the river Ruhr to the river Rhine and look at dune infiltration, used for the city of Amsterdam, it can be said that this method has proved very successful too (Municipal Waterworks, 1977). Due to some additional groundwater and an equalization of fluctuating concentrations, it helps to obtain a better water quality. This can be seen from figure 5-9, showing chloride concentrations in the Rhine and in the drinking water. But here again, infiltration is not the only treatment used, as will be seen from the process scheme shown in figure 5-10.

One can see from the figure that, in addition, flocculation and filtration are used before, while activated carbon and slow sand filtration are used after dune infiltration. Average analytical data for the plant can be seen in table 5-7. By looking at the dissolved organic carbon values in particular, it is obvious that a high removal efficiency cannot be obtained through biological treatment only. The reason for this is that a large part of the organics in the Rhine water are difficult to biodegrade.

Since it was known that this could be changed by employing an oxidation process using ozone, the waterworks in Amsterdam have decided that another process scheme will be put into practice in their newest plant. This scheme is shown in figure 5-11. After flocculation and filtration, ozonation is used, followed by flocculation again and two further filtration steps. It can be expected that this will lead to a far

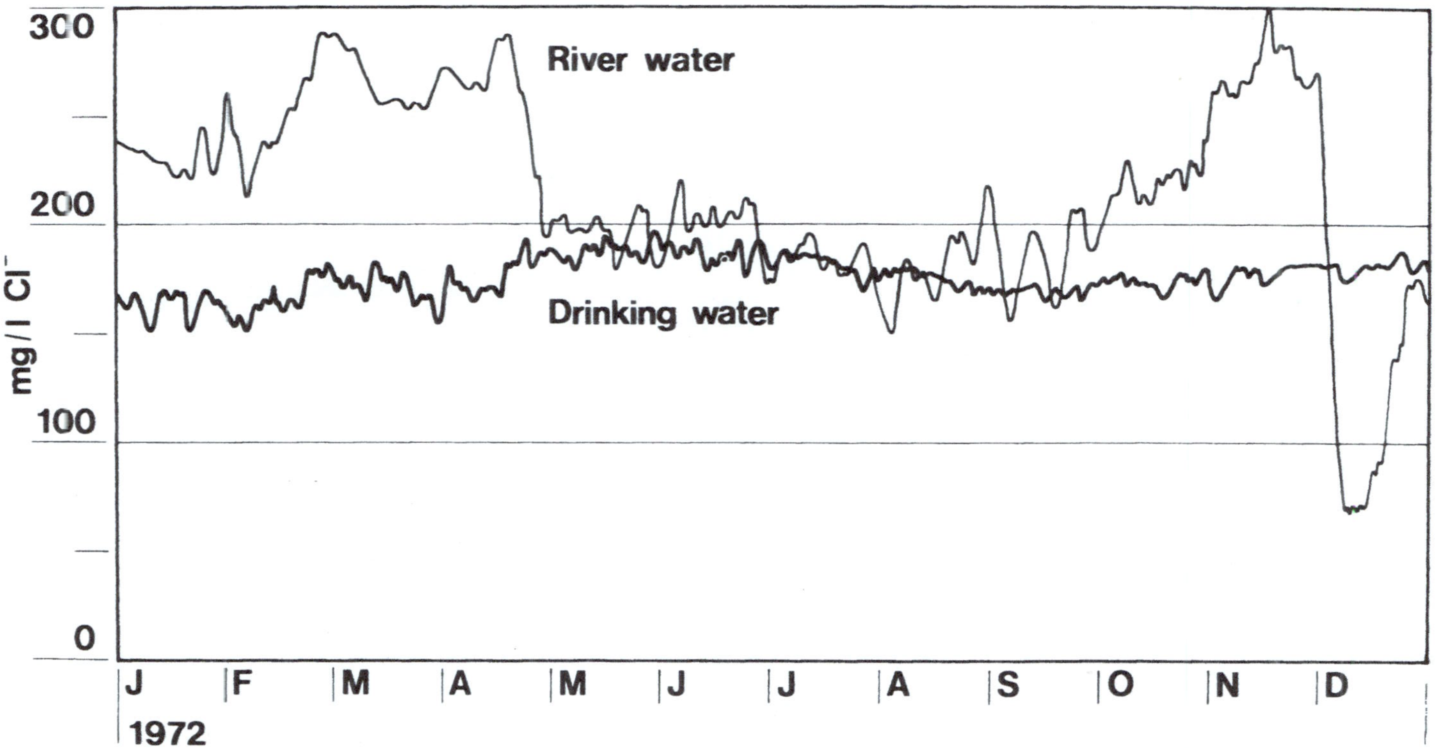

Figure 5-9. Chloride content of water entering and leaving the dune waterworks.

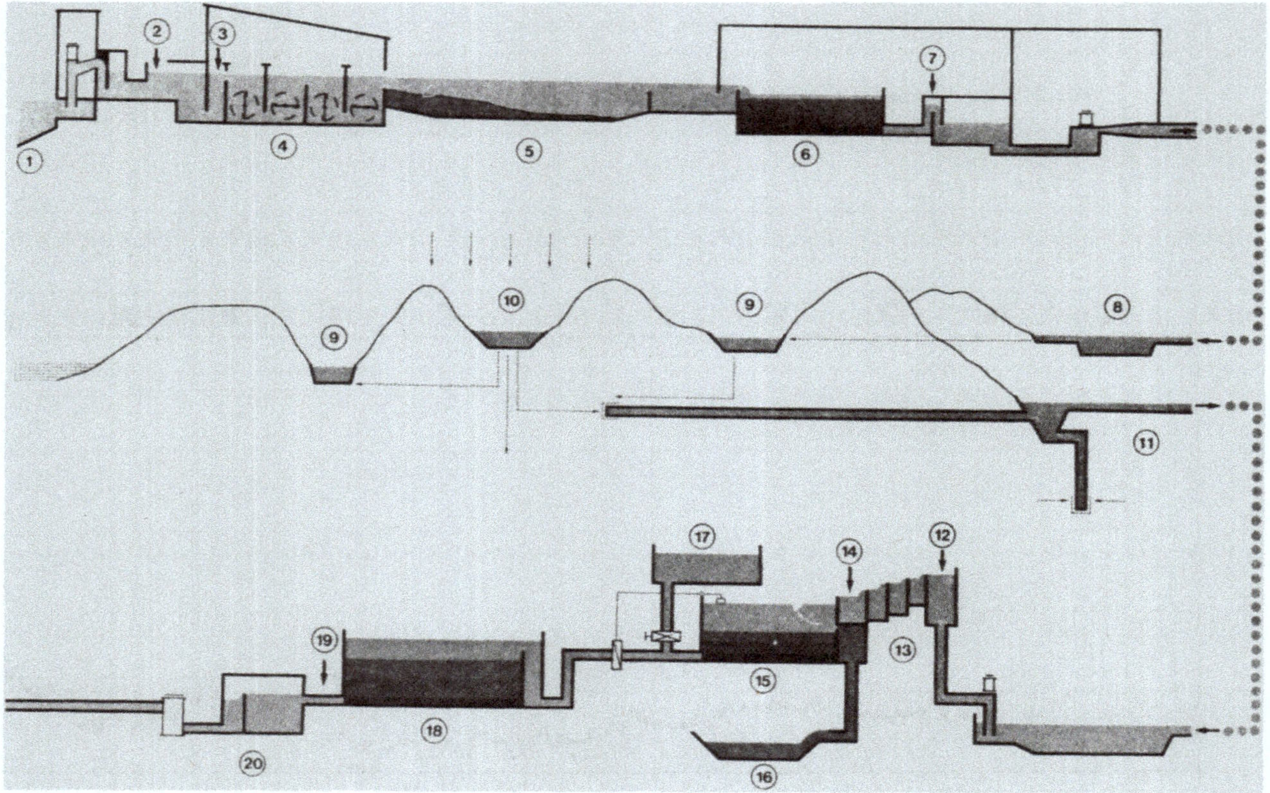

Figure 5-10. Process scheme of dune waterworks, Amsterdam. (1) Lek Canal; (2) $FeCl_3$ dosing; (3) NaOH dosing; (4) Flocculation basin; (5) Sedimentation basin; (6) Rapid filter; (7) Chlorine dosing; (8) Distribution pond; (9) Percolation gullies; (10) Collection and storage canal; (11) Oranje basin; (12) NaOH dosing; (13) Aeration; (14) Activated carbon dosage; (15) Rapid filter; (16) Silt deposit; (17) Wash water reservoir; (18) Slow sand filter; (19) Chlorine dosing; (20) Chlorine contact basin.

Table 5-7. Analytical Data for the Dune Water Treatment, Amsterdam 1976

	River water Rhine	Before dune filtration	After dune filtration	Drinking water
DOC, mg/liter	7.1	5.4	4.1	3.6
Color, mgPt/liter	28.0	15.0	11.0	8.
Threshold number	32.0	5.0	-	0
Hydrocarbons, mg/liter (oil)	0.26	0.03	-	0
Chloride, mg/liter	215.0	227.0	178.0	174.0
Bicarbonate, mg/liter	162.0	158.0	194.0	201.0
Ammonia, mg/liter	1.5	0.2	<0.05	<0.05
Nitrate, mg/liter	17.0	20.0	5.5	5.8
Phosphate, mg/liter	1.2	0.2	0.1	0.1
Iron, μ	1600.0	100.0	50.1	20.0
Chromium, μ	35.0	<1.0	<1.0	<1.0
Copper, μ	22.0	8.0	5.0	5.0
Zinc, μ	150.0	25.0	15.0	10.0
Cadmium, μ	2.9	0.2	0.1	0.1
Mercury, μ	0.5	0.1	<0.1	<0.1
Lead μ	28.0	3.0	3.0	2.0

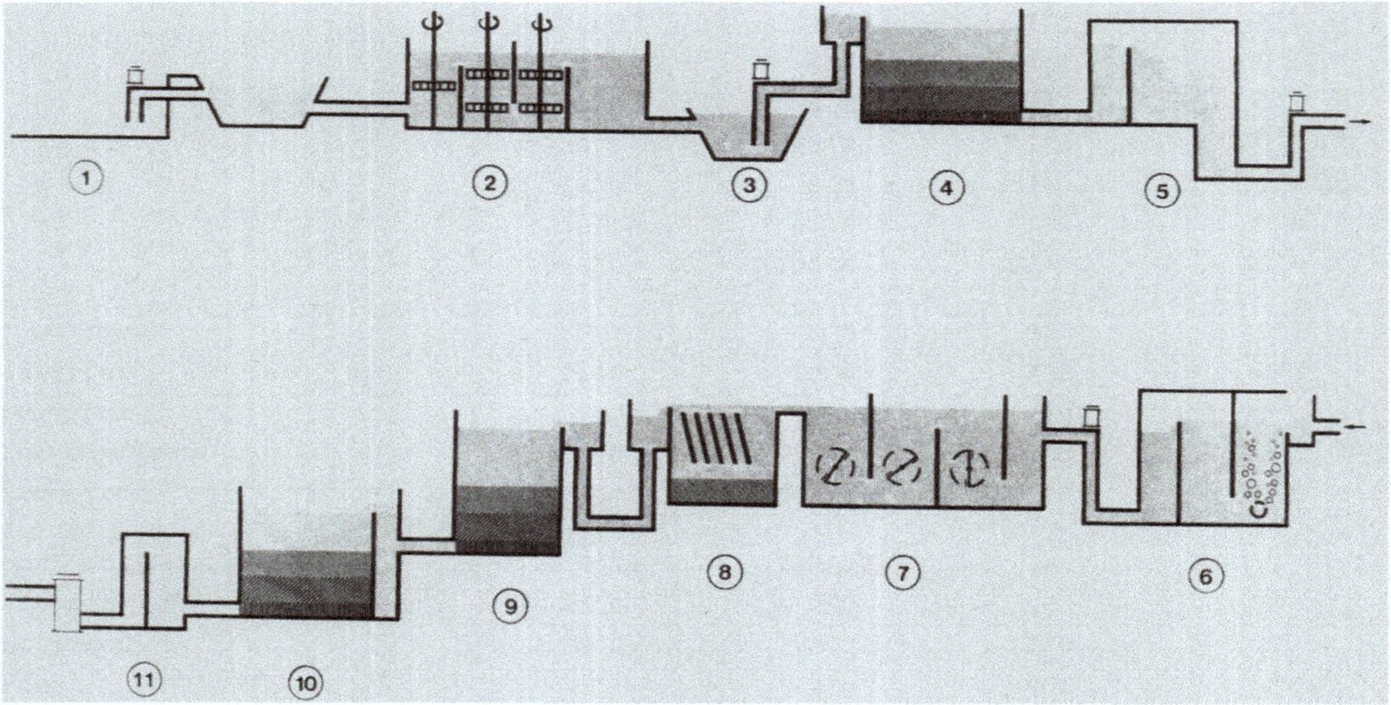

Figure 5-11. Process scheme of lake waterworks, Amsterdam. (1) Bethune polder; (2) Preliminary coagulation; (3) Self-purification basin; (4) Rapid filtration; (5) Filtrate basin; (6) Ozonation; (7) Coagulation; (8) Sedimentation; (9) Rapid filtration; (10) Slow sand filtration; (11) Chlorine contact basin.

more efficient treatment after the necessary optimization of the various treatment steps.

It may be of interest now to look at the waterworks of Zurich, Switzerland, which use good-quality Zurich lake water for their drinking water treatment. Figure 5-12 shows the processes used there twenty years ago, those of more recent times, and those used nowadays.

Filtration was the only treatment step until 1965. It was succeeded by a combination of rapid and slow sand filters using granular activated carbon on top of the slow sand filters. Prechlorination then became necessary to remove the *Dreissenia polymorpha*, but only 0.5 ppm of chlorine proved sufficient (Schalekamp, 1971).

The process scheme used today can be seen in figure 5-13. It includes flocculation and filtration, slow sand filters and ozonation followed by activated carbon filtration, and runs on the same principles as those mentioned for Mülheim and the new Amsterdam plant (Sontheimer, 1975). The results obtained with this method are given in figure 5-14 and show, without going into great detail, the specific effects of so many different treatment steps.

The last example comes from the Wahnbach reservoir waterworks in the Bonn area, which supplies this city with drinking water. The drinking water reservoir is fed by the small Wahnbach river which flows through an agricultural area with a number of villages. Because of the phosphate runoff

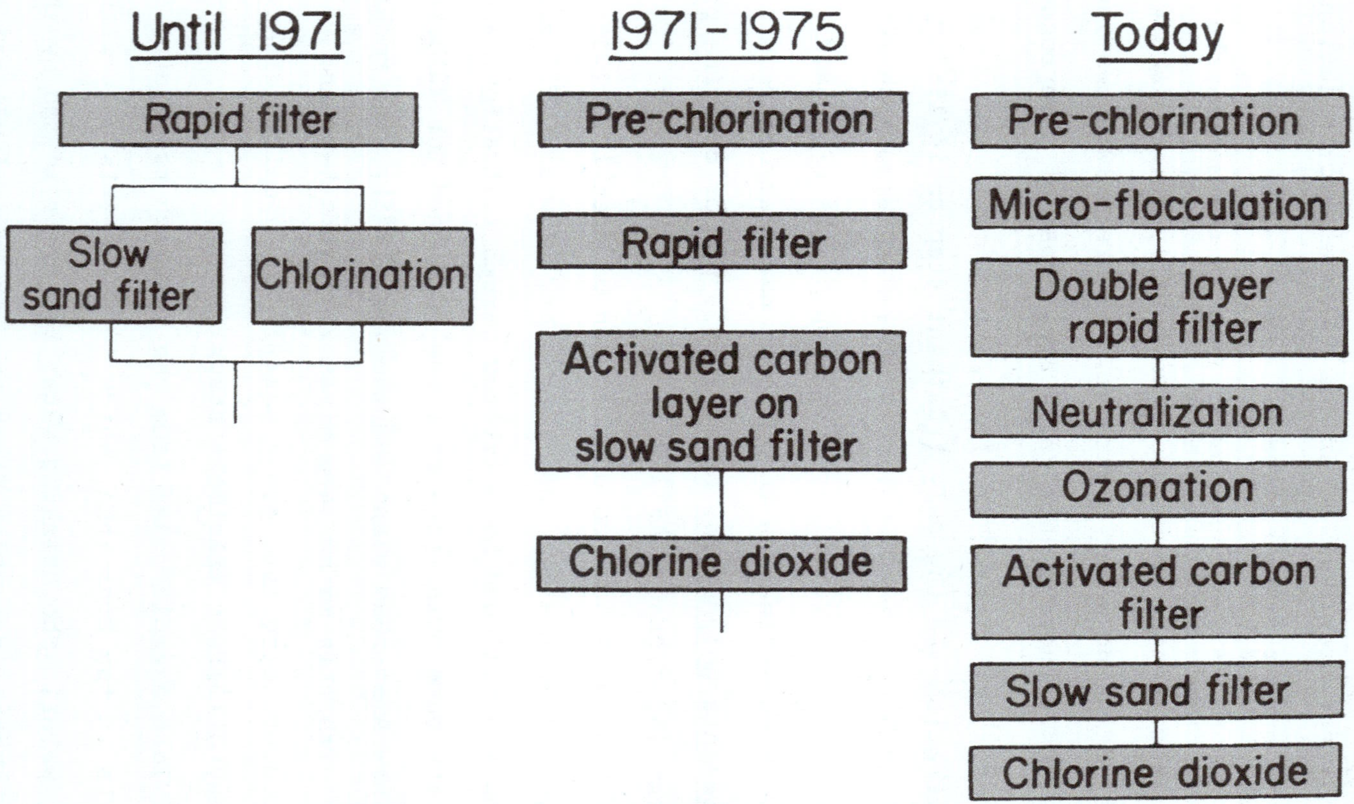

Figure 5-12. Water purification in Zurich.

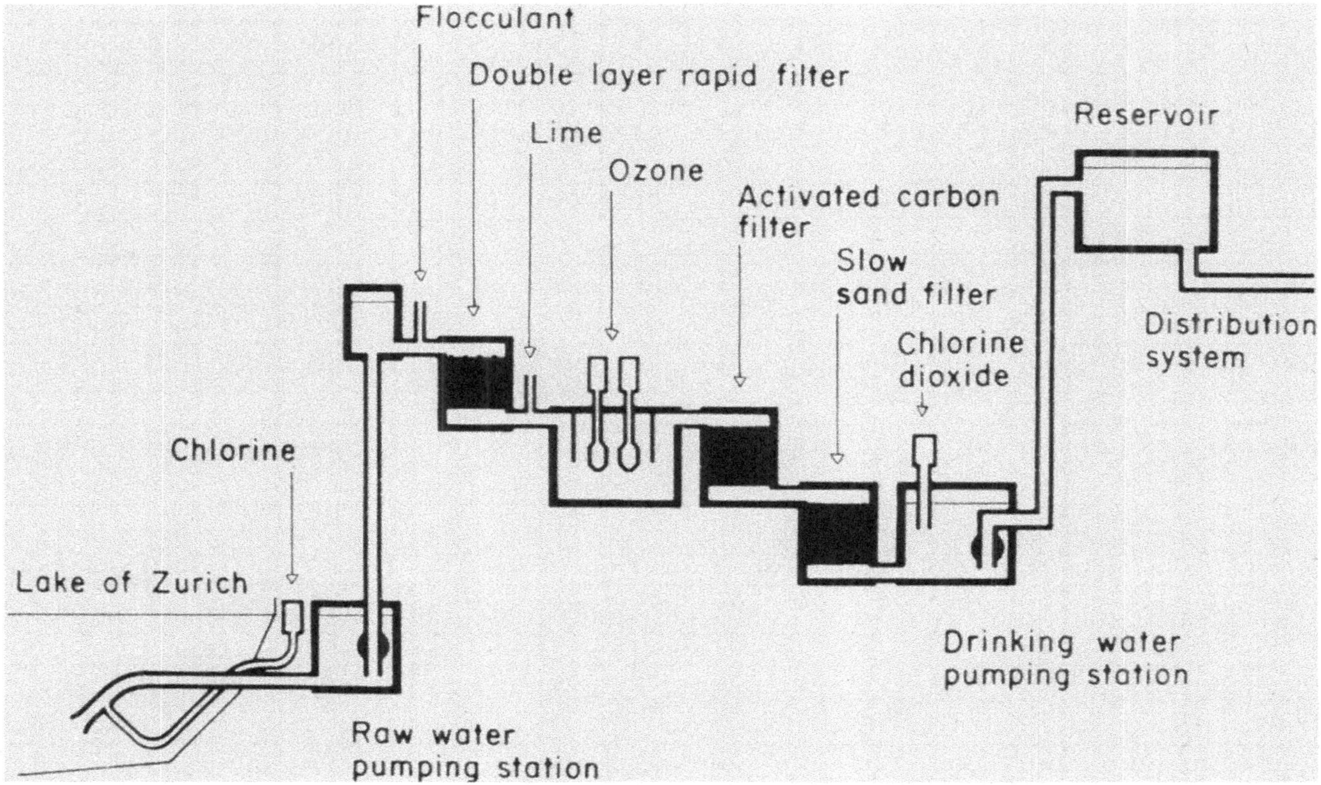

Figure 5-13. Lake water treatment plant of Lengg (Zurich).

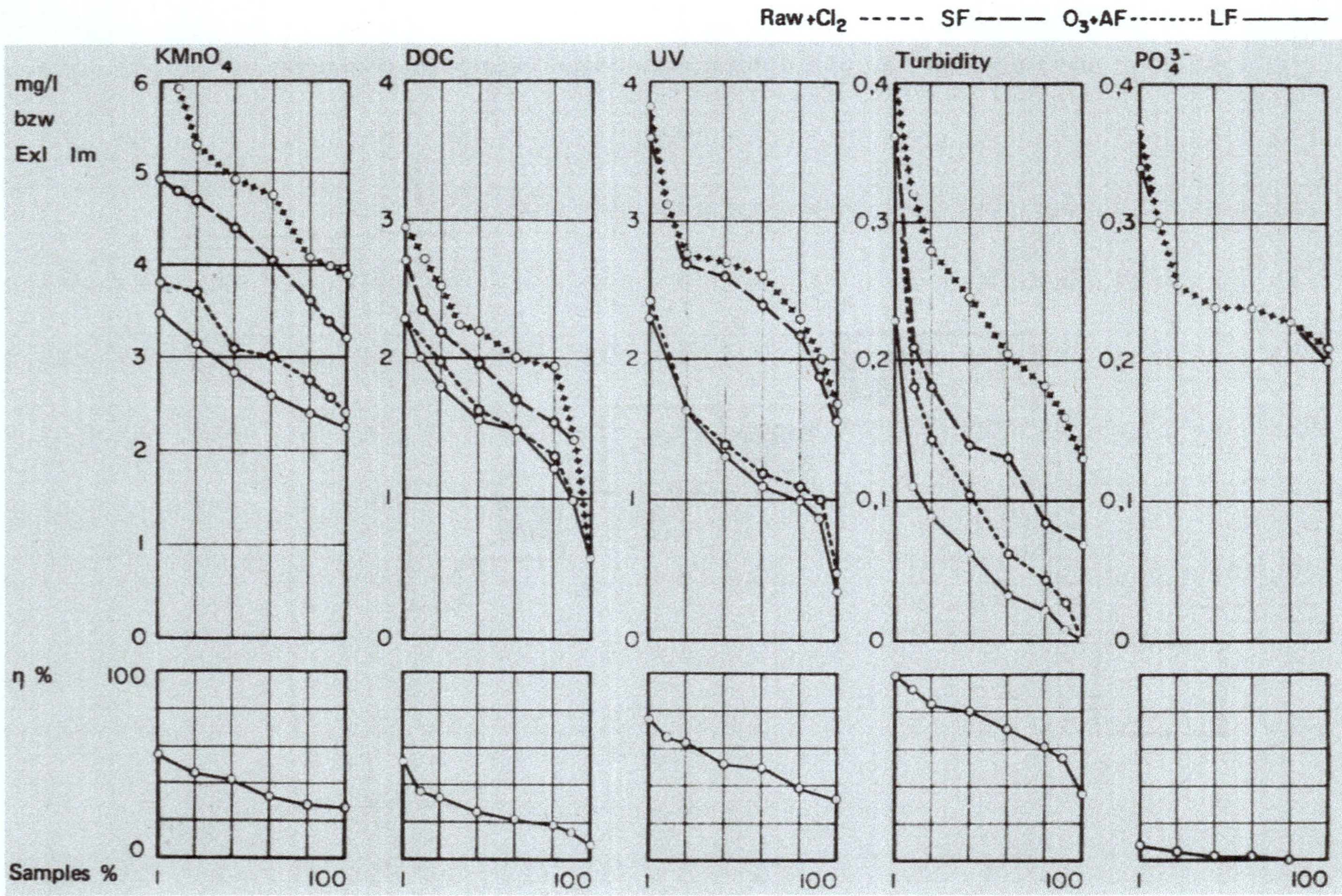

Figure 5-14. Lake waterworks, Lengg, 1974-75.

which entered the reservoir, the water became highly eutrophic, leading to difficulties during treatment and sometimes to a high demand for chlorine. One possible solution to this problem, which may become more difficult in the future, would have been to use a more complicated type of treatment, including ozone, activated carbon, etc., similar to that used in Zurich. The other solution, which has now been put into practice, was based on the general idea of avoiding the negative influence of eutrophication by removing the phosphorus before the water enters the reservoir (Bernhardt, 1976).

To do this in an economical way, extensive studies were made which resulted in the "Wahnbach process," the scheme of which can be seen in figure 5-15. The water is pumped from a primary sampling reservoir to a flocculation- and a three-layer filtration plant, where all suspended and dissolved phosphorus compounds are removed by energy-controlled flocculation, with retention times of a few minutes only. All flocs are removed, with excellent results, by filtration at velocities between 5 and 7.5 gpm/sq. ft. (10 - 17 m/hr). The results obtained at the plant can be seen in table 5-8.

The data show that the plant achieves a reduction of 93 percent and a very low turbidity of only 0.04 formazin units. The other parameters are also reduced to a large extent, and it can be concluded from the data that, after chlorination for safety, this water could be used as drinking

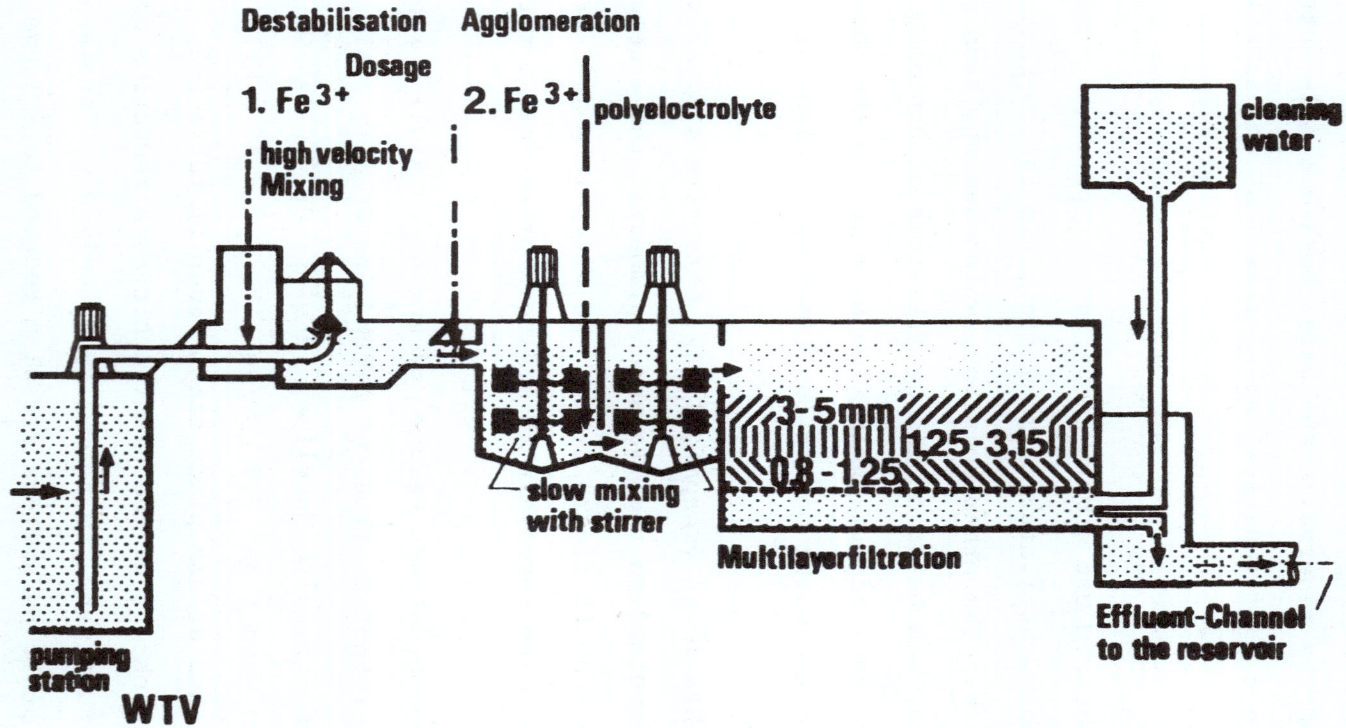

Figure 5-15. Wahnbach system. Flocculation and multilayer filtration with controlled energy input.

Table 5-8. Results of the Wahnbach Water Treatment Plant

		Inlet mean values	Outlet mean values	Elimination mean values
Total phosphorus	µg/liter	72	4.3	93.2
O-PO_3^{-3} phosphorus	µg/liter	20	1.1	94.5
Turbidity TE/F	90°/420 nm	5.9	0.04	99.2
COD	mg/l	8.0	1.9	77.8
Chlorophyll	µg/liter	16.6	1.7	88.7
DOC	mg/l	2.1	1.0	49.5
Bacteria counts	ml^{-1}	8205	214	92.9
Coliforms	ml^{-1}	4435	7	99.7
UV Absorbance at 254 nm	m^{-1}	7.4	2.2	70.9

water without any further treatment. In this particular case, the water is conducted to the reservoir. The good quality of the water should help to avoid further eutrophication and improve the water quality in the reservoir, requiring a simple treatment only with flocculation and filtration.

This process is a good example of how one of our general guidelines on drinking water treatment is implemented. As a general recommendation, all steps should be avoided which lead to a deterioration of water quality. In the example cited, eutrophication would impair the quality of the water, and subjecting the water to oxidation by using chlorine would also lead to deterioration. Organochloro compounds formed through this method are even more difficult to remove than the organics originating from algae, or the algae themselves. So we may learn from this story that it makes sense to avoid all process steps in treating water including raw water storage if necessary, which would lead to a deterioration of the water quality. Using chlorination to prevent the formation of algae, as is done at some waterworks, will only replace one type of deterioration by another one and will not help at all.

The next recommendation regards the use of different treatment steps. The higher the pollution and the higher the amount of possible pollutants, the higher should be the number of treatment steps in order to guarantee complete removal of all undesired substances. If we can do this, if we try to use biological treatment processes for all biodegradable substances and for those which can be made biodegradable by ozonation, and if, finally, we use physicochemical processes such as adsorption for anthropogenic hazardous substances, then we can be sure of obtaining a natural and pure drinking water.

Methods for the Analytical Control of Drinking Water Quality

Recommendations such as those mentioned will only be useful when it is possible to control the relevant parameters. It will not be sufficient to measure the organic pollution of raw water or drinking water by means of the usual summary parameters. A better method would be to have different types of these parameters, such as total organic carbon and dissolved organic carbon, and to correlate these data, but even this method has certain limits.

Another way would be to measure single substances, such as was done for dichlorobenzene in the Rhine river. The data are shown in figure 5-16. In this figure the cumulative frequency distribution for a definite period is plotted, and shows that the maximum values may exceed the mean values for the total period by a factor of 10. So, in order to get useful information on the behavior of single substances, it will

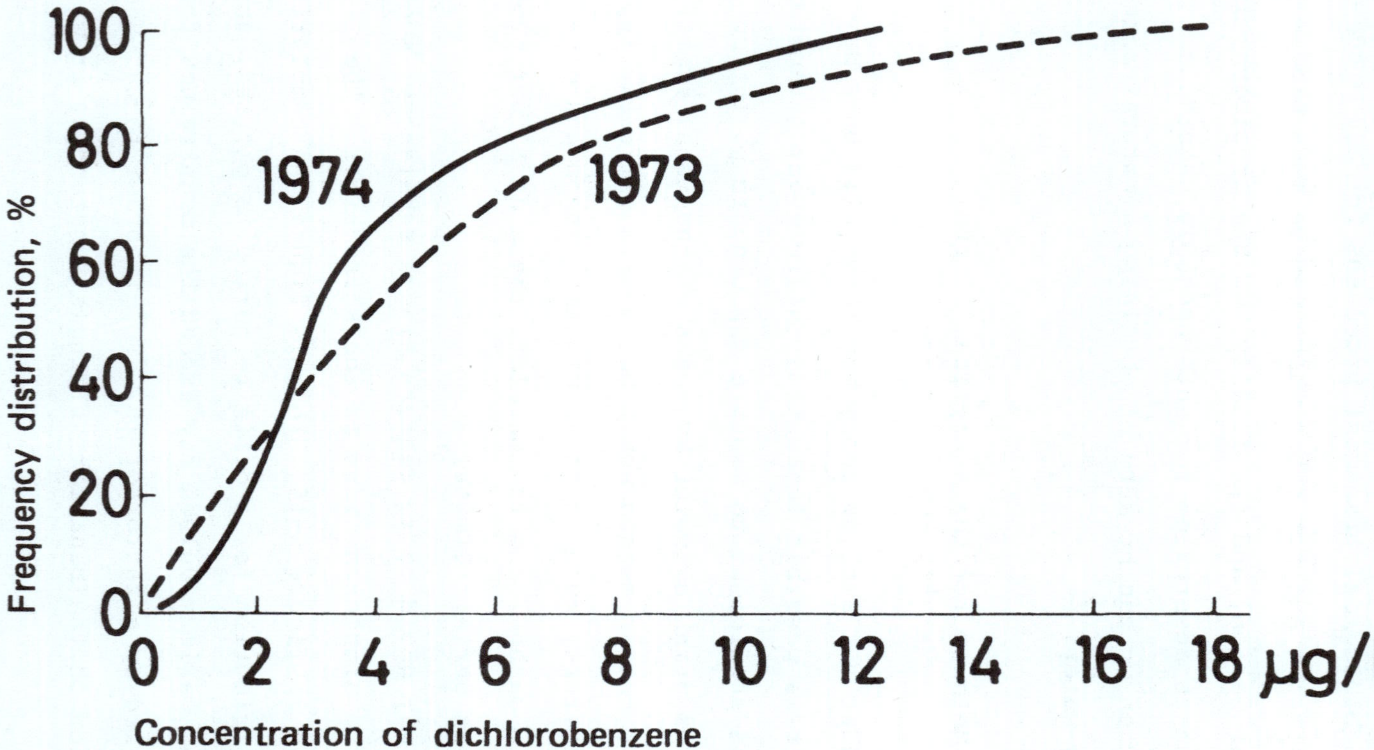

Figure 5-16. Frequency distribution for concentrations of dichlorobenzene at Waal bei Brakel (according to Meijers).

be necessary to carry out a fair number of measurements, which certainly is a drawback.

Another drawback of measuring single substances is shown in table 5-9. Mean data per month are shown for samples taken at three different parts of the Rhine, with special emphasis on the organochloro compounds. They show that in certain parts of the river there are particularly high concentrations of certain organics. However, the total amount of the single solute concentrations measured is still much smaller than that of the total organic chlorine, and the question arises why only single solutes are considered dangerous and not any of the other organochloro compounds, given by the group parameter dissolved organic chlorine (DOCl) (Sontheimer, 1972). The measurement of group parameters does indeed produce fairly good information on the average composition of organics in a river. Figure 5-17 shows the results of such measurements of Rhine water in 1974. The portion of each organic is given in percentage, and it must be remembered that the total concentration will decrease as the river flow increases (Sontheimer and Wagner, 1977). One can see that there are a few large groups of organics which can be measured by group parameters, and also that the concentration of those organics which are not present in natural waters is particularly high at low water flow.

It can be concluded from our measurements that the amount of organochloro and organic sulfur compounds can be taken as

Table 5-9. Measurements for Single Substances and Summary Parameters for Three Places on the Rhine (mg/m^3)

		Basel	Koln	Duisburg
Chloroform		1.1	0.8	1.1
Carbon tetrachloride		0.3	1.4	3.3
Chloroethanes		2.5	0.2	0.6
Trichloroethylene		0.9	0.8	0.6
Tetrachloroethylene		0.3	0.8	1.5
Chlorobutadienes		4.4	0.4	0.3
Chlorobenzenes		0.9	0.6	5.3
Chlorotoluenes		-	0.5	1.3
Benzene		0.2	0.3	0.8
Toulene		0.8	0.7	1.9
$\emptyset - C_2$	Benzene with 2 C-groups	0.2	0.7	1.4
$\emptyset - C_3$	Benzene with 3 C-groups	0.1	0.4	1.0
$\emptyset - C_4$	Benzene with 4 C-groups	-	0.1	0.5
DOCl	Dissolved organic chlorine	55	78	100
DOClN	Nonpolar dissolved organic chlorine	10	5	15
DOC	Dissolved organic carbon	2300	4600	5000

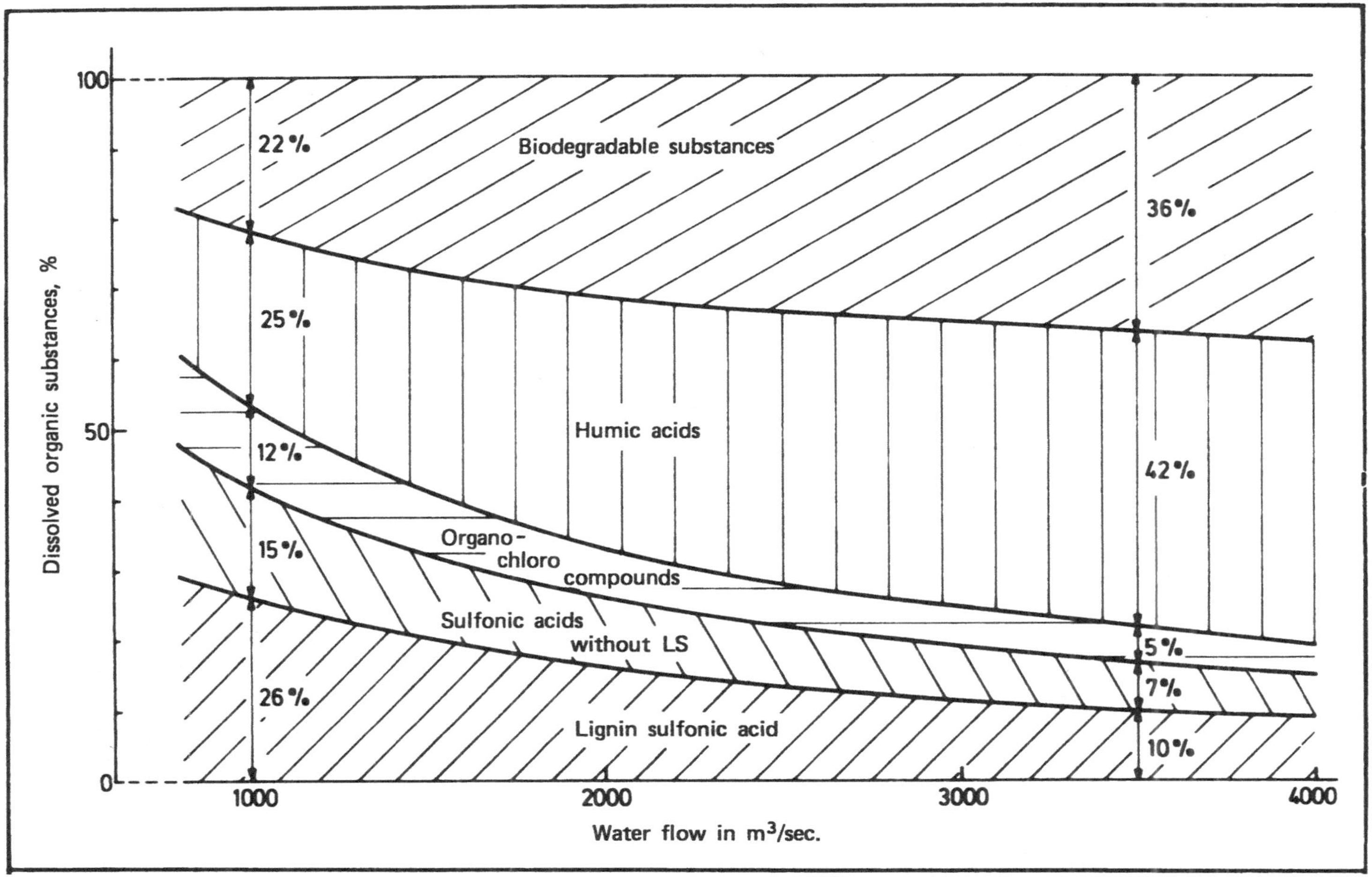

Figure 5-17. Breakdown of dissolved organic substances into principal components for different water flows on the lower Rhine.

a guide number to the degree of anthropogenic pollution of a raw water or drinking water. As table 5-10 shows, quite differing values are obtained from different rivers, which indicates that each river contains different loadings of anthropogenic organics. In order to obtain further information, important for the efficiency of drinking water treatment, special analytical methods have been developed, such as measuring floccability or adsorbability. Figure 5-18 is a drawing of isotherms, measured before and after flocculation, for waters from different rivers. It is evident that the results vary greatly. This can serve as an indication of the appropriate treatment method to be used. Similar tests can be made on oxidation (Rhine Waterworks Association, 1976). These few examples show that there are indeed methods to control water quality and that it is not really all that difficult to put them into practice. The same methods can be used to control the effectiveness of the treatment plant, too.

Conclusions

It is possible to draw some general conclusions on the various points discussed in this paper. Regarding drinking water policy in particular, the following basic points may be of special importance:

Table 5-10. Average Values for Control Filter Analyses (1974)

	DOC (g/m^3)	Adsorbable organics (g/m^3)	Nonpolar organics (g/m^3)	Organic sulfur (mg/m^3)	Organic chlorine (mg/m^3)	Nonpolar organic chlorine (mg/m^3)
Lake Constance	1.4	1.3	0.3	5	6	3
Rhein (Basel)	2.8	2.9	0.4	43	66	18
Neckar (Heilbronn)	4.6	4.6	0.4	53	50	6
Donau (Leipheim)	2.5	2.0	0.4	20	23	5
Lower Rhine (beyond Dusseldorf)	6.0	6.5	1.1	330	160	40

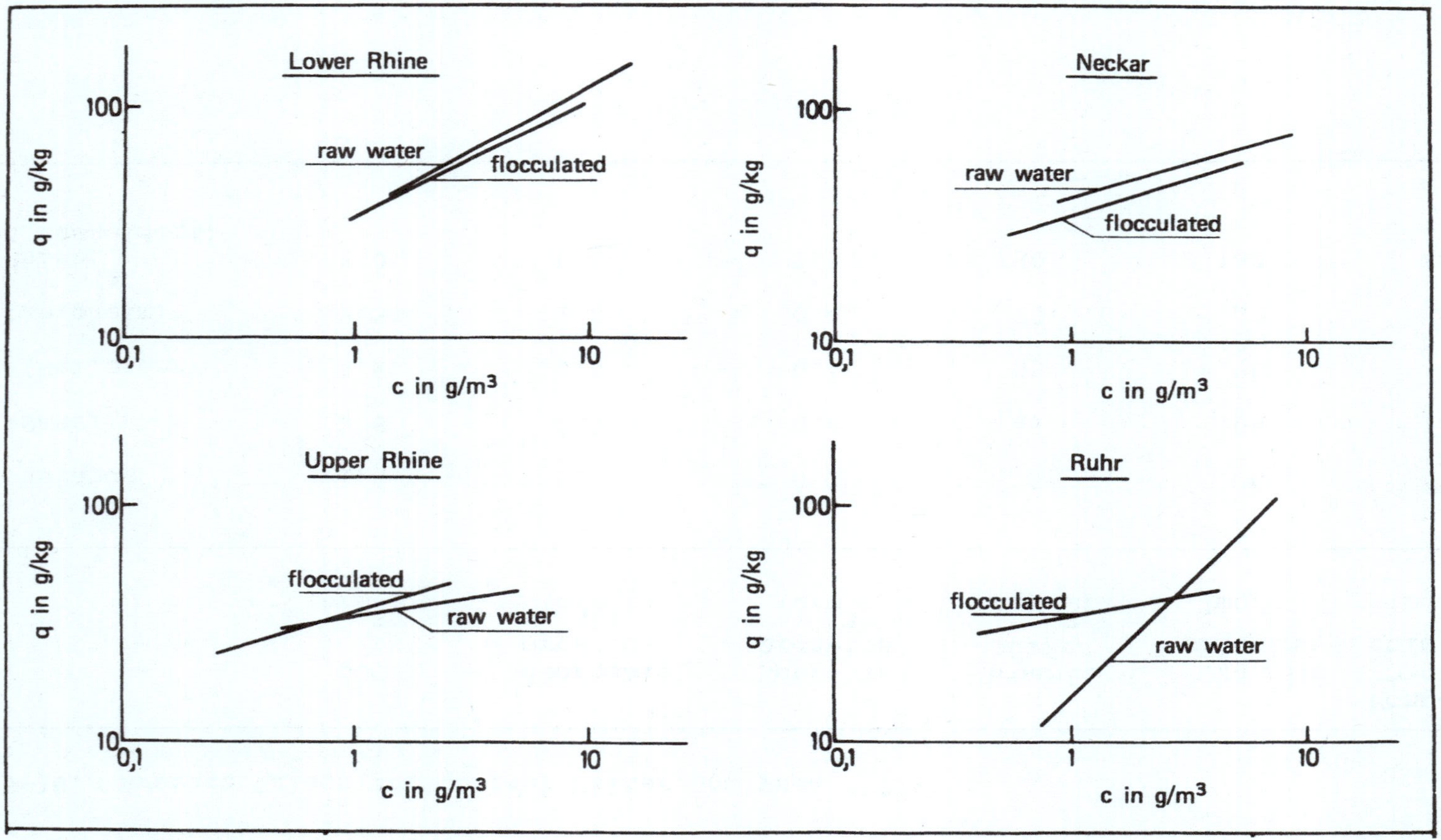

Figure 5-18. Adsorption isotherms of river water before and after flocculation (average values 1976).

1. It seems very difficult to control the quality of drinking water or to direct future developments in this field by laws and regulations only, such as laying down standards, permissible concentrations, etc. The problems arising at different places and with different types of raw water are so complex that it is almost impossible to prescribe general process conditions or data which will be useful in all these cases.

2. Since it will probably be impossible to obtain sufficient information about the effects on human health of all possible water components, standards based thereon will always be limited and very often they will come too late. It is also difficult to derive conclusions on the best process combinations and conditions to be used in water treatment from these data alone.

3. In my opinion, it is far more important and helpful to approach problems on drinking water quality with the view that all human beings have a right to "natural and pure drinking water."

4. This guideline can then be followed in making decisions on the most suitable treatment process or process combinations to be used for any type of raw water, paying attention to the special conditions existing at each individual place.

5. We have to remember that a safe drinking water treatment can be achieved only by using a combination of different processes, and not merely by analytical control. Nevertheless, some of the analytical methods will help to optimize the treatment process, and here group parameter measurements have proved most effective and should be used and developed far more intensely than has been done so far.

6. There are many questions which cannot be answered yet and many problems which still have to be solved if "natural pure water" is to be made available for drinking. This is true for the United States and also for Western Europe. But making use of this general idea will help us to go in the right direction, and this seems worthwhile enough to be considered seriously.

References

Arbeitsgemeinschaft Rheinwasserwerke (ARW) - Jahresbericht. 1976. (Rhine Waterworks Association - Annual Report.)

Bernhardt, H. 1976. "Die technische Konzeption der Massnahmen zur Verhinderung der fortschreitenden Eutrophierung der Wahnbachtalsperre," Limnologica vol. 10, no. 2, pp. 623-635. (The engineering basis of measures for the prevention of the continuous eutrophication of the Wahnbach reservoir.)

Kölle, W., and H. Sontheimer. 1968. "Wassergefährdende Substanzen," Gas-Wasser-Warme (Switzerland) vol. 22, pp. 226-236. (Water-endangering substances, Gas-Water-Heat.)

______, ______, and L. Stieglitz. 1975. "Eignungsprüfung von Wasserwerksaktivkohlen anhand ihrer Adsorptionseigenschaften für organische Chlorverbindungen," Jahrbuch Vom Wasser vol. 44, pp. 203-217. (Suitability test for activated carbon in waterworks on the basis of its adsorption properties for organochloro compounds, Yearbook About Water.)

Kühn, W., F. Fuchs, H. Sontheimer. 1977. "Bestimmung des organisch gebundenen Chlors mit Hilfe eines neuartigen Anreicherungsverfahrens," Zeitung f. Wasser- und Abwasserforschung vol. 6, no. 10. pp. 192-193. (Studies on the determination of chlorine in organic compounds by means of a new enrichment process, Journal for Water and Waste-water Research.)

Park, Y.K., and H. Sontheimer. 1974. "Beitrag zur Untersuchung von unpolaren organischen Chlorverbindungen in Wässern und auf Aktivkohlen," Jahrbuch Vom Wasser vol. 43, pp. 293-313. (Contribution to the analysis of nonpolar organochloro compounds in water and on activated carbon, Yearbook About Water.)

Poggenburg, W., C. Engels, F.J. Weissenhorn, F. Fuchs, H. Sontheimer. 1973. Untersuchungen zur Aktivkohleanwendung bei der Aufbereitung von Rheinuferfiltrat. DGVW-Schriftenreihe Wasser Nr. 7. (Studies on the use of activated carbon in the treatment of Rhine bank filtrate. DVGW-Series Water No. 7.)

Sander, R., W. Kühn, H. Sontheimer. 1977. "Untersuchungen zur Umsetzung von Chlor mit Huminsubstanzen," Zeitung f. Wasserun Abwasserforschung vol. 5, no. 10, pp. 155-150. (Studies on the transformation of chlorine by humic substances, Journal for Water and Waste-water Research.)

Schalekamp, M. 1971. "Warnung vor der Wandermuschel Dreissenia polymorpha Pallas und Bekampfung derselben," Gas-Wasser-Abwasser vol. 3. (Warning concerning Palla's migratory shellfish Dreissenia polymorpha and control measures against it, Gas-Water-Waste Water.)

Sontheimer, H. 1972. "Chemische Parameter der Gewässerverschmutzung," Chemie in unserer Zeit vol. 6, pp. 175-183. (Chemical paramaters of water pollution, Chemistry in our Time.)

______. 1975. "Moderne Verfahrenstechnik der Wasseraufbereitung," Gas-Wasser-Abwasser vol. 9, no. 55, pp. 464-466. (Modern water treatment technology, Gas-Water-Waste Water.)

______. 1977. "Neue Technologien und Trinkwasserqualität," GWF-Wasser/Abwasser, vol. 118, no. 4, pp. 165-173. (New technologies and drinking-water quality, GWF-Water/Waste Water.)

______, R. Gimbel. 1977. "Untersuchungen zur Veränderung der Fracht an organischen Wasserinhaltsstoffen mit der Wasserführung am Beispiel des Rheins," GWF-Wasser/Abwasser vol. 118, no. 4, pp. 165-173. (Studies on the alteration of the load of organic substances contained in water caused by water-flow using the example of the Rhine, GWF-Water/Waste Water.)

______, W. Nissing. 1977. "Anderung der Wasserbeschaffenheit bei der Bodenpassage unter besonder Berücksichtigung der Uferfiltration am Niederrhein," Gas-Wasser-Abwasser vol. 9, no. 57, pp. 639-645. (Alteration of water composition in passage through soil with special attention to the lower Rhine bank filtration, Gas-Water-Waste Water.)

______, I. Wagner. 1977. "Zur Bestimmung von Huminsäuren und Ligninsulfonsäuren aus den UV Spektren," Zeitung f. Wasser- und Abwasserforschung vol. 3/4, no. 10, pp. 77-82. (On the determination of humic acids and lignin sulfonic acids from the UV spectra, Journal for Water and Waste-Water.)

______, E. Heilker, M. Jekel, H. Nolte, F.H. Vollmer. 1978. "The Mühlheim Process." Experience with a new process scheme for treating polluted surface waters. To be published in the Journal of American Waterworks Association.

Chapter 6

BRITISH EXPERIENCES ON PROBLEMS OF DRINKING WATER QUALITY

John W. Atkinson

I should like to highlight some of the problems that we face in England, and initially it is necessary for me to set the scene of the water supplies in the United Kingdom. However, may I say at the outset that there appears to be only one major difference between the problems of the United States and ours, and that is waterway size and its resultant effect on residual concentrations.

In 1975 the actual annual withdrawal of water for use in treated public water supplies in the United Kingdom was about 5.6×10^9 m^3 per year, of which roughly one-third was groundwater, one-third upland or impounded waters, and one-third was river-derived water. At present our total consumption of treated water is fairly stable at the above value; industrial demand has eased, partly because of reuse and recirculation, but domestic demand continues to increase because of changes in living styles.

Water Sources and Associated Difficulties

Groundwater sources usually present few problems: iron and manganese may be present and hardness may be high, but, in the majority of cases, disinfection is the only treatment necessary. The iron and manganese levels can be reduced by simple chemical oxidation methods, and hard waters may be softened, although there is a strong feeling in the United Kingdom that, because of the apparent link between cardiovascular disease and soft waters (Shaper, Clayton, and Morris, 1974), softening should no longer be performed as part of water treatment. It is not, therefore, the practice to carry out softening on any new sources of potable water, but if extensions are being made to existing treatment plants, the full former treatment may be applied if it is required.

In a paper presented at a meeting in 1976 (Robertson, 1975), it was stated that there was a correlation between the sodium levels of waters and the medically unexplained deaths of infants while asleep in their beds. This statement caused some concern for a while but no new evidence to substantiate the hypothesis has been produced, and attempts to correlate sodium levels of blood with those in water appear to have failed.

Some groundwaters may contain high levels of nitrate and aggressive carbon dioxide, thus leading to corrosion difficulties. These phenomena will be discussed later.

Many of the United Kingdom's impounded waters are situated in remote, barren areas, their inaccessibility at the time of reservoir construction being regarded as a means of ensuring a water quality free from human pollution. However such waters tend to be extremely soft, of very low turbidity, and of high color (due to the humic acids present in peaty vegetation); pH values can be as low as 3.0, and it is not unknown for iron, manganese, and aluminum to be naturally present in these waters. The treatment difficulties are obvious.

River-derived waters cause the most concern. Apart from the sewage effluent discharges in the rivers, pollution may also occur because of industrial discharges and agricultural drainage, and all too frequent accidental spillages, for example, from road accidents. It must be remembered that since most of our rivers are small compared with those in the United States, a spill into a river used as a source for potable water in the United Kingdom could be a major disaster, whereas it would be barely noticeable in the United States and Canada. Compare, for example, the river Dee in Chester, with an average residual flow of about 675 x 10^3 m^3 per day (180 x 10^6 U.S. gal. per day) after total abstractions of a similar amount have been made, and that of the Ottawa River in Canada with an average flow of 300 x 10^6 m^3 per day (78,000 x 10^6 U.S. gal. per day) and an abstraction of less than 450 x 10^3 m^3 per day (120 x 10^6 U.S. gal. per day) for treatment for water supply. The UK rivers may contain up to 35 percent sewage effluent at average flows, and it must be noted what that proportion could become, given less than average flows. Results for 1975 have indicated that about 45 percent of the total abstractions had a less than 5 percent effluent while 30 percent had a 15 to 20 percent effluent. Land drainage from agricultural areas presents its own problems: nitrates, phosphorus, and herbicides and other pesticides can be more than a nuisance. Industrial discharges are a constant headache, though we now have more power to regulate the quantity and quality of such discharges (Control of Pollution Act, 1974).

In the United Kingdom, there are, in all, 137 abstraction sites, not necessarily intakes, for potable supplies from 89 rivers and canals. Another 11 sites are either under construction or are planned. At present there are 151 treatment plants providing physical treatment, chemical treatment; and disinfection, and of these, 67 have provision for activated carbon treatment though, normally, carbon is used only for taste and odor troubles and to remove color. It has been shown that activated carbon is relatively inefficient in removing trihalomethanes, and this confirms Dr. Sontheimer's findings.

One hundred thirty-two of these treatment plants, a majority, use coagulation processes prior to filtration--in nineteen plants, slow sand filtration is employed, and in eighteen plants slow sand filtration is carried out without coagulation.

Slow sand filtration is not normally a method of treatment at the more modern plants, partly because of costs, so that rapid filtration and pressure filtration are the most favored processes. Chlorination is used for disinfection on the majority of the treatment plants, but chlorine dioxide is used on seven and ozone on three. Let me note in passing that ozone is used on one of the large supplies from the Lake District for color removal.

Storage is employed on about half the plants, varying from a half day to several years: the longer storage time is related more to water conservation than to water quality; the shorter periods are intended to cater for intermittantly poor water quality by facilitating the bypassing of the contaminated water. Some years ago a report (First Annual Report of the Steering Committee on Water Quality, 1971) was published in which a recommendation was made that all river abstractions should incorporate seven-day storage reservoirs. This particular recommendation is now being restudied in the light of further experience, for it has been noted that even if the bacterial counts in the water decrease, other aspects of water quality may show a deterioration. On the mechanical side, it is debatable whether pure plug flows are possible or whether complete mixing ever takes place. The importance of adequate monitoring of water quality at all times should not be underestimated.

Standards

A few comments about water quality standards are required. The bacteriological standards adopted are those set down in "The Bacteriological Examination of Water Supplies" (1969), first published in the 1930s and now, after several revisions, being rewritten. This booklet also recommends the methods of examination and includes guidance on sampling frequencies and techniques. Chemical and radiochemical standards are usually those laid down by the World Health Organization (1971) although recommendations from the American Public Health Association and others are used as guidelines.

In 1975 the United Kingdom formally became a member of the European Economic Community (EEC). Automatically it became a member of EUREAU (European Water--"EAU"), a group of experts involved in the scientific aspects of water supply in the EEC countries. Unfortunately, by that time the first environmental program had already been planned and the influence of the United Kingdom was not as great as it could have been. Nevertheless, there is generally agreement and acceptance of the directives produced by the EEC although the final directive may only appear after many hours of discussions. The directives themselves are designed to improve the environment, that is, they have similar aims to the directives of the EPA and, so far, three directives relating to water quality have been implemented.

One directive concerns the quality of surface water intended for abstraction for drinking water (EEC, 1975). It assesses waters on the basis of their treatment requirements but does allow countries to apply for waivers where they cannot meet the required standards. These waivers allow for the difficulties where natural "pollution" has created problems, for example, where iron, manganese, and color are higher, or pH levels are lower, than the obligatory levels.

Another is devoted to the quality of bathing waters (EEC, 1976a). This directive not only covers bathing beaches but also stretches of rivers and lakes where people congregate to bathe. Swimming pools have had to be excluded.

The last concerns pollution caused by certain dangerous substances discharged into the aquatic environment of the Community (EEC, 1976b). This seeks to control the emission of specified substances into the aquatic environment by means of emissions standards and environmental quality objectives in the receiving water. A "black" list and a "gray" list are included, and among the chemicals listed are heavy metals, for example, mercury and cadmium; and organohalides, for example, dieldrin, aldrin, and endrin.

Further proposals for the water directives include the dumping of wastes at sea; the quality of water for human consumption; the quality requirements for waters capable of supporting freshwater fish; the quality requirements of waters capable of supporting shellfish growth; the pollution caused by certain dangerous substances discharged into groundwater; and the quality of water used for agriculture.

Particular Problems in the United Kingdom

I would now like to comment specifically on the topics which are of present concern in the United Kingdom.

Lead in Drinking Water

The WHO limits for lead in drinking water have for years been set at 0.1 mg per liter in the running water and 0.3 mg per liter in the water after it stands overnight. (To many scientists, it is impossible to define overnight so that the value of 0.3 mg per liter becomes almost meaningless.) The EEC recommendation is to reduce the level to 0.05, and the recent EPA value is 0.025 mg per liter. After discussions with WHO experts in London, in September 1977, it was decided that for the time being 0.1 mg per liter would remain accepted as the standard.

In 1976 a national Lead in Drinking Water survey (Department of Environment, 1977) was carried out in the United Kingdom. Some 2,600 statistically selected samples were taken throughout England, Scotland, and Wales in an attempt to assess the size of the lead distribution and to assess the costs of replacing lead piping should it be necessary. The survey produced expected and many unexpected results. It demolished the long-held belief of many that plumbosolvency was only related to soft waters, but it also highlighted the importance of standardizing sampling techniques and analytical methods. Further work on medical aspects has been performed; a duplicate diet study has recently been completed and attempts have been made to correlate lead in water with lead levels in blood. (These results will be published when completed.) Meanwhile, individual water authorities continue to analyze for lead as part of the normal sampling programs. Where lead values appear to be high, the initial reaction is to try to ensure that treatment procedures are correct, although in certain old distribution areas, correct treatment to modern standards can result in the stripping of "protective coatings" laid down after many years of pipe service, thus resulting in "dirty water" problems and where the "dirt" has extremely high lead values.

Should treatment alterations fail to remove the high values of soluble lead, complete pipe replacement may be the only answer. It is necessary here to stress the word <u>complete</u>. Further work is being carried out by the Water Research Centre in conjunction with the water authorities in England on:

1. Identifying priority areas within distribution systems
2. The influence on plumbosolvency of variations in type of pipe
3. Seasonal variations (including thermal effects) in water quality and also the effects of mixing waters

4. Corrosion inhibitors and the effects of removing humic substances

5. Monitoring prior to and after the introduction of control measures

6. The economic aspects of plumbosolvency control.

Nitrates in Drinking Water

Over thirty years ago, Comly (1945) made the first positive attempts to relate the nitrate content of drinking water to methemoglobinemia for he was the scientist who described "well-water cyanosis," having noted that the illness tended to occur among bottle-fed infants where well waters, rich in nitrates, were used to make up the infant's formulas. The cause of the trouble is not really nitrate but nitrite, the compound produced by nitrate-reducing bacteria in the child's gastrointestinal tract. The nitrite then oxidizes hemoglobin to methemoglobin which itself is unable to carry oxygen to the tissues, and the child can succumb to cellular anoxia.

It is worth noting that, in adults as well as young children, nitrites can react with secondary and tertiary amines in the food to form nitrosamines which are potentially carcinogenic, teratogenic, and mutagenic. In the light of evidence which became available, limits of nitrate in drinking water were fixed. The 1971 WHO International Standards for Drinking Water recommended a maximum level of 45 mg per liter of NO_3; above this level, infants less than one year old were considered to be at risk. The first edition of WHO European Standards gave a value of 50 mg NO_3 per liter, but the second edition revised this to: recommended level, <50 mg per liter; acceptable level, 50 to 100 mg per liter; and unacceptable level, >100 mg per liter. These values are generally in accordance with British experience, for no cases of methemoglobinemia have been reported where the nitrate content was <100 mg per liter of NO_3.

Monitoring of water sources has indicated that many UK rivers have shown a steady increase in nitrate over a period of years, and the drought of 1976 caused a sharp upturn in the curve. The same effects are now being noted in some groundwaters. Because of these increases in particular localities, for example, the Thames Water Authority and Anglian Water Authority, other sources of water had to be considered. The overall national policy of action now adopted is based on the WHO European Standards; that is, for 0-50 mg per liter of NO_3, no action is to be taken; for 51 - 100 mg per liter of NO_3, a warning is to be given to environmental officers and medical officers of health; and for >100 mg per liter of NO_3, bottled water is to be supplied to infants as required. Because of the 1976 drought, serious consideration was given to the problems of bottling waters, and the distribution and delivery of such water to those families requiring them. In addition, the Water Research Centre is investigating means of nitrate removal, including specific ion-exchange resins and the use of methanol as a substrate for bacteria culture.

Micropollutants in Water

Some micropollutants were mentioned earlier in this chapter, in conjunction with the EEC directives. They are derived mainly from agricultural use or from manufacturing industries and thus arrive in the river by natural drainage or discharge or after passage through a sewage water-treatment plant. Most water authorities in the United Kingdom are able to carry out the analytical work required to monitor these problems, but there is little detailed information available, mainly because they have not been considered very important until recently. For example, estimates of organic micropollutants extractable in pentane were made on samples of river Dee water before and after treatment: the raw water showed no individual compound greater than 1 mg per liter and the levels in the final water going into public supply were all <0.2 mg per liter. The removal of organics from water is another project being undertaken by the Water Research Centre, and the Joint Committee on Medical Aspects of Water Quality is considering the associated medical and toxicological problems.

Haloforms

Basically haloforms fall into two categories: those which are formed from naturally occurring compounds such as humic and fulvic acids; and those which are derived from sewage works effluents, that is, mainly man-made. Recent work by Rook (1977) has suggested that the chloroform concentrations produced by treating colored, upland surface waters may be higher than the chloroform concentration derived from the more polluted lowland (river) waters. Our own work has not proved his suggestions, but a few of the results that we have obtained are given below in milligrams per liter.

Source	Chloroform $CHCl_3$	Bromodichloromethane $CHBrCl_2$	Others
Thirlmere, treated	21	8	<1
River Dee, raw	3	<1	<1
River Dee, treated (including prechlorination)	22 - 55	10 - 20	<1

The Thirlmere water is soft, colored, unpolluted upland water, whereas the river Dee exhibits the typical lowland qualities of a polluted river. In passing, it should be remembered that it is not UK practice to chlorinate sewage effluents, but since our sewage effluents are about four times stronger than those in the United States and since the effects on UK rivers can be much more pronounced, we are now concerned that by prechlorination, including breakpoint chlorination of some of our river sources, we could be creating further long-term problems for ourselves. We have to compare the benefits of prechlorination as an aid to water treatment against the risks of forming unwanted, and possibly dangerous, chemicals.

It is known that in general the formation of chloroform is favored by: (1) high organic content of the water; (2) high pH; (3) high water temperature; (4) high chlorine doses; and (5) long contact with chlorine. Only the first two are likely to be controllable without impairing the essential purposes of chlorination, but possibly by adjusting other treatment stages--that is, by removing prechlorination and leaving chlorination until a later treatment stage--the need for items 4 and 5 may be overcome. As mentioned earlier, we have already demonstrated the limited value of activated carbon for trihalomethane removal.

The fact that chlorine itself contains impurities should not be ignored, and there is a need for further investigation directed to finding more suitable disinfectants. Ingols and others are examining chlorine dioxide and ozone, although very little is known about the reaction products of these materials. It is believed that trihalomethanes are not produced when chlorine dioxide is used, but there is a probability that aldehydes, carboxylic acids, ketones, and quinones are formed. Similarly, ozone cannot produce trihalomethanes, but epoxides are possible. Unfortunately, some quinones and expoxides are of questionable safety.

Doubts have also been expressed about chlorine dioxide in that the chlorite ion may be released, and comparisons have been made with nitrite; could the chemical reactions within the digestive system be similar to nitrite reactions? For this reason, some European countries have set maximum limits for chlorine dioxide doses.

A process called Photozone has recently been brought to our attention. It has been claimed that it is valuable as a disinfectant and for color removal. So far very little information has been published, but this product could be of considerable interest to the water supply experts.

Taste and Odor Problems

It is not unknown for normal treatment processes, particularly chlorination, to aggravate tastes and odors. As is well known, straight chlorination of waters containing minute concentrations of phenols produces the objectionable taste of trichlorophenoxyacetic acid (TCP). Activated carbon is a very useful antidote for this, but in some of the treatment plants, chlorine dioxide or ozone is used instead of chlorine for disinfection so that no odoriferous phenol intermediates are formed. Unfortunately, these chemicals only react with certain types of tastes and odors; those caused by microbiological growths, for example, are not always amenable to straight oxidation and may have to be adsorbed on granular or powdered carbon. Typical of this type are tastes caused by the organism, actinomycetes, which produces the chemicals geosmin and methyl isoborneol, giving an unmistakable earthy, musty odor.

Asbestos

There are difficulties in assessing the effects of asbestos on health, and investigations are being made on techniques for counting asbestos fibers in water. This work is being carried out independently by the Water Research Centre and by some of the manufacturers of asbestos and asbestos/ cement pipes. As it is known that inhaled fibers are dangerous, strict health and safety regulations are laid down for anyone handling or cutting asbestos materials.

Waste Disposal

Waste disposal is not strictly related to water supply, but as runoff from waste dumps enters the hydrological cycle, it is a topic in which we have an interest.

The particular topic of the moment is what happens to salmonellae in sewage sludge when it is spread on agricultural land. We have regulations on land usage where such materials have been spread, but very little monitoring has yet been carried out on salmonellae survival on various types of land. The North West Water Authority is actively involved in such monitoring work, and some interesting results are now being produced.

The above are only a few of the items of interest and concern within our water industry. Work continues on many other fronts, including the extremely important ones of sampling techniques, sampling frequencies, analytical quality control, interpretation of results and their application to day-to-day operations. I hope that the foregoing remarks indicate how close our two countries' problems actually are.

Conclusion

In the past, the basic requirements of the water supply industry were to produce and supply a water which was pure and wholesome. That is, the water on tap should be odorless, colorless, and tasteless, and it should not contain harmful bacteria or harmful chemicals. The first three criteria are relatively simple to satisfy, but the last two are more difficult. Adequate disinfection removes harmful bacteria, but it should not be forgotten that as our analytical techniques improve, we are able to determine and monitor for smaller and smaller amounts of more and more chemicals--at what level can it be said that a water contains no harmful chemicals? Obviously, it is imperative that there is a very close working relationship between toxicologists and analysts. It seems to me, therefore, that we should not be striving to lay down inflexible rules for treatment methods of water quality. There is no off-the-shelf standard treatment for all waters; and, if possible, allowances should be made in all legislations and regulations for local conditions. The water supply industry is <u>not</u> irresponsible.

References

Bacteriological Examination of Water Supplies. 1969. Pamphlet No. 71 (London, Her Majesty's Stationery Office).

Comley, H. H. 1945. "Cyanosis in Infants Caused by Activates in Well Waters," *Journal of the American Medical Association* vol. 129, p. 122.

Control of Pollution Act. 1974. (London, Her Majesty's Stationery Office).

Department of Environment. 1977. *Lead in Drinking Water Survey, 1975-1976*, Pollution Paper No. 12 (London, Her Majesty's Stationery Office).

European Economic Community. 1975. "Quality Required of Surface Water Intended for Abstraction for Drinking Water," 75/440/EEC/OJ 25.7.75. No.L 194 (Brussels, EEC).

______. 1976a. "Quality of Bathing Waters," 76/160/EEC/OJ 5.2.76. No.L31/1 (Brussels, EEC).

______. 1976b. "Pollution Caused by Certain Dangerous Substances Discharged into the Aquatic Environment of the Community," 76/464/EEC/OJ 18.5.76. No.L129 (Brussels, EEC).

Robertson, J. S. 1975. "Sodium in Drinking Water," *Drinking Water Quality and Public Health Colloquium* (Medmenham, Buckinghamshire, Water Research Centre).

Rook, J. J. 1977. "Chlorination Reactions of Fulvic Acids in National Waters," *Environmental Science Technology* vol. 11, pp. 478-482.

Shaper, A. G.; D. G. Clayton; and J. N. Morris. 1974. "The Hardness of Water Supplies and Cardiovascular Disease" (London, International Water Supply Association).

Steering Committee on Water Quality. 1971. *First Annual Report* (London, Department of Environment).

World Health Organization. 1971. *European Standards for Drinking Water* (Copenhagen, WHO).

Chapter 7

WATER AND THE FOOD INDUSTRIES

Emil M. Mrak

I have been pointing out to various segments of the food industry that it should be well informed about its water supply, its purity, and possible sources of contamination. I was well aware that there would be changes in laws, rules, regulations, and philosophies, and that these in turn would affect the food industries. Unfortunately, there were too many other problems to think about so that, until very recently, little consideration was given to this matter. Today, however, as I see it, the picture is entirely different.

For several years there have been strong indications of interest on the part of government agencies and Congress in the quality of our water supply, whether it be used for drinking, in food preparation, or for other purposes. This has been especially true of the Environmental Protection Agency (EPA) since its creation. Prior to that, of course, there was the Public Health Service and the Food and Drug Administration (FDA).

During the past several years, there have been changes in our water supply. Industry has grown and has inadvertently contaminated with organics, as well as inorganics, supplies taken from streams and even underground water.

Recently, the use of chlorination has been questioned, even though it has solved many of our public health problems, because chemical reactions seem to be taking place between chlorine and the organics in the water. These latter might be the result of industrial contamination, or may be naturally occurring substances, such as humic acid.

Then there has been considerable use of water softeners, and now it is said that, in some instances, these may have a questionable effect on drinking water. There seems to be some evidence to indicate that soft water tends to remove from the pipes and conveying systems such metals as zinc, cadmium, lead, and copper, and that these are suspected of having an adverse effect on the human circulatory system.

All these observations have added to the concerns about the safety of our water supply, especially during the last few years. The evolution of these developments should have been foreseen by the food and beverage industries, but apparently was overlooked by them.

Water, of course, is an important component of our diet, and is taken into our bodies several times a day. As a matter of fact, it is for all intents and purposes the largest single component of our diet. We ingest six to eight cups, or about the equivalent of 1.5 quarts, of water a day, as a liquid in one form or another--from such sources as tap water, bottled water, coffee, tea, soft beverages, beer, milk, canned goods, soups, fruits, and so on. A vast majority of our processed foods contain substantial quantities of water.

I would like to give a few statistics, which although crude, do give some indication of the amount of water used by the food industries, and what new regulations may mean to these industries. These estimates were obtained from Dr. John Birdsall of the American Meat Institute, Norman Olson of the National Canners Association, Dr. Jerry Hutton of the Foremost Foods Company, and from various journals.

First, a little should be said about the fruit-and-vegetable products industry in the United States. This includes the canning, freezing, dehydrating, and pickling processers. It is estimated that about 100 billion gallons of water per year are utilized by these sectors of the industry. It is further estimated that one-quarter of the 100 billion gallons is used for cooling and condensing, and possibly 5 to 10 billion is used in products such as syrups and brines. There is an additional unknown amount that does not come in contact with the product, for it is used in plant clean-up, but this is probably a relatively small figure.

It is estimated that the meat industry uses, for slaughtering and processing, about 24.4 billion gallons each year. As an example of this use, about 187 gallons are used in slaughtering one hog, but not all of this water comes in direct contact with the carcass. Even more important is the water added to meat products: it is estimated that the industry produces about 8.5 billion lb. of meat products to which water is added. Some of the water is cooked out during processing, although about 10 percent remains in the product.

Canned meat products include such items as chili, beef stew, hash, and corned beef. About 2.5 billion lb. of such products are produced each year, and approximately 10 percent of this, or 250 million lb., is added water. Overall the water added directly to animal products is estimated to be about 1 billion lb. per year.

For the dairy industry, it is estimated that the annual use of water amounts to more than 6 billion gallons per year. Of this, 5 billion gallons are used to wash equipment, thereby coming indirectly into contact with food. About 193 million gallons are used as an ingredient in fruit drinks, ice cream, ice milk, sherbets, water ices, and similar products. In addition, an estimated 700 million gallons are used to wash cottage cheese curd.

According to figures reported in the publication "Beverage World," the yearly U.S. consumption is more than 6 billion gallons of bottled, canned, or in-bulk soft drinks. It is estimated that U.S. beer and ale consumption tops 4.6 billion gallons of water a year. These are indeed significant figures, and one can understand the need for serious consideration and perhaps, in many cases, concern about the water used by the food and beverage industries.

If one considers the farm areas of milk production, on the average 0.8 of a gallon of hot water per cow per day is used in modern milking parlors, for cleaning and sanitation of pipeline milk systems and the related equipment. This figure, however, varies from 0.31 to 1 gallon per cow per day. If this is multiplied by 365 days, and then by the number of cows, it may be seen that the amount of water coming into direct contact with equipment which comes into contact with milk is indeed considerable.

Sources of water on the farms are usually wells rather than city systems. The food industries in general, however, obtain much of their water from city systems, although many have their own wells and supplies.

While all these figures are, at best, estimates, they do give an indication of the importance of the water supply and of the quality of the water supply to the food industries.

Now, let us return to the regulatory situation. Because of the increasing interest in water and its quality, there have been studies, meetings, discussions, publications, laws, rules, regulations, judgments, and standards relating to water pollution and, of course, drinking water.

While most of these actions have been recent, some of them go back a number of years. The U.S. Public Health Service (1962), for example, published a manual for evaluating public drinking water. This applied only to water moving in interstate commerce so that, in general, it has been overlooked. The document mentioned bacteriological, physical, and chemical quality and included lists of chemicals and the then estimated human tolerances for them. In 1962 the list consisted mostly of metals, ranging from arsenic to zinc, although detergents, defined as methylene blue-active substances, were also included. Today, of course, many more have been added. Even at that early date, however, the list included fourteen pesticides, three of which were herbicides. Also discussed were the physical properties of water, important to its potability, that is, those involving color, odor, and turbidity.

As interest in the safety, purity, and drinkability of water increased, the National Academy of Sciences-National Research Council (1973) was called upon to make a study of water quality. The study pointed out that comprehensive analytical screens must be developed for the isolation and identification of all classes of organic materials in aquatic environments. This has been done, to a large extent, and is one reason for increased interest in our water supply.

The report also indicated that systematic studies were needed of the origins of various classifications of organic compounds that occur in water, their concentrations, fate, and persistence in the aquatic environment. It also pointed out that the relation of specific compounds to tastes and odors should be identified. It stated further that low concentrations of the organics and heavy metals such as barium, cadmium, chromium, silver, selenium, lead, arsenic, and mercury, have a physiological significance for humans.

Increased knowledge and changed thinking about water have developed a recognition that what is considered safe today may not be considered safe a few years from now. Scientific advances and forthcoming new information relating to safety can change the thinking of those establishing rules and regulations and the chances are that they will. The purveyors of potable water and the food industries using water should, therefore, be alert to new developments.

The Safe Drinking Water Act came into being in 1974, and this certainly stimulated an enormous amount of interest in water--what it is, how safe it is, and what its standards should be. The act focuses on water at the tap--as provided by all water systems serving more than twenty-five people. This definition not only includes the community systems serving residents and populations, but also self-supplied noncommunity systems, such as parks, residences, and motels serving the traveling public.

The Food and Drug Administration (FDA) also has a legal responsibility for water, since it is food and drink for man and animals. The question then arises as to whether water is legally a food and which agency ultimately has responsibility for its regulation. I will come back to this, but first, I would like to say more about the Safe Drinking Water Act which directs the administrator of the EPA to make a comprehensive study of public water supplies and drinking water sources and report the findings. This study involves the determination of the nature, extent, sources, and means of control of contamination by chemicals, or of other substances suspected of being carcinogenic. This is, indeed, a large order, and what is meant by "suspected" I cannot say. In any event, the EPA has given this law a great deal of consideration.

Above all, it requires the administrator of the EPA to prescribe national regulations for contaminants in drinking water which may adversely affect public health, and requires public water systems to deliver, monitor, and test water, and to operate and maintain their systems in compliance with the drinking water regulations. It requires drinking water purveyors who fail to comply to notify their customers of that failure, and it requires the administrator of the EPA to encourage maximum levels of state involvement, and accordingly, authorizes states to adopt and implement adequate standards and enforce them.

It also establishes state and federal programs to protect underground sources of drinking water and authorizes the EPA administrator, on a temporary basis, to certify the need of chlorine or, perhaps, other materials. This is indeed a large order, and one can but wonder if such an all-inclusive law is essential, and whether the present community water supplies are adequate to meet the new requirements. As for the food industries, the question is, Where do they stand in this matter?

The EPA was asked what its jurisdiction might be for water safety. It was pointed out that the EPA administrator is required to develop drinking water standards at the tap and to set minimum requirements for the supervision of public water systems. The states are assigned the responsibilities from then on. The EPA is also charged with the development of regulations to minimize the pollution of groundwater and aquifers, which serve as sources for drinking water for 50 percent of the nation's population.

In brief, the EPA's regulatory effects are supported by activities aimed at (1) determining the current state of knowledge as to analytical methods, health effects, control technology, and related economic effects as a prerequisite to drinking water standards; (2) establishing and implementing needed research; and (3) developing public information.

The question has arisen as to the FDA's jurisdiction relating to water safety. It has been pointed out that FDA has always been concerned with water, having issued regulations and advisory opinions relating to the use of water in food, drug, and cosmetic manufacture, and having also been responsible for bottled water. Some of the food industries may seek refuge in the reference to water from the tap, but

I am certain that much, if not most of the water used in the food industries, does come from a tap.

I may say at this point, however, that steps have been taken by EPA Administrator Costle to bring together various regulatory groups, with common interests, in an attempt to develop common approaches and regulatory procedures. The agencies involved are the EPA, the FDA, the Occupational Safety and Health Administration (OSHA), and the Consumer Product Safety Commission. I should point out that this last agency is also concerned with water from the standpoint of the use of purifiers that might be attached to the tap, or used otherwise to improve the water supply in the home. Costle deserves a great deal of credit for displaying such leadership. It constitutes one of the most constructive steps in government that I have observed in a long time, and I have high hopes for this collaborative procedure.

Nevertheless, in the end, it may be necessary for the Congress to clarify the situation, for individuals are involved in making such an arrangement work, and the present leaders will not always be there, and may well change with a change in administration. In any event, the collaborative efforts and hopes of Costle are timely, for the last paragraph of the Safe Drinking Water Act makes it clear that when the EPA prescribes interim or revised primary drinking water regulations, the secretary of Health, Education and Welfare shall consult with the EPA administrator and, within 180 days, adopt these regulations or publish in the Federal Register HEW's reasons for not making amendments to the FDA's bottled water standards. The second hope for clarification naturally lies with the Congress and the possibility of changing the law to clarify responsibilities.

The question of definition has come up any number of times. This very question was asked Rep. Paul G. Rogers, who was one of the chief architects of the Safe Drinking Water Act. He was asked whether water is to be considered a food. Representative Rogers indicated that when Congress thinks of water for human consumption, it classifies water as a food. Rogers also pointed out that many food processors today use municipal water, and the question has arisen as to whether it would be possible to do so any longer. Furthermore, the question was raised whether a food processor in effect would be allowed to follow the FDA additive regulations with respect to water. Rogers replied that there may be occasions when the FDA would feel it significant enough to apply the additive provisions, and he would not want to rule out the FDA.

Rogers was also asked whether, if drinking water requirements are tightened, the food processors will be affected. His answer was essentially that if the water is not safe but is polluted, it would adversely affect the food. He also indicated that he thinks food processors should be particularly careful and anxious to make sure that their water is clean and safe. The FDA might even be concerned about the mixing of water in foods as an additive, while at the same time, the EPA sets general standards for water quality. This makes it very clear to me that the food industry should be genuinely interested in what is going on with respect to water.*

During the last few years, a great deal of concern has been caused by the revelation of the harmful effect of asbestos inhaled by workers in factories, especially those who smoke. This has given rise to the question of what effect asbestos has if it occurs in drinking water. The National Research Council study on water indicates that there is no cause for concern about asbestos fibers occurring in water. This finding should be of interest to food and beverage industries that may still use asbestos filters.

Quite a few food industries are located along the Mississippi River and, accordingly, use the river as a source of water. Since so much as been said about contaminants in this water, it should be of concern to the food industries using it.

De Rouen and Diem (1975) published a paper in which they pointed out that there was considerable inconsistency in the evidence supporting the hypothesis of water as a causative agent of cancer, as reported in the Panel Discussion by Harris of the Environmental Defense Fund. They also pointed

---

*Ed. Note: This question of FDA/EPA jurisdictional overlap is discussed by others in this volume, most importantly by Harker.

out, considering white males only, the group on which the original report was based, that some contradictions to the theory that water was causing cancer were overlooked. One such was that four of the nine parishes included in the study apparently were rural and not using the Mississippi River as a source of their water, but no mention was made of this. De Rouen and Diem (1975) also commented, "Whether the drinking water in New Orleans is contributing to cancer mortality is an extremely difficult question to answer in the presence of many other suspected and known causes of cancer."*

In an editorial Dr. William H. Stewart, former Surgeon General of the U.S., pointed out that the defensive posture has become more commonplace for public health officials in the last few decades: "This is because of our ability to detect foreign substances in the environment in increasingly smaller amounts, while at the same time, our ability to understand the biological consequences of exposure to these small quantities over long periods of time has advanced slowly and with great uncertainty." This is, indeed, the situation today, and judgments unfortunately can only be based on theory rather than on fact.

In any event, subsequent studies made throughout the country indicated that there are well over 700 different

---

*Ed. Note: See Schneiderman's paper, however, for a discussion of later work by these authors which tends to confirm the Harris et al., results.

chemicals occurring in drinking water, although not all at the same place, and some of which result from use of the chlorination process. If chlorination is not used, then we may be confronted with waterborne diseases. If we continue to use chlorination, then there is the question about carcinogenic compounds, such as chloroform and the other members of the trihalomethane group.

In 1975 the EPA assistant administrator responsible for water programs asked the Science Advisory Board of the EPA to study the matter and to advise him on the significance in terms of health risk of the observed occurrence of certain organics in drinking water. The committee recommended that further epidemiological studies be made on populations exposed to known common patterns of water contamination, and on situations in which water supplies of major population groups are changed through the introduction of new sources or of new processing facilities.

The study group also concluded that current methods of extraction, identification, and measurement of drinking water contaminants are adequate for field surveys. It went on to point out, however, that it is unlikely that the majority of the drinking water purveyors would have available the sophisticated equipment and trained personnel to provide monitoring of individual contaminants on a routine basis. In this connection, I cannot help but wonder about the food and beverage industries. I am certain that large organizations with adequate facilities can do this without much difficulty, but small operations will find it a difficult, if not impossible, task.

I have already indicated the concern about the effect of trihalomethanes on health. In this connection, the committee of the Science Board concluded that there may be some cancer risk associated with the consumption of chloroform in drinking water. The level of risk estimated from consideration of the worst expected cancer site, the liver, might be extrapolated to account for up to 40 percent of the observed cancer incidence rate. It went on to say, however, that a more reasonable assumption, based on current water quality data, would show much lower levels than the worst case in the majority of the U.S. drinking water supplies, and would place the additional risk of <u>hepatic</u> cancer much lower, possibly nil. Furthermore, the committee stated, "Epidemiological studies do not thus far support the conclusion of increased risk of liver cancer, although hypothesis-formulating studies in southern Louisiana suggest the possibility of an association with contaminated water and overall high cancer incidence."

The committee made an interesting comment with respect to the food industries. It pointed out that attention has been focused on the contaminants in drinking water itself, while a complete analysis of the problem would require analytical chemical data on exposure to these chemicals by ingestion of foods and beverages processed with contaminated water. In addition, other possible exposures, resulting indirectly from environmental redistribution and possible biomagnification of chemicals in plants and animals used for food, should also be considered.

In this connection, and with respect to the organics in drinking water, Junk (1977), who did so much of the recent analytical work, stated:

> The data from eighty different water sites can be used to reliably predict the distribution and the amounts of six halogenated hydrocarbons. Thus, chloroform is 100 percent distributed in amounts varying from 0.2 to 311 micrograms per liter; bromodichloromethane is 95 percent distributed in amounts from 0.3 to 116 micrograms per liter; and dibromochloromethane is 89 percent distributed to the extent of 0.4 to 100 micrograms per liter. The other three halocarbons, bromoform, 2-dichloroethane and carbon tetrachloride, are much less widely distributed and are present in smaller amounts.

Much has been said about nitrosamines in foods, but less about their occurrence in water. They may arise from natural sources or from the use of agricultural chemicals. In this connection, a recent EPA (1977) publication indicates that nitrates, nitrites, and amines, which are precursors in the formation of nitrosamines, are chemical compounds that occur naturally in food, water, soil, and air, and, unlike many recently produced manmade chemical compounds currently causing environmental problems, nitrates, nitrites, and amines have been in existence throughout the evolutionary history of man.

The report goes on to say, "Nitrites are in themselves toxic to man, other animals, and plants. However, the nitrosamines that may be formed in soil, water, air, food, and in the gastrointestinal tract through the chemical and microbial nitrosation of secondary and tertiary amines are the greater hazard to man."

I cannot refrain from mentioning this, for here again is something that may be on the horizon. What it all means, I do not know, but I think the food industry should be aware of the existence of this type of statement.

At an international water conference held under the auspices of the International Academy of Environmental Safety (Junk, 1977) it was indicated that there are probably three great concerns with respect to water standards and requirements. One, of course, relates to the occurrence of infectious bacteria and viruses in water, and it was and is in this area that scientists from Pan-America and underdeveloped countries expressed most concern. On the other hand, the U.S. representatives seemed to be more concerned about the occurrence of toxic organics, which I have already discussed.

I found it of particular interest that the representative from Finland stressed the importance of naturally occurring inorganic trace elements as water pollutants and their implications for the health of man and the aquatic biota. There seemed to be general concern about naturally occurring chemicals in water, or perhaps about the water being too soft and thereby permitting contamination with metals such as lead, zinc, cadmium, and copper that are picked up by the soft water during its passage through the pipelines.

Conclusions

Where the food industry fits into this picture at present, and in the future, remains to be seen, but it should be well aware of the situation, and the standards and guidelines and the philosophy behind them. In any event, it is clear to me that the food industries use an enormous amount of water, directly in the foods, in contact with the foods, and in contact with equipment that comes into direct contact with foods. Earlier in the paper I gave estimates of what this might be, but these estimates were for only a few of the food industries. For example, I did not say anything about fish products, condiments, baking industries, candy, milling cereals, the sugar industry, and others. The question is, "How good is the water that is being used, and if it is not satisfactory, then what must be done to make it satisfactory?" This is something that the food industries should think about seriously. Indeed, I would recommend that the industry embark on a formal program to see just where it stands.

Finally, there is potential for conflicts between the EPA and the FDA in the regulation of suspected carcinogens. The FDA's mandate, as I see it, calls for a "zero tolerance," while the EPA is given congressional authority to set non-zero or maximum contaminant levels (MCL). I can foresee difficulties arising, for example, with plastics used to store, pipe, pump, or package water. Polyvinyl chloride and acrylonitrile are two such examples. Another potential problem concerns polycarbonate plastic, which is known to contain trace amounts of chloroform. In water, the EPA has proposed an MCL of 100 parts per billion for chloroform, but the FDA proposes to prohibit chloroform in food packaging. At this time, polycarbonate, which may contain chloroform, is used commerically to package bottled water. It appears to me this gives the food industries not only food for thought, but water for thought.

References

De Rouen, T. A., and J. E. Diem. 1975. "The New Orleans Drinking Water Controversy: A Statistical Perspective," _American Journal of Public Health_ vol. 65, no. 10, pp. 1060-1063.

Fairbanks, W. C.; R. Eide; H. S. Etchegary; and G. C. Gurtie. 1978. "Hot Water Consumption in Milking Parlors," _California Agriculture_ vol. 32, no. 1, pp. 7-9.

Federal Register. 1978. p. 5761.

Junk, G. A. 1977. "Organic Contaminants in Water," in F. Coulston and E. Mrak, eds., _Water Quality, Proceedings of an International Forum_ (New York, Academic Press) pp. 137-141.

Kimm, V. J., and R. Angelloti. 1977. "Who Safeguards Your Water?" _Professional Nutritionist_ (Winter) pp. 1-5.

Mrak, E. M. 1977. "Water Quality?" _Professional Nutritionist_ (Winter) pp. 3-9.

National Academy of Sciences-National Research Council, Environmental Studies Board. 1973. Water Quality Criteria 1972, EPA Report, EPA-R3-73-033 (Washington, D.C.).

Rogers, P. G. 1977. "Is Water a Food?" Professional Nutritionist (Fall), pp. 4-6.

Stewart, W. H. 1975. "Editorial: Facts and Fancy About Drinking Water," American Journal of Public Health vol. 65, no. 10, p. 1111.

U.S. Environmental Protection Agency. 1975. Assessment of Health Risks from Organics in Drinking Water, Report of an Ad Hoc Committee of the Hazardous Materials Advisory Committee of the Science Advisory Board of the EPA, April 30 (Washington, EPA).

______. 1977. "Scientific and Technical Assessment of Nitrosamines," in Star Series EPA-600/6-77-001 (Washington, EPA), pp. 1-4 to 9-1.

U.S. Public Health Service. 1969. Manual for Evaluating Public Drinking Water Supplies, U.S. Bureau of Water Hygiene, PHS Pub. 1820 (Washington, D.C.).

Discussion by John T. Quigley

First, let me compliment Dr. Mrak on his thoughtful analysis of the complex issues relative to the special needs of industrial and commercial water users in the food and related industries. Now, I do not suggest that I can speak for specific industrial or commerical users, but only make comments from my understanding of these concerns.

I wish to apologize also for any apparent duplication that may occur, since so many related topics have been mentioned in other papers and discussions. However, my objective is to emphasize certain aspects that seem to fit from an industrial user's point of view. My remarks are divided into (1) direct reactions to Dr. Mrak's paper; and (2) comments relative to these problems.

Direct Reactions

As Dr. Mrak correctly observed "Small companies have a problem similar to that of small (water) utilities. . . ." Certainly, the costs of monitoring the control will weigh most heavily on the smaller manufacturers. Scaling factors are especially important in this area.

Referring to an earlier committee report, Mrak says that it is important to emphasize that "A complete analysis of the problem would require analytical chemical data on exposure to these chemicals by ingestion of foods and beverages processed with contaminated water. In addition, other possible exposures, resulting indirectly from environmental redistribution and possible biomagnification of chemicals in plants and animals used for food, should also be considered."

It should be noted that in the food-processing industry water quality traditionally is a prime consideration during site selection together with access to transportation, adequate power, suitable labor, and so forth. For this reason, accommodation to stringent new regulations may be a problem, even for the large firm, because of the fact that site commitments may have to be reviewed. Water quality is such a fundamental consideration that it is easy to overlook when placed in competition with other more dynamic day-to-day concerns.

Routine input monitoring is costly and apt to be ineffective. In fact, usually a first step to problem solving is to institute a thorough water quality-monitoring program as a means of problem delineation and definition. Often "the problem" may change or diminish when carefully scrutinized.

Further Comments

Industrial Attitude. Industrial managers tend to think most naturally in terms of output parameters, that is, product quality and system performance of efficacy. Subtle or seemingly minor changes in raw materials or process inputs tend to go unassessed until a problem appears in one of these output parameters. Then, as a recent visitor characterized their situation, "All ---- broke loose when a shipment was stopped by QA!!"

Therefore, I suggest that the FDA's participation in this area is still the most certain and probably the most economical way of getting at the sum total of all these problems. Let the food processor be free to solve his own problems. He may simply treat the water, or in certain cases he may modify the overall process moderately in order to achieve stated product quality.

Let us simplify the FDA's involvement and coordinate but not unnecessarily throw the baby out with, you guessed it, the water.

Agency Responsibility. It is encouraging that the regulatory agencies, notably the EPA, the FDA, the USDA, and the Consumer Product Safety Commission, have moved to coordinate their efforts, but it remains for the EPA to reduce ambient levels of trihalomethanes within the water utility system at the point of end use; and through stringent treatment regulations, determined by discharge permits, to reduce input loads of other organics. It is important to note that for those chlorinated organics used by industry, agency decisions may act as economic allocators among users and those saddled with contaminated raw materials. That is, these decisions will determine who has to pass these costs along to the public. Finally, the relationship between food product specifications and tap water quality will determine the extent of treatment required by the food processor.

Economic Factors

Water treatment is relatively easy to provide for the purpose of maintaining product quality control. The industrial or commercial user assumes responsibility to take whatever water quality he receives and to improve it sufficiently to meet product requirements. Generally, he would rather perform this treatment in the plant, first, to assure control and, second, to reduce overall cost. Here I refer to the fact that the food processor's water use may represent only a small fraction of the system total. Yet, if all utility water is treated to the highest levels, then total cost for water supplied to the food processor will rise in proportion to the charges that accrue for treated water expended in non-consumptive uses. It is hoped that he can then pass these costs (whatever they may be) along through his customers. The water purveyor's situation is similar to the user's in that the purveyor wishes to deliver a certain quality product for a certain price.

One final comment that cannot be over-emphasized: Whatever rules might be adopted should be consistent and promulgated without later amendment. Uncertainty and frequent change carry with them the threat that venture capital will be lost or wasted without satisfying the need.

On-site Solution

Regarding the matter of small-system and individual supplies, I cannot help noting what appears to me to be an analogous problem to that now being addressed quite realistically by the "dirty" water side of the EPA.

Some seven or eight years ago I created the Small Scale Waste Management Program at the University of Wisconsin for the purpose of improving wastewater treatment and disposal practice--mainly for individual and small cluster residential systems. With increasing acceptance of this work (and that of others) in this area, these small-scale systems are now being used, together with innovative collector systems, for small or areawide distributed services. Of course, the institutional arrangements are critical to assure proper supervision and maintenance, as well as to provide funding.

The apparent analog that I perceive has to do with the "point of use" treatment approach to meeting the most stringent drinking water standards. While such units may appear to be costly and, perhaps, inconvenient to service, these factors alone should not cause this avenue to be dismissed without a thorough and careful analysis. From comments made at this meeting, the relative cost of complying with the Safe Drinking Water Act seems to be uncertain and may vary greatly from system to system. This household point-of-use service option should be completely researched and available "on the shelf" for systems where service costs might rise toward the higher levels mentioned here yesterday.

Another reason for following up this approach would be the possibility of such an at-the-tap solution being applied as a companion for these noncentral waste disposal systems in rural areas.

In time the "dirty" water system inspector may put on his white gloves and check the tap <u>first</u>! Actually, where costs permit, an information and educational effort can be directed toward having the community decide through public participation whether they want central treatment or treatment at the tap. The service can be offered either under the auspices of the water utility, as a service from a private supplier, or as a joint effort.

Where hard waters prevail and softening is common, another delinquency could be corrected as well. In training courses, I often explain that for properly plumbed homes, it is likely that only the kitchen tap does not present a high sodium problem. And, many who are in the water supply business seem to be surprised.

This approach by water utilities would have the added advantage of keeping down cost for industrial and commercial users by stressing drinking water quality only at the point of use.

Part II

Costs and Financing for Compliance with Safe Drinking Water Act

Chapter 8

MEETING THE DRINKING WATER STANDARDS: THE PRICE OF REGULATION

Robert M. Clark and Richard G. Stevie*

Introduction

For the water supply industry, the Safe Drinking Water Act of 1974 (PL 93-523) may be one of the most important laws ever passed. This act will bring about a fundamental examination of the way water is handled before it is delivered to the customer.

The U.S. Environmental Protection Agency (EPA) has been designated as the agent responsible for implementing the act. EPA is required to issue regulations, state agencies are to develop water programs, and local governments are to pursue them. Because the law will have a significant economic impact on water utilities, the act strongly emphasizes the need to consider all aspects of the economics of water delivery.

---

*Robert M. Clark is a Systems Analyst and Richard G. Stevie is a Research Economist at the Water Supply Research Division, Municipal Environmental Research Laboratory, EPA, Cincinnati, Ohio. The authors wish to acknowledge the assistance of Gregory Trygg, Janine Hessoun, and Daniel Guttman in preparing this manuscript. They also wish to thank Harry von Huben of EPA Region V, Robert Bridges of MERL, Cincinnati, and C. Dameron, Office of Water Supply, for the data they provided.

In 1912, the first federal water quality-related regulation was adopted--the prohibition of the common drinking cup on interstate carriers. Although the use of a common drinking cup on interstate carriers is almost unthinkable by today's standards, the prohibition met strong opposition. The theory was that if the cups became unavailable for any reason, the crowd so deprived of drinking facilities would have their health seriously endangered and might even be forced to succumb to the obvious temptation of alcoholic beverages (Symons, 1976).

Though today's society may be more sophisticated and interdependent than that of ancient Rome or the early twentieth century, basic human concerns have remained much the same. We are concerned, as we should be, over the proper investment of public funds and the benefits derived from them.

This paper will briefly describe the water supply industry of today. Then some of the economic consequences of the Safe Drinking Water Act and some of the central economic issues surrounding implementation of the act will be discussed.

The Water Supply Industry Today

The water supply industry in the United States is a complex and interdependent one and accounts for an annual expenditure of nearly $6 billion (Bureau of Census 1971-76). Currently, estimates show more than 240,000 public water supplies in the United States (Federal Register, 1975), including systems that serve the traveling public.

Public systems can be categorized as either community or noncommunity. A community system is defined as one that serves at least fifteen service connections used by year-round residents, or one that regularly serves at least thirty-five nonresidents. The noncommunity category includes systems that serve a transient population. At present there are an estimated 40,000 community systems and 200,000 noncommunity systems.

Based on data contained in the EPA Inventory of Public Water Supplies (Federal Register, 1975, p. 59582), approximately 177 million persons are served by community water supplies. Table 8-1 shows the distribution of community systems by population served. Most of these water systems are small. More than 90 percent serve fewer than 10,000 persons, but they provide water to less than 25 percent of the total population served by all community systems.

A variety of treatment processes are employed by these systems. The EPA Inventory of Public Water Supplies indicates that the most prevalent treatment processes are used to control bacteriological contamination and turbidity (Federal Register, 1975, p. 59582). Table 8-2 shows the percentage of systems employing treatment processes.

Community water systems may be publicly or privately owned. The majority (58 percent of the 40,000 community water

Table 8-1. Distribution of Community Water Systems

System size (persons served)	Number of water systems	Total population served (in thousands)	Percent of total population served
25 to 00	7,008	420	0.2
100 to 9,999	30,150	36,816	20.8
10,000 to 99,999	2,599	61,423	34.6
100,000 and over	243	78,800	44.4
Total	40,000	177,459	100.00

supplies) are publicly owned, and these systems supply 88 percent of the total U.S. drinking water production (Federal Register, 1975, p. 59583). Most of the 200,000 public, noncommunity water supply systems are privately owned. These noncommunity systems may be found in service stations, motels, restaurants, rest areas, campgrounds, state parks, beaches, national parks, national forests, dams, reservoirs, and other locations frequented by the traveling public. Some schools and industries are also included in this category. Data on these systems are sparse, and only rough estimates of their number can be made. The sector of the water supply industry considered in this paper includes all the systems that supply water for primarily residential, commercial, and municipal use. An approximate national average allocation of water use by category of user is as follows (Federal Register, 1975,

Table 8-2. Treatment Processes Employed and Objectives Pursued by Community Water Systems

Treatment processes	Percent of systems employing process[a]
Aeration	6.6
Prechlorination	7.8
Coagulation	11.3
Sedimentation	8.9
Filtration	12.8
Ammoniation	0.9
Fluoride Adjustment	8.5

Treatment objective	Percent of systems pursuing objective[a]
Softening	4.9
Taste and odor control	3.4
Iron removal	5.7
Disinfection	35.2[b]

[a]Percentages do not add to 100 because some systems have multiple processes and objectives while others employ no treatment.

[b]Ed. Note. One reviewer has suggested that as much as 90 percent of the population served by public water systems may receive chlorinated water. Assuming this is even roughly true, it becomes apparent that tabulation by number of systems is quite misleading in this field. Bigger systems tend to have more extensive treatment.

p. 59583): residential, 63 percent of total use; commercial, 11 percent; industrial, 21 percent; and public, 5 percent.

Cost and Economic Impacts of the Safe Drinking Water Act

Throughout the Safe Drinking Water Act, emphasis is placed on the need to consider economics. The principal purposes of this paper are to provide estimates of the cost of complying with the act, and to translate these costs into customer and national aggregates. The areas considered are research and regulation, cost function characteristics, cost of interim standards, cost of the organic standards, regional solutions, and individual and national economic impacts.

Cost of Research and Regulation

To carry out the provisions of the Safe Drinking Water Act, it is necessary to establish a bureaucratic structure at the federal and state levels. EPA's Office of Water Supply and Water Supply Research Division constitute the regulatory and research arm of EPA. Figure 8-1 shows the federal expenditure for water supply research and regulation from 1972 through 1978. As can be seen, the total federal budget for water supply increased from approximately $4 million in 1972 to over $55 million in 1978, with the biggest increase coming after passage of the act.

Much of the responsibility for carrying out the major regulatory provisions of the act lies with the various state

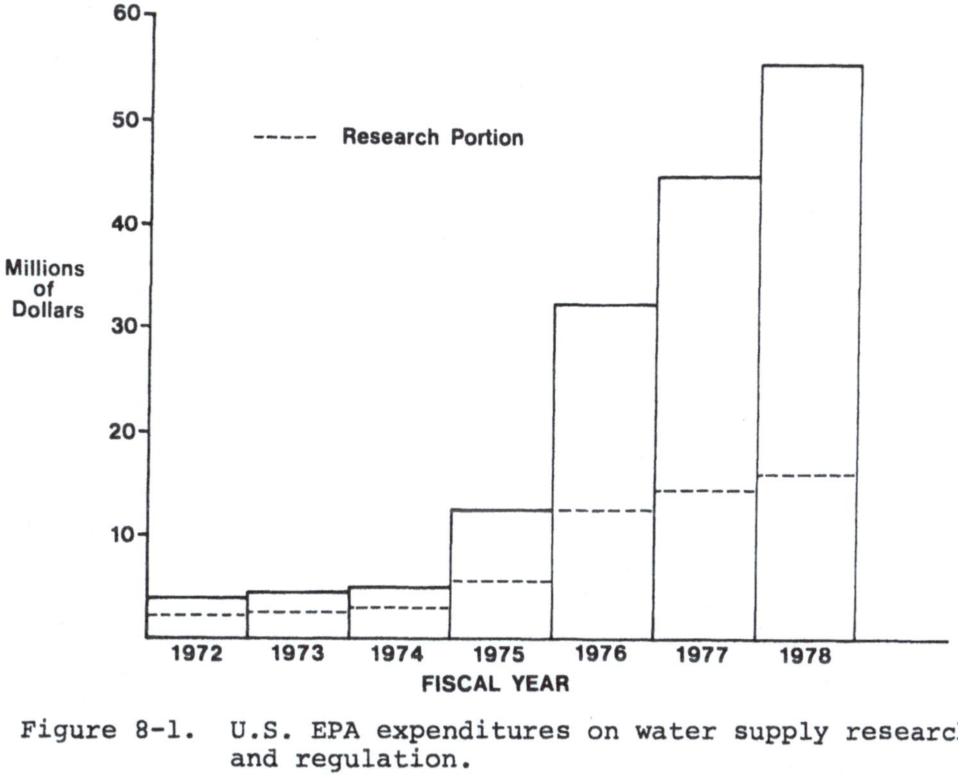

Figure 8-1. U.S. EPA expenditures on water supply research and regulation.

Source: Expenditures for water supply research and regulation compiled by C. Dameron, U.S. EPA Office of Water Supply and by R. Bridges, U.S. EPA, MERL, Cincinnati.

level agencies. At present, there is little comprehensive information on state level expenditures for water supply, but figure 8-2 shows expenditures for EPA's region V and the state of Ohio in 1975 and 1978. The budget for both Ohio and all of region V has obviously increased over the past three years, but by next year there should be a dramatic increase in state expenditures for water supply as the individual states take primacy under the act.

Water Supply Cost Characteristics

Before one can understand the potential cost impacts of the Safe Drinking Water Act, it is necessary to understand the economic characteristics of water supply technology and systems. A pervasive characteristic is the existence of economies of scale (increasing unit costs with decreasing system size). For example, Orlob and Lindorf (1958, p. 45) examined treatment cost to determine its relationship to the cost of surface water importation, reclamation of wastewaters, groundwater recharge, and other alternatives available for increasing water supply to California. Construction cost was hypothesized to be a function of the explanatory variable, design capacity, of the form:

$$C_c = \alpha Q_n^{\beta} \tag{1}$$

where C_c = total capital cost of a complete water treatment facility

Q_n = the design capacity of the plant in millions of gallons per day (mgd), and

α = constant

β = scaling factor

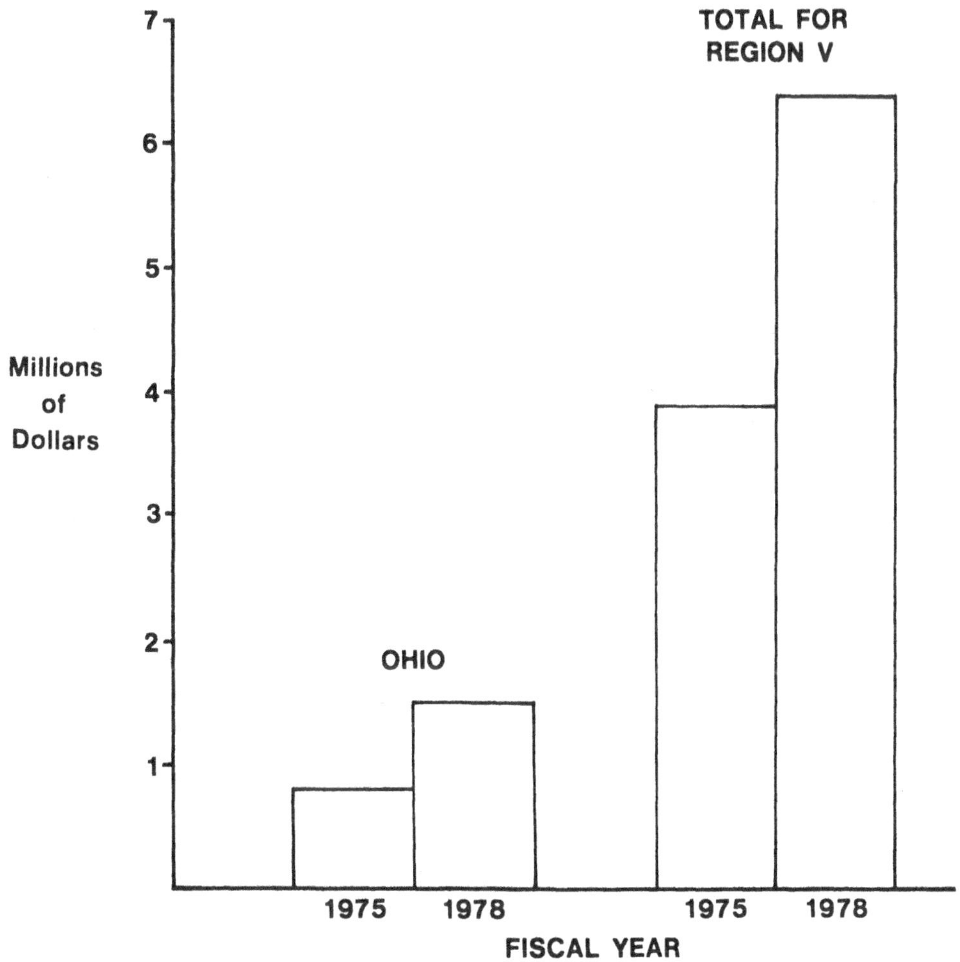

Figure 8-2. State expenditures for water supply regulation and research.

Source: State and regional expenditures were compiled by H. von Huben of EPA's Region 5 office.

Table 8-3. Costs for Large and Small Water Supplies by Function

Utility size	Revenue-producing water (mgd)	Cents/1,000 gal					
		Support services	Acquisition	Treatment	Distribution	Interest	Total
Large	86	9.9	5.4	4.8	12.7	8.7	41.5
Small	1.7	15.0	12.9	8.8	35.3	12.9	85.4

Estimation of the equation generated the following relationship:

$$C_c = 257\ Q_n^{0.67} \tag{2}$$

The value of $\beta = 0.67$ implies that economies of scale exist in treatment plant construction. Dividing both sides of equation (2) by Q_n yields:

$$C_c/Q_n = 257\ Q_n^{-0.33} \tag{3}$$

The term C_c/Q_n is the unit cost of capacity, and the exponent of Q_n is less than 0. Equation (3) is continuously differentiable and monotonically decreasing. According to equation (3), unit cost therefore decreases with increasing Q_n or capacity, illustrating economies of scale. Figure 8-3 shows the results from a detailed study of twelve water utilities geographically scattered across the United States (Clark, Gillean, and Adams, 1977, p. 117). Both total supply cost and total treatment costs are shown.

Table 8-3 summarizes data collected by the Water Supply Research Division on large and small systems (Clark, 1978). As can be seen, the average cost of twelve large water supply systems (averaging approximately 85 mgd) is 42¢/1,000 gal. The average cost for thirty small supplies (averaging approximately 1.7 mgd) is 85¢/1,000 gal.

As shown by table 8-3, the impact of the regulations will be amplified for small systems because of economies of scale. Installation, operation, and maintenance of sophisticated control processes, together with monitoring requirements, may result in substantial cost increases for small systems. In addition, many large systems already have sophisticated

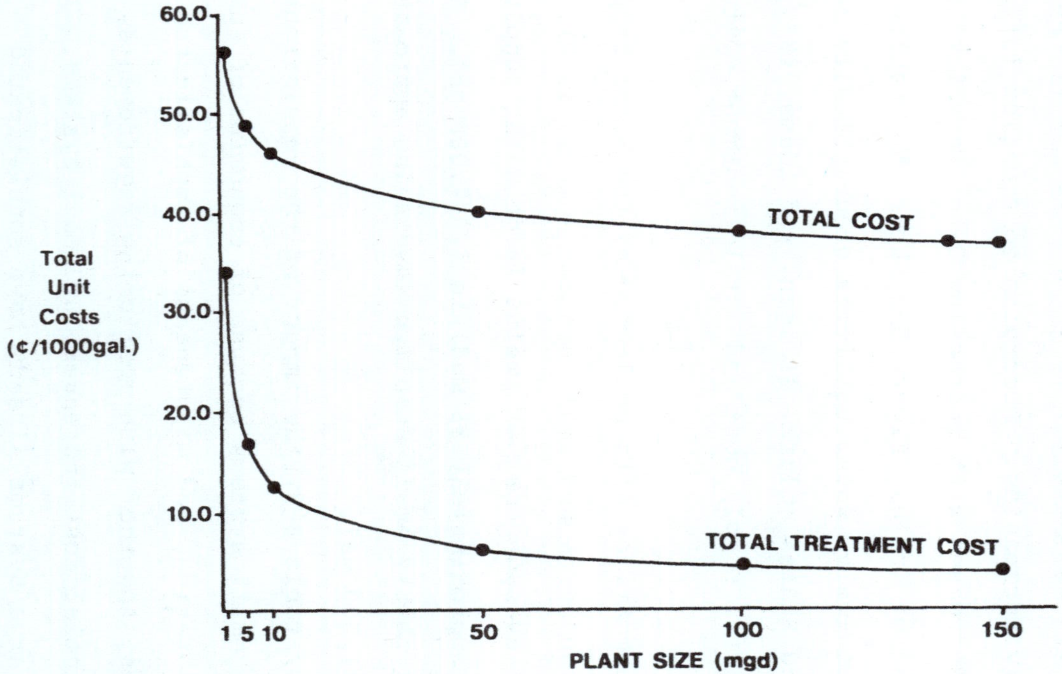

Figure 8-3. Total unit costs for water supply systems and treatment.

Source: Clark and coauthors, 1977, p. 114.

treatment, but many small systems do not. The absolute incremental costs to a small system may be substantial, and they will be compounded on a unit cost basis by the diseconomies of small scale.

It is obvious, therefore, that we are discussing two different conditions for the application of the interim primary standards. One condition applies to small water systems (taken here to be 10 mgd and less), and the other to larger systems (larger than 10 mgd). The choice of 10 mgd as a cutoff point is arbitrary, but it is a convenient level that is close to the point at which the unit cost for most types of technology turns rapidly upward.

Cost of Meeting Existing Interim Primary Standards

As shown in table 8-2, water utilities employ a wide variety of treatment processes, but the most common are disinfection and turbidity control. All utilities will be required to monitor their product to meet the requirements of the act. The cost of routine monitoring will therefore be discussed in the following sections, along with costs for package plant technology, disinfection, and conventional treatment. Package plant technologies provide a possible alternative to conventional treatment for small systems. Disinfection and conventional treatment represent the extremes in cost for complying with the Interim Primary Standards.

Monitoring Costs. Monitoring costs vary with the size of the water system involved. The required number of samples for routine bacteriological monitoring is a function of the number of persons served (Federal Register, 1975, p. 59585). For community supplies, this can range from a minimum of one sample per quarter for systems serving 1,000 persons or less to a maximum of 500 samples per month for systems serving more than 4.7 million persons. For noncommunity supplies, only one sample per quarter is required.

In general, the annual impact of routine chemical monitoring varies according to the frequency of sampling rather than the number of samples. The frequency of sampling depends on the system type--groundwater versus surface water, and community system versus noncommunity. Figure 8-4 shows the costs for monitoring requirements by system size. Costs are very high for small systems, but they drop off rapidly for larger systems.

Cost for Disinfection. As shown in table 8-2, disinfection is the most common treatment process presently in use in the United States. Indeed, in many water supplies, chlorination is often the only treatment process used. When other treatment processes are used, disinfecting chlorine may be added to the raw water (prechlorination), the partially treated water, or the finished water (postchlorination). Where the distribution system contains open reservoirs, the treated water may be rechlorinated in the distribution system.

The cost of chlorination is presented here as representative of disinfection costs. Because the formation of

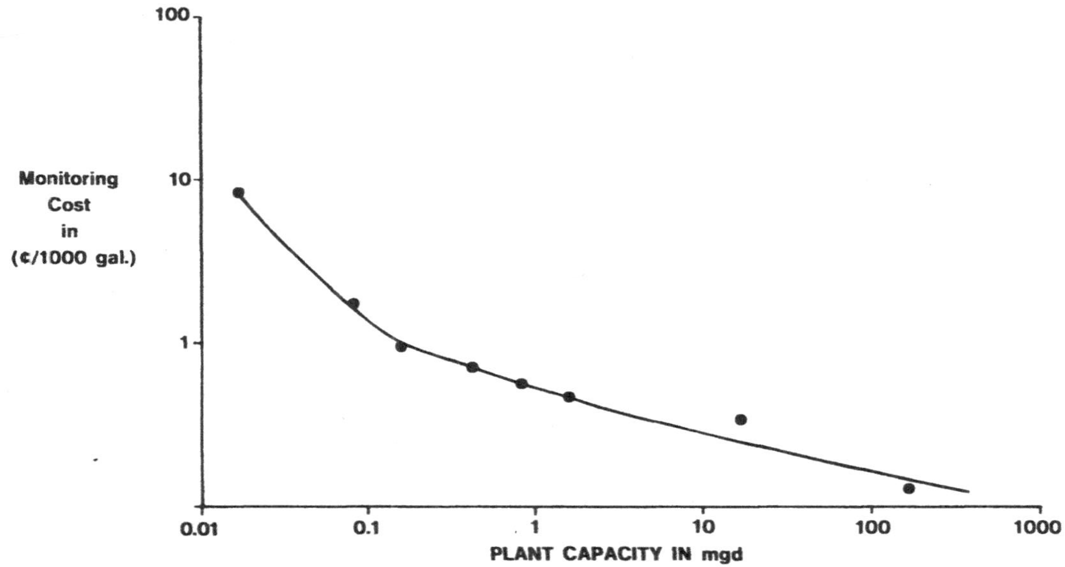

Figure 8-4. Monitoring costs for water supply.

Source: Federal Register 1975.

trihalomethanes is associated with chlorination, several options can be considered in lieu of chlorination (Clark, Guttman, Crawford, and Machisko, 1977, p. 4). The benefits as well as the problems associated with changing to some alternative must be weighed when choosing another form of disinfection. These problems will be considered in more detail below. Figure 8-5 shows the cost for chlorination in relation to system size (at 70 percent of capacity).

Package Treatment Plants. As the name implies, package plants are water treatment plants in which the major components are prefabricated. The components are transported to a given site and assembled into a working water treatment plant. The theory behind the cost savings associated with the package plant is that it is factory-built in an efficient atmosphere where costs can be closely controlled. For some package plant designs, minimal on-site assembly and installation are required, a feature that amplifies the economies realized with the package plant concept (Hansen and Conley, 1976).

So-called package water treatment plants have been manufactured for many years. Treatment concepts employed in these package plants have by and large paralleled standard recognized water treatment plant design criteria.

A study of package treatment plants conducted by the Water Supply Research Division (unpublished results from ongoing EPA field research) has documented the fact that, when properly operated, these plants can achieve compliance with

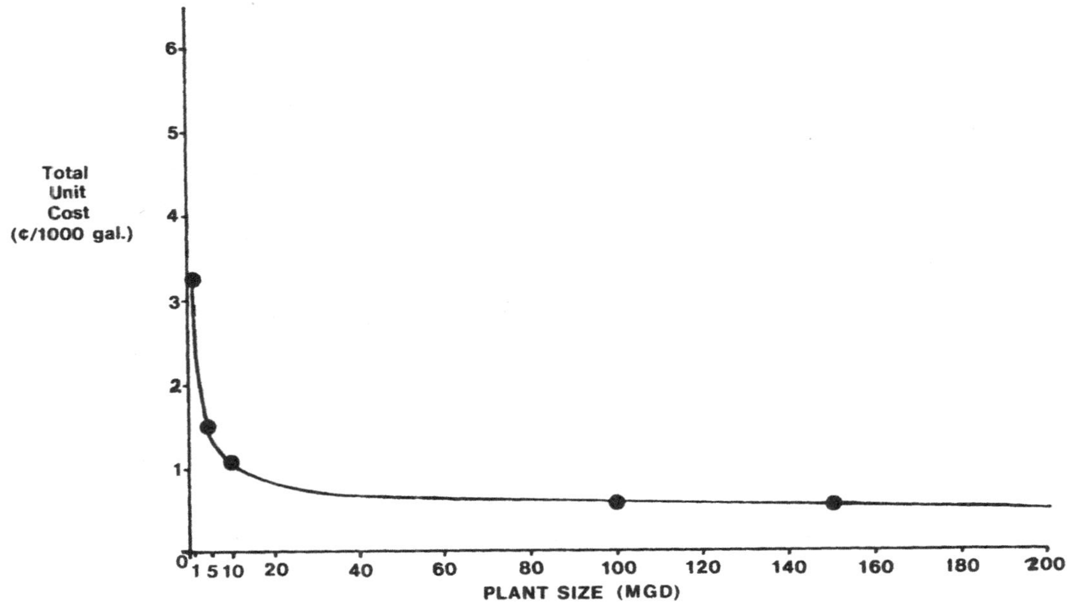

Figure 8-5. Unit cost for chlorination versus plant size.

Source: Clark and coauthors, 1977, p. 4.

the Interim Primary Drinking Water Standards. Figure 8-6 shows the unit cost of water treatment for small systems included in this study (at 70 percent of capacity).

Conventional Treatment for Turbidity Removal. A maximum contaminant limit (MCL) has been established for turbidity because certain types of turbidity-causing solids (such as organic matter) can interfere with disinfection or microbiological determinations and can also prevent maintenance of an effective disinfectant throughout the distribution system. Suspended solids that cause turbidity can be removed from water by coagulation, sedimentation, and filtration (Logsdon, 1977, p. 43). In addition to preparing the raw water for disinfection, sedimentation and filtration offer some fringe benefits. Very clear waters are more esthetically appealing to consumers. Suspended matter removed by filtration cannot settle in dead ends in the distribution system and cause problems with chlorine demand, microbiological growths, or taste and odor. Also, removal of suspended matter can result in removal of contaminants (for example, heavy metals, pesticides, and asbestos adsorbed or attached to the suspended matter). Sedimentation and filtration can also remove precursor substances that could form trihalomethanes by combination with free residual chlorine. The MCL for turbidity is 1 turbidity unit (tu), but under certain circumstances, it can be 5 tu. Figure 8-7 shows the total production cost for conventional treatment, which includes settling and filtration (at 70 percent of capacity).

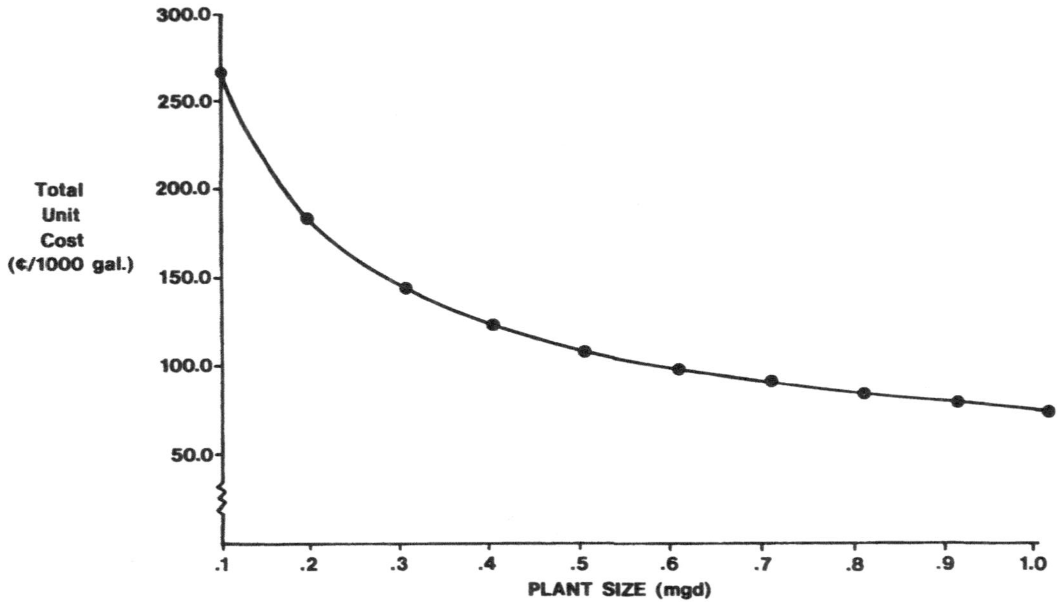

Figure 8-6. Package plant total unit treatment cost.

Source: Inhouse EPA Research.

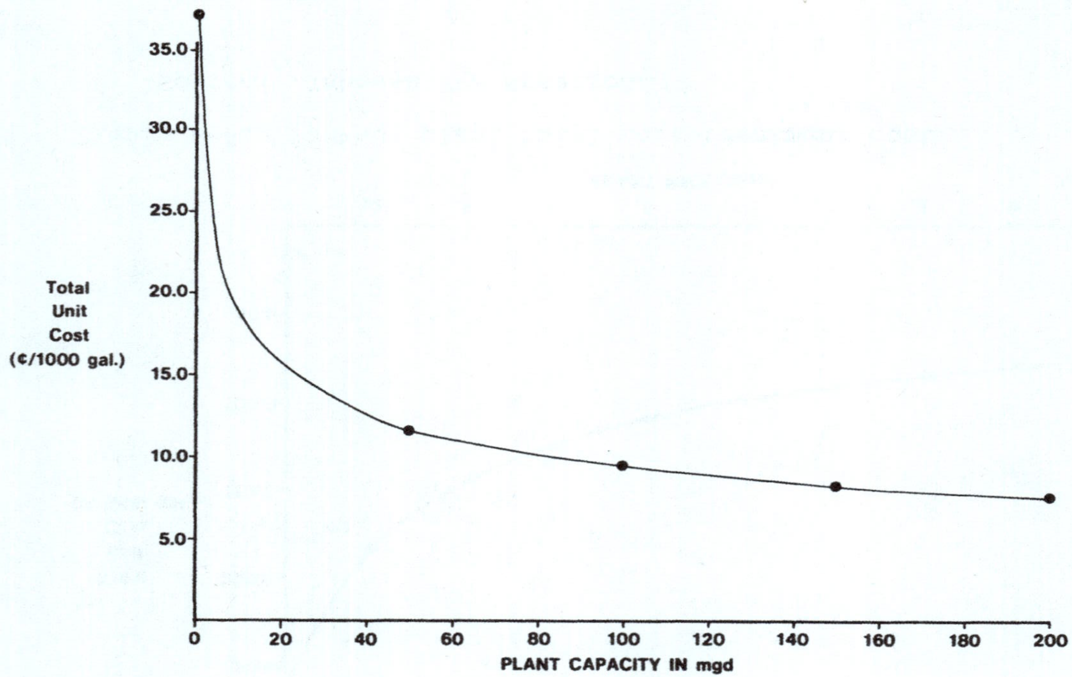

Figure 8-7. Total unit costs for conventional treatment plants with settling.

Source: Logsdon, 1977.

For very small systems, package treatment is cheaper than conventional treatment. However, as system size increases, conventional treatment becomes the cheaper alternative, and, as figures 8-6 and 8-7 show, by 1 or 2 mgd size, conventional treatment is already cheaper.

Costs for Meeting the Proposed Organic Standards

Two aspects must be considered in meeting the requirements of the newly proposed organic standards. The first relates to the formation of trihalomethanes in drinking water. Trihalomethanes in general, and chloroform in particular, have recently been determined to be carcinogens (National Cancer Institute, 1976). Trihalomethanes are found in drinking water as a direct consequence of chlorination; chlorine has probably been reacting with certain organic materials to produce chloroform and related organic by-products since this process was initiated (Symons, 1976, p. 1). These compounds have previously escaped detection in drinking water because they are low in concentration and possess low boiling points, and thus were lost during analysis.

The second aspect to consider relates to synthetic organics in drinking water, which occur as a result of pollution from upstream point or nonpoint discharges.

Several approaches are available for minimizing or eliminating the trihalomethane problem (Symons and Clark, 1978). Possible solutions include changing disinfectants or using granular activated carbon (GAC) as a means of removing trihalomethanes once they have been formed, or for removing their precursors. Synthetic organics are probably best removed by granulated activated carbon. [Ed. Note. See Sontheimer paper in this volume for considerable evidence relevant to the options.]

In the previous section, the costs of chlorine disinfection were presented as typical of disinfection costs in general. In this section, costs for chlorine, alternative disinfection, and granulated activated carbon will be presented.

Trihalomethane Removal and Control. An approach that may be followed in minimizing the trihalomethane problem is to use a disinfectant other than chlorine. In this subsection, the costs associated with various disinfectants are presented, including the cost of chlorination (Clark, Guttman, Crawford, and Machisko, 1977, p. 4).

Actual costs can vary, due to local conditions. For example, operation and maintenance costs can be reduced if the delivered costs for chemicals are less than the costs upon which the estimates were based. A cost index is used to provide a baseline for projecting costs and for estimating escalation because of inflation.

The following design parameters were assumed constant for all disinfectants considered: electric power cost, \$0.03/kWh; EPA construction cost index, 273.8; wholesale price index, 194.6; direct hourly wage rate, \$5.58/hr.; amortization interest rate, 7 percent; amortization period, 20 years; and capacity factor, 70 percent.

All cost computations have been made assuming design flow rates of 1, 5, 10, 100, and 150 mgd, and the required dose is that calculated to have disinfecting effect equivalent to the chlorine dose of 2 mg/liter with 20 minute contact time.

1. Chlorine. The design variables unique to the cost computations for chlorination are as follows: chlorine dose, 2 mg/liter; contact time, 20 minutes; and cost, $300/ton. Unit costs for chlorination, using both the specific and general design parameters, are given in table 8-4.

Table 8-4. Unit Chlorination Costs (cents/1,000 gal)

	Design capacity (mgd)				
Item	1	5	10	100	150
Capital cost	2.19	0.88	0.62	0.26	0.24
Operating cost	1.06	0.56	0.46	0.32	0.31
Total cost	3.25	1.44	1.08	0.58	0.55

2. Ozone. Specific design parameters and levels for ozonation are as follows (Clark, Guttman, Crawford, and Machisko, 1977, p. 19): ozone dosage, 1 mg/liter; contact time, 20 minutes; cost, $0.046/lb. Unit costs for disinfection with ozone using both air and oxygen are shown in tables 8-5 and 8-6.

Table 8-5. Unit Costs for Disinfection with Ozone--Air (cents/1,000 gal)

Item	Design capacity (mgd)				
	1	5	10	100	150
Capital cost	2.90	1.36	1.11	0.76	0.73
Operating cost	2.75	1.05	0.77	0.40	0.38
Total cost	5.65	2.41	1.88	1.16	1.11

Table 8-6. Unit Costs for Disinfection with Ozone--Oxygen (cents/1,000 gal)

Item	Design capacity (mgd)				
	1	5	10	100	150
Capital cost	4.46	1.50	1.08	0.61	0.58
Operating cost	2.87	1.17	0.88	0.52	0.49
Total cost	7.33	2.67	1.96	1.13	1.07

3. Chlorine Dioxide. In this cost analysis, the dose of chlorine dioxide was assumed to be half the dose of chlorine to achieve equivalent disinfection results (Clark, Guttman, Crawford, and Machisko, 1977, p. 8). If 0.5 mg/liter of chlorine is combined with 1.6 mg/liter of technical grade sodium chlorite, 1 mg/liter of chlorine dioxide will result.

The equation below shows this relationship (assuming 80 percent pure $NaClO_2$ yields 1.3 mg/liter of reactive material):

$$2\ NaClO_2 + Cl_2 \longrightarrow ClO_2 + NaCl \tag{4}$$

mol. wt. (181) (71) (135) (117)

1.3 mg/liter 0.5 mg/liter 0.9 mg/liter 0.8 mg/liter

The chlorine feed system and contact basins were therefore estimated for a 0.5 mg/liter dose of chlorine with the rest of the standardized values fixed at the following levels: chlorine dose, 0.5 mg/liter; contact time, 20 minutes; cost of sodium chlorite, \$0.70/lb; chlorine cost, \$300/ton. The ratio of sodium chlorite concentration to chlorine concentration is 3.2. Costs for chlorine dioxide are shown in table 8-7.

Table 8-7. Unit Cost of Disinfection with Chlorine Dioxide (cents/1,000 gal)

	Design capacity (mgd)				
Item	1	5	10	100	150
Capital cost	1.90	0.76	0.51	0.22	0.20
Operating cost	1.55	1.18	1.12	1.03	1.02
Total cost	3.45	1.94	1.63	1.25	1.22

Unit costs for four disinfection processes have thus been developed and are summarized in table 8-8.

Table 8-8. Summary of Unit Costs for Disinfectants (cents/1,000 gal)

Item	Design capacity (mgd)				
	1	5	10	100	150
Chlorination 2 mg/liter	3.25	1.44	1.08	0.58	0.55
Cost of ozone--air 1 mg/liter	5.65	2.41	1.88	1.16	1.11
Cost of ozone--oxygen 1 mg/liter	7.33	2.67	1.96	1.13	1.07
Chlorine dioxide 1 mg/liter	3.45	1.90	1.63	1.25	1.22

Granular Activated Carbon (GAC) Costs. For this analysis, two types of GAC systems are considered. One system uses carbon in separate contactors after sand filters (hereafter called postfilter adsorbers) and the other uses the GAC as a replacement for the sand in existing filter beds (hereafter called sand replacement). Both systems will be considered with on-site reactivation (Clark, Guttman, Crawford, and Machisko, 1977, p. 26 and following). The economics of using truck transport combined with centralized reactivation furnaces will also be discussed briefly.

The need to consider the cost of separate GAC contactors is eliminated if it is assumed that GAC will replace sand in existing filters. For purposes of the sand replacement analysis, a water treatment plant is assumed to consist of an integral number of 1-mgd filters. Design parameters assumed for the sand replacement systems are shown in table 8-9 and design assumptions for postfilter adsorber systems are shown in tables 8-9 and 8-10.

Note that for sand replacement, a GAC loss of 10 percent per reactivation cycle is assumed, but a GAC loss of only 5 percent per cycle is assumed for postfilter adsorbers. These two assumptions are intended to reflect differences in the operation of the two systems. Sand replacement systems are labor intensive and increase the possibility of GAC loss because the activated carbon is changed manually. In postfiltration systems, the activated carbon is assumed to be changed hydraulically, leading to fewer possibilities for handling losses. Table 8-11 shows the unit costs for both types of systems for 1, 5, 10, 100, and 150 mgd plants.

Several economic relationships should be explored with regard to GAC. One is the effect of reactivation frequency. Figure 8-8 shows the cost of GAC for a 100-mgd sand replacement system in relation to the period between reactivations. As can be seen from the figure, costs escalate rapidly when the period between reactivations is shortened.

Another important economic option, particularly for small systems, is that of banding together, either as a group of small systems or as a larger system to minimize the cost

Table 8-9. Design Parameters for Granular Activated Carbon

Design parameter	
Sand replacement:	
Activated carbon cost	\$.55/lb
Activated carbon loss per reactivation cycle	10%
Fuel cost	\$1.50/million Btu
Construction cost index	273.8
Wholesale price index	194.6
Direct hourly wage rate	\$5.58/hr
Amortization rate	7%
Amortization period	20 years
Volume per filter	856 ft^3
Loss in adsorptive capacity	0%
Design capacity	70%
Empty bed contact time	13.4 min
Reactivation frequency	Every 1.2 months
Postfilter adsorption:	
Activated carbon cost	\$.55/lb
Activated carbon loss per reactivation cycle	5%
Fuel cost	\$1.50/million Btu
Construction cost index	273.8
Wholesale price index	194.6
Direct hourly wage rate	\$5.58/hr
Amortization rate	7%
Amortization period	20 years
Loss in adsorptive capacity	0%
Design capacity	70%
Empty bed contact time	18 min
Reactivation frequency	Every 2.4 months

Table 8-10. Assumptions for Separate Postfiltration Systems

Item	Plant capacity (mgd)				
	1	5	10	100	150
Number of contactors	3	6	12	40	60
Diameter of contactors (ft)	8	12	12	20	20
Depth of contacts (ft)	18	13	13	14	14
Volume of GAC per contactor (ft^3)	653.1	1469.5	1469.5	4369.0	4369.0
Apparent contact time (min)	18	18	18	18	18

Table 8-11. Amortized Capital and Operating and Maintenance Costs for GAC Systems

(cents/1,000 gal)

Cost	Plant capacity (mgd)				
	1	5	10	100	150
Amortized capital cost					
Sand replacement	24.1	5.8	4.4	2.0	1.7
Filter adsorber	40.6	15.1	13.5	6.9	5.5
Operating and maintenance cost					
Sand replacement	26.3	13.8	11.1	6.6	6.4
Filter adsorber	28.3	13.5	10.5	5.8	5.4
Total cost					
Sand replacement	50.4	19.6	15.5	8.6	8.1
Filter adsorber	68.9	28.5	24.0	12.7	10.9

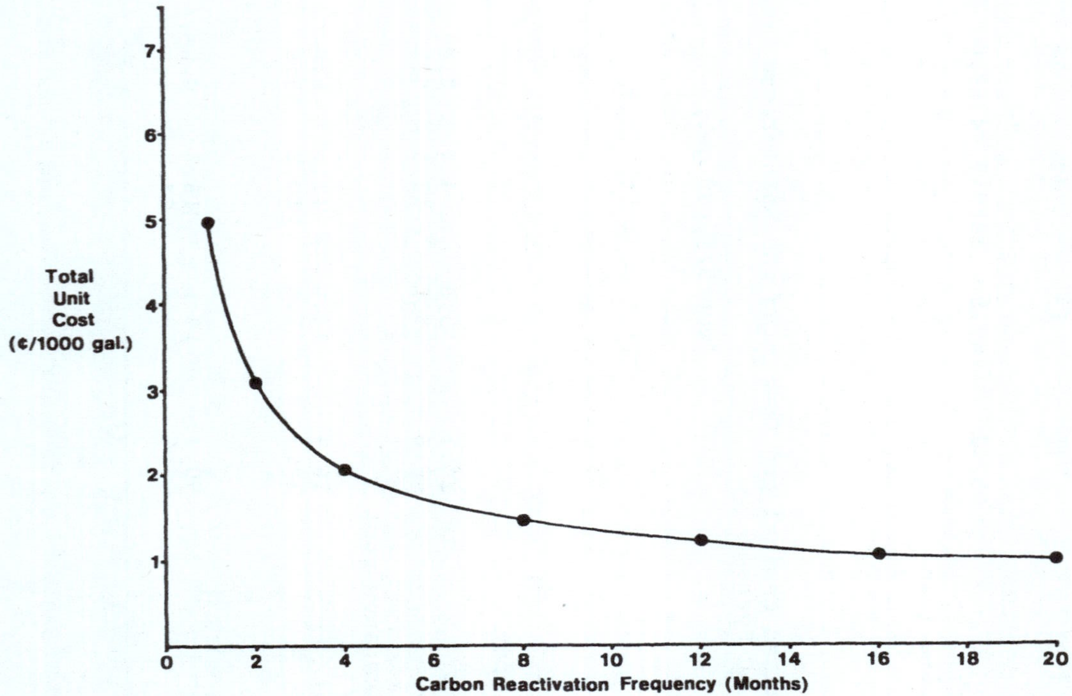

Figure 8-8. Total unit cost of GAC for a 100 mgd sand replacement plant in relation to the time between reactivations.

Source: Clark and coauthors, 1977, p. 52.

of regeneration. Figure 8-9 shows the tradeoffs that exist between hauling carbon to a central regeneration furnace and on-site regeneration. In this case, it is hypothesized that a group of ten 10-mgd utilities have joined together to provide regional regeneration services.

Regional Solutions

The Safe Drinking Water Act, along with other environmental legislation, encourages the development of regional systems for managing water supplies. Regionalization has great appeal to regulatory groups, too, because it reduces the number of individual entities with which they must work. Tables 8-3, 8-4, and 8-5 demonstrate that small systems tend to have more quality control problems than large systems. A common argument goes that regional systems exhibit economies of scale, thereby minimizing costs to the consumer (Kneese, 1972, p. 21-44).

The argument for economies of scale holds true for a limited set of circumstances, as can be demonstrated by the following data. Figure 8-10 shows unit costs of water treatment for the Fairfax County Water Authority (Virginia) uncorrected for inflation over a ten-year period (Clark, Gillean, and Adams, 1977, p. 89), and four other utilities. The Fairfax County Water Authority was developed by acquiring individual water service areas around the original Alexandria Water Company, and those data seem to support an argument for economies of scale. Compared with the other nonregional

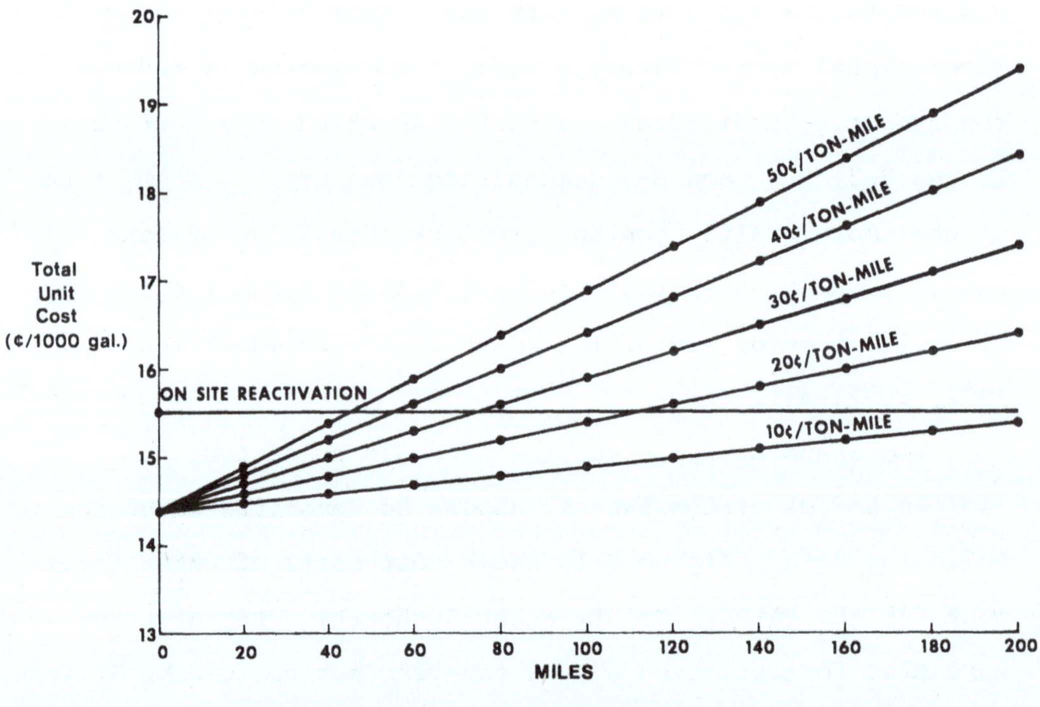

Figure 8-9. The sensitivity of reactivation costs to transportation cost variations.

Source: Clark and coauthors, 1977, p. 86.

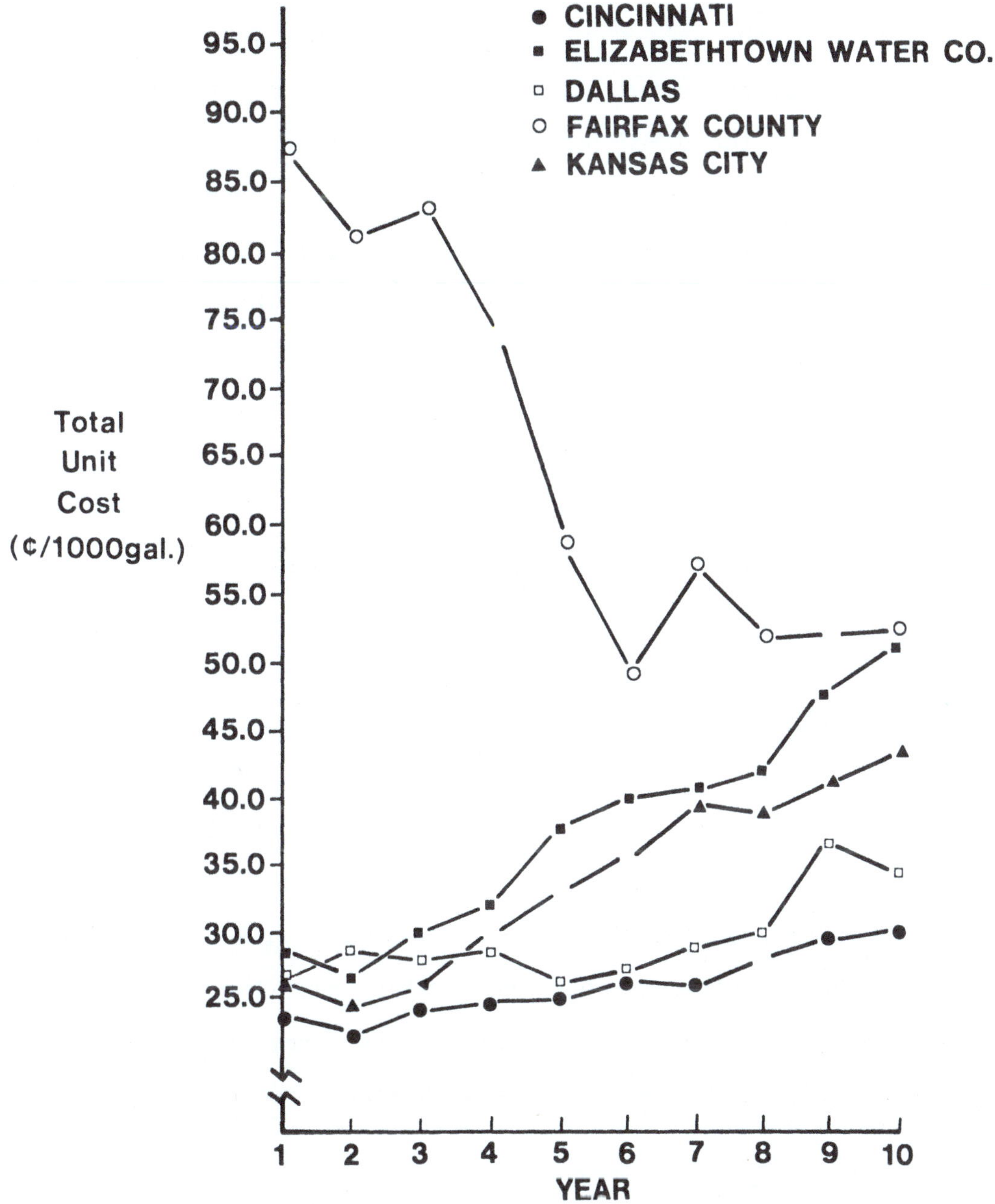

Figure 8-10. Total unit costs for five utilities.

Source: Clark and coauthors, 1977, p. 89.

utilities, the unit costs for the Fairfax Authority continuously declined between years 1 and 5, the periods of maximum growth. On the other hand, research conducted by the authors indicates that there may be limits to these economies of scale (results from in-house EPA research). The relationship between total costs and treatment and distribution was estimated to be:

$$UC = 512\ Q^{-0.23} + 990.2\ Q^{-0.39} d^{0.53} \quad (r^2 = 0.68) \qquad (5)$$

where UC = unit cost in \$/million gal per yr

Q = million gal/yr

d = distance in miles from treatment plant

When graphed for population density distributions typical for an urban area and for a system size of 235 mgd, the equation yields an average cost curve (figure 8-11). As can be seen, the unit cost first drops and then begins to rise between 7 and 8 miles, indicating that the economies of scale obtained by treating more water may ultimately be lost in transmission and distribution. Caution should therefore be used when recommending regional solutions. The apparent economies of scale and efficiencies in a regional system may ultimately be lost when the system grows too large.

Individual Economic Impacts

When the Safe Drinking Water Act was conceived, it was intended that direct costs of control under the act be assessed against the consumer. This section calculates hypothetical cost impacts for various combinations of treatment

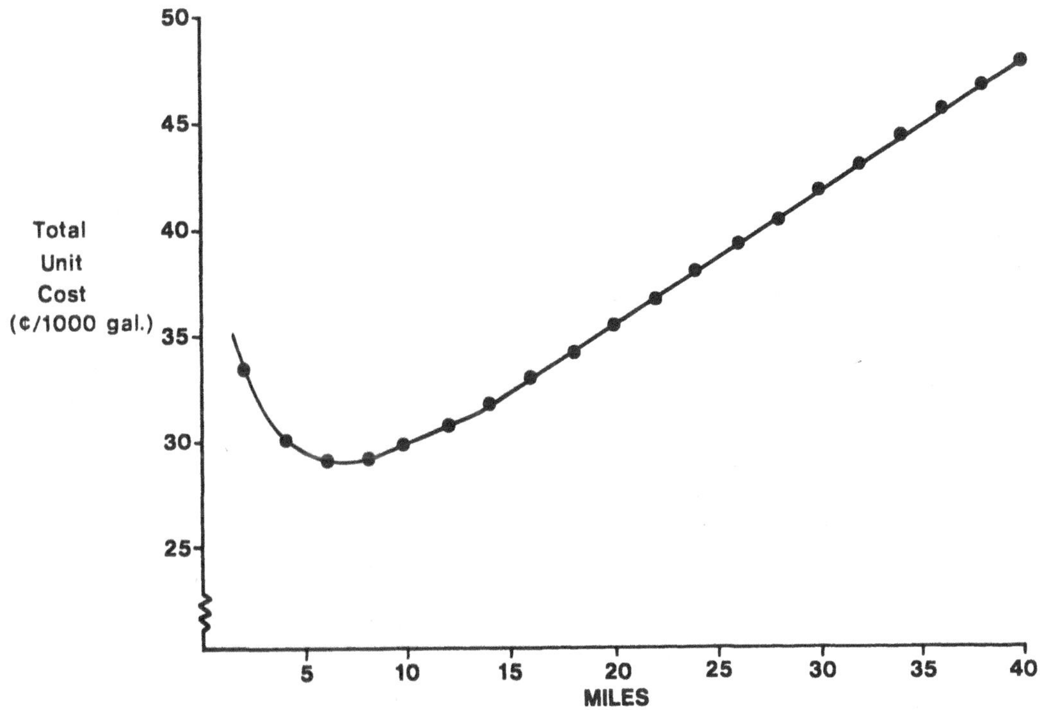

Figure 8-11. Unit costs versus miles from treatment plant.

Source: Unpublished EPA research.

options for 1-mgd and 100-mgd systems. The costs presented in table 8-6 will be used as the basis for the impact analysis.

For the first analysis, it was assumed that the large and small utilities would be required to add monitoring capability only (table 8-12).

Table 8-12. Monitoring Cost Impact on 1-mgd and 100-mgd Plants

Item	1-mgd plant		100-mgd plant	
	Cost in ¢/1,000 gal	% Increase	Cost in ¢/1,000 gal	% Increase
Current cost	85.4	-	41.5	-
Cost with monitoring	86.0	1	41.6	<1

If the data presented in figure 8-4 are used, table 8-12 shows that compliance monitoring will have little impact on the cost of water supply, except for extremely small systems.

Another more interesting impact will be on supplies that currently have no treatment capability (table 8-13). Again, this table uses data from table 8-3 and figure 8-4 but it subtracts the treatment costs from the total unit costs in table 8-3.

The water bill increases resulting with added treatment are easy to calculate if a family of four is assumed to use 117,000 gal of water per year (80 gallons per day per capita).

Table 8-13. Treatment Requirement and Monitoring Cost Impact on Water Supplies with No Treatment Capability

Item	1-mgd plant		100-mgd plant	
	Cost in ¢/1,000 gal	% Increase	Cost in ¢/1,000 gal	% Increase
Current cost (without treatment)	75.6	-	36.7	-
Cost with monitoring	76.2	1	36.8	< 1
Cost with monitoring and disinfection	79.5	5	37.4	2
Cost with monitoring and conventional treatment	114.2	39	46.3	26

Ed. Note. It has been asserted by one reviewer that there exist no 100-mgd systems without any treatment. This does not destroy the illustrative value of the comparison between the cost effects at the two system sizes.

Other impacts to be considered are those related to the cost of complying with the proposed organic standards. The first to be considered will be that of alternative disinfectants. Table 8-14 summarizes the costs incurred when switching disinfectants. The current costs for the 1- and 100-mgd systems are assumed to include chlorination.

Another impact that will be significant in terms of cost is the use of GAC for removing organics. Table 8-15 shows the effect on an existing system (1 and 100 mgd), assuming the existing costs in table 8-3.

Table 8-14. Cost of Trihalomethane Control Alternatives

	1-mgd plant			100-mgd plant		
	Unit cost for disinfection (¢/1,000 gal)	Current cost (¢/1,000 gal)	increase	Unit cost for disinfection (¢/1,000 gal)	Current cost (¢/1,000 gal)	increase
Chlorination, 2 mg/liter	3.25	85.4	---	0.58	41.5	---
Ozone--air, 1 mg/liter	5.65	85.4	2.8	1.16	41.5	1.4
Ozone--oxygen, 1 mg/liter	7.33	85.4	4.7	1.13	41.5	1.3
Chlorine dioxide, 1 mg/l liter	3.45	85.4	0.2	1.25	41.5	1.6

Table 8-15. Granular Activated Carbon Cost

Treatment alternative	1-mgd plant			100-mgd plant		
	Unit cost for for GAC (¢/1,000 gal)	Current cost (¢/1,000 gal)	% Increase with GAC	Unit cost for for GAC (¢/1,000 gal)	Current cost (¢/1,000 gal)	% Increase with GAC
Sand replacement	50.4	85.4	59	8.6	41.5	21
Postfilter adsorber	68.9	85.4	81	12.7	41.5	31

Another impact is on system capital requirements. The total capital cost for a 1-mgd replacement system is $655,000; for 100-mgd, $5,440,000. For a 1-mgd postfilter adsorber system, the cost is $1,100,000, and for 100-mgd, $18,800,000.

National Economic Impact

An analysis of economic impact would not be complete without examining estimates for the total annual cost of capital, operations and maintenance, and monitoring necessary to comply with the Interim Primary Drinking Water Standards (Energy Resources Co., Inc., 1975, p. 14). The estimated annual costs of complying with the standards are shown in table 8-16.

Table 8-16. Estimated Capital and Annual Costs of Implementing the Interim Primary Drinking Water Regulations for Public Water Supply

Item	Cost (in millions of 1975 dollars)
Total capital cost	1071 - 1800
Annual capital[a]	146 - 247
Annual operations and maintenance	263
Annual monitoring (routine only)	17 - 35
Total annual	426 - 545

[a]Assumes capital costs amortized over 15 years at 7 percent interest.

The proposed trihalomethane and synthetic organic regulations will affect a limited number of water supplies. For those systems affected by both regulations, it appears that only the use of GAC will allow compliance, and the operating parameters of contact time and reactivation frequency will have to be chosen to achieve both sets of requirements. Accordingly, some systems may be forced to operate with longer contact times or shorter reactivation periods than would be possible were they affected by only one of the regulations. In combination, the two regulations are expected to require sixty-one to seventy-eight large water systems to utilize GAC (or an equivalent carbon treatment) solely to comply with the trihalomethane regulation. Thirty-five systems will install it solely to meet the organic requirements, and fifteen systems will use it to meet both requirements (Temple, Barker and Sloane, Inc., 1977). The national costs for both trihalomethane and general organics control are given in table 8-17.

Table 8-17. Combined Costs for Meeting the Trihalomethane and General Organics Regulations

Item	Cost (in millions of 1976 dollars)
Annual capital expenditures[a]	23 - 53
Annual O & M expenditures	34 - 92
Total	57 - 145

[a]Amortized at 8 percent over forty years.

Summary and Conclusions

The Safe Drinking Water Act reflects a growing national concern over the safety of public drinking water systems. Many surface and some groundwater supplies are vulnerable to pollution by upstream point and nonpoint discharges. The economic impact of the act will be felt at the national, state, and local level. Although implementation of the National Interim Primary Drinking Water Standards will have limited economic impact, the most expensive control technologies for meeting the proposed organic standards may increase the cost of water supply to the individual affected utilities by 20 percent or more. Even though the economic costs are significant, they still amount to a relatively small percentage of total annual expenditures on a national basis. For example, assuming that total expenditures by both public and private water supplies are $7 billion, and that annual implementation costs are $700 million, the impact will be less than 10 percent of national costs. Nevertheless, there is currently a high level of controversy over the benefits to be obtained from implementing the act, and in particular the proposed organic standards.

Part of the controversy over the act centers around the question of who should pay for pollution control, the consumer or the polluter. Currently, as the act is written, the price for control technology is placed directly on the user and is not spread across the population through a general federal support program. Congress intended that the consumer would pay for his protection directly. By so doing, the consumer may, for the first time, begin to understand what the cost of pollution control is to him as an individual. The act may, therefore, in a very real sense "protect" public water supplies by forcing a closer examination of upstream discharges into vulnerable water supply sources.

The act, as with much recent federal legislation, encourages the development of regional compacts. Data in this paper show that there are limits to the application of the regional concept. Regionalization can lead to lowered unit costs by taking advantage of the inherent economies of scale in water delivery systems. At some point, however, the distribution system becomes so large that the diseconomies of transporting water overwhelm the economies of production. Utility managers and planners must be careful not to become trapped in the "bigger is better" philosophy that tends to be embodied in the engineering profession. Rather, each situation must be evaluated separately with regard to the benefits and costs of regionalization.

Implementation of the act raises many important questions relative to social values. The basis for environmental control and sanitary engineering is preventive public health practice. Applying cost-benefit analysis in this area is a difficult exercise, because it goes to the heart of human values. How much are we willing to pay for prevention of a small number of cancer deaths or to prevent the anxiety associated with industrial spills? These are fundamental questions in human values and add an extra dimension to the argument. In a recent situation in which Cincinnati, Ohio, was faced with unexpected contamination of its water system by an industrial spill of carbon tetrachloride, there was near panic in the city. The city council, when presented with some of the cost data contained in this paper, agreed that indeed the benefits exceeded the costs.

The act can therefore be thought of as an insurance policy and, as with all insurance policies, we must ask if the benefits are worth the cost. Is the policy worth the price? Ultimately, the American public must decide the answers to these questions, but in the authors' opinion the answers are affirmative.

References

Ciriacy-Wantrup, S. V.; Wells A. Hutchins; Clyde O. Martz; Sho Sato; and Albert W. Stone. 1967. Water and Water Rights: A Treatise on the Laws of Waters and Allied Problems: Eastern, Western, Federal. vol. 1 (Indianapolis, Ind., Allen Smith Publishers).

Clark, Robert M. 1978. "The Safe Drinking Water Act: Its Implications for Planning," in D. Holz and S. Sebastian, eds., Municipal Water Systems: A Challenge for Urban Resource Management. (Bloomington, Indiana University Press).

______; James I. Gillean; and W. Kyle Adams. 1977a. The Cost of Water Supply and Water Utility Management. vols. I and II. Water Supply Research Division, Municipal Environmental Research Laboratory, Cincinnati, Ohio.

______; D. L. Guttman; J. L. Crawford; and J. A. Machisko. 1977b. The Cost of Removing Chloroform and Other Trihalomethanes from Drinking Water Supplies. U.S. Environmental Protection Agency, Cincinnati, Ohio.

Energy Resources Co. 1975. Economic Evaluation of the Proposed Interim Primary Drinking Water Regulations (Washington, U.S. Environmental Protection Agency, Office of Water Supply).

Fair, G. M., and J. C. Geyer. 1954. Elements of Water Supply and Waste Water Disposal (New York, Wiley).

Federal Register. 1975. National Interim Primary Drinking Water Standards. vol. 40, no. 248, December 24.

Hansen, S. P., and W. R. Conley. 1976. A Decade of Experience with Package Water Treatment Plants. (Corvallis, Ore., Neptune Microfloc).

Herschell, Clemens. 1973. Trans. 1973. The Two Books on the Water Supply of the City of Rome. New England Water Works Association.

Hines, N. William. 1971. Public Regulation of Water Quality in the United States. U.S. Department of Commerce, Legal Study 18. Prepared for National Water Commission, Department of Commerce (NTIS PB-208 309).

Kneese, Allen V. 1972. "Analysis of Environmental Pollution," in Robert Dorfman and Nancy Dorfman, eds., Economics of the Environment. (New York, W. W. Norton)

Logsdon, Gary S. 1977. Treatment Techniques for the Removal of Turbidity from Drinking Water. Manual of Treatment Techniques for Meeting the Interim Primary Drinking Water Regulations. EPA-600/8-77-005, U.S. Environmental Protection Agency, Cincinnati, Ohio.

McDermott, James H. 1973. "Federal Drinking Water Standards--Past, Present, Future," Journal of the Environmental Engineering Division, ASCE vol. 99, no. EE4, Proc. Paper 9924.

National Cancer Institute. 1976. Report on the Carcinogenesis Bioassay of Chloroform. Carcinogen Bioassay and Program Resources Branch, Carcinogenesis Program Division of Cancer Cause and Prevention, National Cancer Institute, Bethesda, Md.

Orlob, G. T., and M. R. Lindorf. 1958. "Cost of Water Treatment in California," Journal of the American Water Works Association vol. 50 (January).

Symons, James M. 1976. Interim Treatment Guide for the Control of Chloroform and Other Trihalomethanes. U.S. Environmental Protection Agency, Cincinnati, Ohio.

______ and Robert M. Clark. 1978. Interim Treatment Guide for Controlling Organic Contaminants in Drinking Water Using Granular Activated Carbon. U.S. Environmental Protection Agency, Cincinnati, Ohio.

Temple, Barker, & Sloane, Inc. 1977. Unpublished Economic Analysis of Proposed Regulations on Organic Contaminants in Drinking Water, submitted to EPA's Office of Water Supply.

U.S. Bureau of the Census. 1975. 1972 Census of Manufacturers, Area Series: Alabama, Georgia, Illinois, Indiana, Kentucky, Maryland, Mississippi, New York, North Carolina, Ohio, Pennsylvania, Tennessee, Virginia, West Virginia. (Washington, GPO).

______. 1977. 1974 Census of Agriculture, Vol. 1, State Reports: Colorado, Iowa, Kansas, Minnesota, Missouri, Montana, Nebraska, North Dakota, South Dakota, Wyoming. (Washington, GPO).

______. Governmental Finances in 1971-72, 72-73, 73-74, 74-75, 75-76. Government Finance Scores: 2466-2.6, Table 11. (Washington, GPO).

U.S. Department of Health, Education, and Welfare. 1970. Community Water Supply Study: Analysis of National Survey Findings. Public Health Service, Environmental Health Service, Bureau of Water Hygiene.

U.S. Environmental Protection Agency. 1978. Statement of Basis and Purpose for an Amendment to the National Interim Primary Drinking Water Regulations on a Treatment Technique for Synthetic Organic Chemicals. Office of Water Supply, Criteria and Standards Division, U.S. Environmental Protection Agency, Washington, D.C.

Discussion by Walter B. Sinnott

The authors should be complimented for the excellent compilation of capital and operating cost data on water treatment, disinfection, and monitoring. My comments will be directed toward the costs of meeting the interim drinking water standards, particularly for those large municipalities in the Northeast which currently do not have filtration plants. To cite just a few of these systems, this category includes:

	Peak daily demand (mgd)
New York City, N.Y.	1,800
Bridgeport Hydraulic Co.	60
Utica, N.Y.	40
Newark, N.J. (Pequannock Supply)	90
Boston, Mass.	800

Several other large systems in the Northeast, such as the North Jersey Water District and Jersey City, have filtration plants under construction. The capacities are 100 and 80 mgd, respectively.

All of the systems noted secure water from impounded, upland supplies. Some of the watersheds are, of course, more protected than others, but none can be characterized as highly developed. What development there is usually consists of small, residential communities, scattered residences, and agriculture.

In addition to the two New Jersey treatment plants under construction, New York City and Newark, N.J., are in various stages of planning for filtration of some of their surface supplies. The filtration plants are being planned primarily to meet the interim standards for turbidity. Actual construction costs for the two plants in New Jersey and estimated construction costs for the others are as follows:

	Peak capacity (mgd)	Type of treat-ment	Estimated cost, 1978 Total (millions)	Per mgd	Remarks
New York City Croton supply	270	D.F.	$108	$400,000	(1)
North Jersey Water District	100	C.F.	26	260,000	
Newark Pequannock supply	94	D.F.	22	234,000	
Jersey City	80	C.F.	31	387,000	(2)

D.F. - Direct filtration without settling

C.F. - Conventional filtration with settling

(1) - Ozonation and diatomaceous earth filtration was found to be more economical but will require further full-scale testing program before being selected.

(2) - Bid in 1975. Costs escalated by 25 percent to reflect 1978 conditions.

The relatively high cost for the proposed New York City treatment plant is attributable to the high labor costs in the city and a costly site at Jerome Park Reservoir in the Bronx.

From the foregoing, one can conclude that a 100-mgd plant will cost on the order of $250,000 to $300,000 per mgd of capacity for suburban northeast locations and may be considerably more expensive for in-city sites. In most large water systems in the Northeast, peak daily demands are between 1.3 and 1.5 times the average daily demand. Therefore, a 100 mgd plant would be required to serve a system with an average demand of about 67 mgd of demand. Using these figures and assuming that capital costs are financed by twenty-five year bonds at 7.0 percent rate of interest and level payments over the twenty-five year period, the actual capital cost per million gallons of water treated is estimated as follows:

1. For a 100-mgd conventional treatment plant, capital cost = 100 x $300,000/mgd	$30,000,000
2. Annual debt service for twenty-five year bonds at 7.0 percent interest, assuming equal annual payment - 0.08581 x $30,000,000	$ 2,574,000
3. Cost per million gallon treated assuming average demand is 67 mgd = $2,574,000 ÷ (67 x 365)	$ 105/per million gallons

In addition to the capital cost of filtration plant, there are the annual costs for labor, chemicals, power, fuel, replacement of parts, sludge disposal, etc. Our experience indicates that for a 100-mgd plant these costs will be between $50 and $60 per million gallons of water treated. Adding this to the capital costs gives a total cost of about $155 to $165 per million gallons.

Let us compare this with the figures presented by the authors in figure 8-7 for a 100-mgd conventional plant. Figure 8-7 indicates about 9¢ per 1,000 gallons or $90 per million gallons. Since this estimate seems to have been based on an EPA cost index of 273.8 and the present index for the New York metropolitan area is 300±, the figure 8-7 estimate should be adjusted upward by about 10 percent. The $90 estimate would become about $100 per million gallons. This is still considerably below our estimates and may be attributable to our average peaking factor of 1.5 included in our analysis. Although the authors indicate a capacity factor of 70 percent was used in the analysis of the cost of the carbon filters, there is no indication of any such factor for the conventional treatment plants.

In the large northeast cities, the average family of four uses about 70 gallons per capita per day, or a total of about 280 gallons per day. (In some housing projects, however, we have found per capita consumption exceeding 125 gallons per day.) To supply 280 gallons per day at the consumer's meter, most systems must produce 20 percent more at the treatment plant with the difference accounted for in system leaks, water used for fire-fighting purposes, street flushing, etc. Therefore, the consumer's cost will be based on producing 1.2 x 280, or 336 gallons per day. At $160 per million gallons, the new conventional treatment facilities would require an additional $20 per year in revenue.

In smaller communities, this figure would increase drastically because of several factors:

> The unit capital and operating costs for small treatment plants would be much higher.
>
> The peaking factors in small systems can be as high as 2.5 to 3.0 times average daily demands.
>
> Smaller systems may have significant growth rates compared with the large, old cities, which are barely holding their own. Most small plants are built to provide capacity to meeting projected demands ten to twenty years hence. Therefore, present users pay for the extra capacity.

Similar analyses for smaller municipalities have indicated that the additional costs for treatment facilities will be on the order of $100 per family per year.

In the large municipalities such as Newark, which will be building water plants on the order of 100 mgd in capacity, we have found that the average domestic consumer is currently paying about $50 per year. Therefore, the additional $20 for treatment will require at least a 40 percent increase in water rates. For the smaller municipalities, present rates are on the order of $100 per year and a $100 per year increase for conventional treatment facilities will mean at least a 100 percent increase in rates.

I have not examined the estimates by the authors for granular activated carbon filters in the same detail as I have for the conventional filtration, but would raise similar questions. Is the peaking factor adequately provided for in the analysis? Does the analysis provide sufficient redundancy for having some carbon reactors out of service for regeneration during peak demands? How much extra capacity to handle growth is included in the analysis? (In systems exhibiting growth, present users will have to finance extra capacity until the growth develops; should the growth not materialize this will be a permanent burden.)

Fortunately, most of the impounded supplies in the Northeast do not have problems with organics. Assuming that these municipalities will receive variances from the proposed EPA requirement to install GAC filters for systems serving more than 75,000 people, the municipalities will only be faced with adding filtration to meet turbidity requirements. As a practical matter, the major portion of the New York City supply, the Catskill-Delaware systems, can meet the interim requirements for turbidity without filtration and the city has no plans for filtering this supply.

In summary, municipalities in the Northeast that currently do not filter their upland supplies will have to build filtration plants to meet the turbidity requirements of the Interim Primary Drinking Water Standards. It is estimated that the larger municipalities will require rate increases of at least 40 percent to finance these plants while some of the smaller municipalities will require increases of 100 percent or more. Obviously, this impact will be substantial compared with the impacts on municipalities that presently have filter plants and which may not require GAC filters because of size of system, source of supply, etc. For cities like New York and Newark, the financial situation is so critical that it is doubtful that the cities can sell bonds to finance the initial capital costs of filter plants. While a 40 percent increase in water rates may not, by itself, seem too high, taken together with all of the other problems of the northeastern cities, outside assistance will undoubtedly be required. With all of the money being spent under the water pollution control act amendments of 1972 and 1977 to clean up our rivers and estuaries, there is no question that we can and should afford financial aid to help protect our water supplies. I don't think we should go overboard on treatment, but there are definite needs for filtration in the Northeast. In most of these areas, federal assistance for the filter plants will have far greater direct benefits for every citizen than does the aid for water pollution control plants. The inner city citizens rarely get a chance to use rivers and estuaries for recreation and whether a waste treatment plant provides 50 or 90 percent BOD removal will have no measurable effect on them. However, since they all drink the water, there is no better way of reaching the masses than providing excellent water. Let's recognize that the cost will be substantial, that considerable aid will be required, and that the benefits from such an aid program would be much more people-oriented than the present water pollution control program.

Chapter 9

FINANCING SYSTEM CHANGES

Henry Graeser

The preceding speakers on this program have presented the facts and figures, and the whys and wherefores of the problem. Here, at last, comes the $64 question: Who pays, and how?

My premise is that the local citizen should pay directly through a fair and equitable user charge to a local or regional water utility operation, either public or privately owned. I base this opinion on my previous experience as a manager of the water utility at Dallas, Texas, which has been operated as a "sacred fund" enterprise function of local government since 1936. The department operated on user charges prior to 1936; however, diversion of revenues to other city operations had mired the utility in debt. At that time, a charter reform, which instituted the city-manager form of government, assured protected funds to the water department. Since that time, the water utility has operated in the black, without any subsidy from local, state, or federal government.

As an outspoken advocate of locally self-sustained operation of water utilities, I have consistently advised against the federal grant process or federal financing of local operations of this type. Also, I am convinced that local government should not engage in enterprise endeavors for any reason other than to render services <u>at cost</u> and at the lowest possible price. The foundation of this premise, however, rests upon availability of revenues protected from political diversion. The charge for service should not be diluted or increased by transfer. Likewise, the investor-owned utility should not be used as a tax collector. Properly regulated, a rate of return is not profit.

Dallas has financed development of water resources for a year 2000 growth potential and a 1985-plus system of purification, production, and distribution, capable of supplying the anticipated growth of the central city and some twenty-two satellite municipalities. This was accomplished by issuance of twenty-year revenue bonds with a 1.5 coverage requirement. Cities which pioneered in revenue bond financing were often saddled with a rather severe bond indenture. The Dallas bond indenture required that there be sufficient income after operating expenses to pay the outstanding debt service in addition to debt service on any proposed bond issue, plus 50 percent of the highest year's debt service thus incurred. So, with $5 million as debt service outstanding, and, if additional bonds are issued, incurring $2 million per year of additional debt service, a 1.5 coverage would require an annual income sufficient to provide $7 million, plus an additional $3.5 million per year over and above all operating expenses. Even under these severe financial criteria, however, the city was able to build, not only for the central city's future, but to prefinance a reasonable growth expectation for the surburban communities.

Local financing of water-system improvements is not only possible, but desirable. Dallas was able to accomplish its program largely because of continued growth, financial stability guaranteed by a city charter protecting the water utility's revenues from diversion, a well-designed master plan, a citizenry willing to invest in the future by accepting rate increases, and an orderly plan of implementation. I want to emphasize this last point. The rate at which financial demands are placed on an operating entity must bear some relationship to its ability to secure additional income. It sounds elemental; but it is easier said than done. Water rates were raised rather consistently throughout my twenty-five-year association with the Dallas system, but even so, rate increases occurred only in 1951, 1955, 1959, 1969, and 1976. Considering the inflation occurring during these years, it is rather remarkable that rate increases of about 20 to 30 percent each time were able to support a program of construction and development of water resources for both current and future growth. Dallas built two major reservoirs in the fifties which were jointly used with the Corps of Engineers. Admittedly, these were quite a bargain, but Dallas financed its share of the cost with twenty-year revenue bonds instead of accepting the normal

Corps payout of fifty years. In other words, the city paid cash to the government for its share of the reservoir construction. Two other major reservoirs, transmission pipelines, and water-production facilities were developed by the city in the sixties totaling more than $80 million. These, too, were financed with twenty-year revenue bonds supported entirely by the income of the utility. Thus, a major portion of the system has been built since the fifties. During this period, no federal grant or loan funds were utilized until about 1974, when $1 million in funds from the Department of Housing and Urban Development (HUD) were employed for extension of water services to newly annexed areas which had been developed with a substandard distribution system outside the city limits.

The history of Dallas' financing has been related simply in order to emphasize that significant local financing is possible and desirable. Whereas Dallas rates were higher than those of most cities in the fifties and sixties, its water rates today are not necessarily out of line with other communities, and the gap will continue to narrow as growth stabilizes and twenty-year bonds are paid off. In 1974, for example, the average of some sixty U.S. cities, for the most part of 100,000 population and above, was about 51¢ per 1,000 gal for 7,500 gal of water per month. The Dallas water rate at that time was about 70¢ per 1,000 gal for 7,500 gal per month, about 30 percent above the average of those sixty cities. This is not exorbitant. Today, after the last increase, the

Dallas water rate for a 7,500 gal-per-month customer is 76¢ for 1,000 gal. The average income from sale of water, however, is about 45¢ for 1,000 gal, based on a billed consumption of 77.380 billion gal of water per year, in fiscal year 1977, an income of $35 million which includes sales "at the gate" to the twenty-two customer cities. Lost water in Dallas is about 12 percent currently, so that the gross output by the filter plants was in the order of 87.93 billion gal.

In order to illustrate the impact of the safe drinking water regulations on a water utility which is locally financed, let us assume Dallas would be required to install granular activated carbon (GAC) in order to comply with the proposed organic standard. Based on studies for the past four years, Dallas has no concern about trihalomethanes and synthetic organics; however, for the sake of illustration, let us assume that studies indicated carbon contactors with regeneration facilities would be the most cost-effective approach to GAC application. What will be the impact on costs and capital structure?

Previous speakers have quoted price estimates, but I will use cost estimates prepared by Black & Veatch and the Environmental Protection Agency's Water Supply Research Division, in Cincinnati, Ohio (Symons, 1977), of a typical plant using the design and operating and maintenance (D & M) parameters shown in table 9-1.

Table 9-1. Design Parameters, Granular Activated Carbon, and Postfilter Adsorption for Three Plants

	Black & Veatch plant	EPA plant	Dallas actual
Design capacity in mgd	150	100	500
Average production	100	70	212
Utilization, in %	66	70	42
Carbon cost per lb.	57	45	
Carbon loss per cycle, %	10	4	
Amortization rate, %	7	7	
Amortization period, in yr.	20	20	
Carbon contactors			
No. of filters (three reserve)	24	28	
Diameter, in ft	30	20	
Volume per filter, in ft^3	10,602	4,396	
Empty bed contact, in minutes	16	19	
Regeneration period, in days	60	60	
Surface loading, in gpm* per ft^2	7	5.5	
GAC bed depth, in ft	15	14	
Loading intensity, in gpm* per ft^3	0.47	0.39	
Bulk density carbon, in lb per ft^3	26	30	

*gallons per minute

Table 9-1. Continued.

	Black and Veatch plant	EPA plant	Dallas actual
Labor rate, per hr	8.00	5.58	
Fuel cost, in $ per million BTU	$2.70	$1.50	
EPA index	287	273.8	
Contingencies, engineering & administration, %	25	15	

The respective cost estimates are shown in table 9-2. Annual income must be based on billed consumption, therefore, 70 mgd average production with 12 percent system loss means that the unit price for water to the customer must be based on a billed consumption of only 88 percent of production, or 62 million gallons per day (mgd) for the EPA plant and 88 mgd for the Black & Veatch plant. The unit price for billed consumption will then be 20¢ per 1,000 gal according to the Black & Veatch estimates, and about 10¢ per 1,000 gal for the EPA's figures.

We can safely say that the cost of GAC contactors will lie somewhere between these figures, probably about 15¢ per 1,000 gal based on billed consumption. Based on the 45¢ per-1,000-gal average income and ignoring indenture requirements, which will require an additional 50¢ on every dollar of annual amortization, GAC facilities at the 15¢-per-1,000-gal figure would require a 33 percent overall rate increase. If the 20¢

Table 9-2. Capital and Operating Expenses for Granular Activated Carbon Systems in Two Plants

Costs and operating expenses	Black and Veatch plant	EPA plant
Capacity, maximum, in mgd	150	100
Capacity, average in mgd	100	70
Billed throughput mgd	88	62
Capital cost total	$28,900,000	$12,599,000
Capital cost per mgd	$192,700	$126,000
Annual capital cost[a]	$2,728,000	$1,189,000
Annual O&M cost	$3,800,000	$1,024,550
Total annual cost	$6,528,000	$2,213,550
Unit price (per 1,000 gal. billed)	.20	.098

[a]Amortized over twenty years at 7 percent.

figure is correct, the increased income required will be 44 percent, or if the 9.8¢ figure is correct, it will be 22 percent. In either case, the amount is significant, especially when the total capital investment in a major surface water system is considered. For example, capital investment in the Dallas system is shown in table 9-3.

As you can see, the treatment plant represents only 10 percent of the capital cost in the system, so that a 44 percent increase in revenue, necessitated by one system element and one water quality standard, is quite significant, and could easily be preemptive of all other capital requirements.

Table 9-3. Dallas Water Utilities' Water Plant Investment, as of September 30, 1977 (in thousands of dollars)

Function	Original cost Amount	%	Replacement cost Amount	%
Raw water supply and transmission	90,095	29	248,791	30
Purification	31,405	10	86,003	10
Transmission and distribution	167,684	54	449,661	55
Meters	10,860	4	16,074	2
Fire hydrants	5,476	2	14,736	2
Engineering, administrative, and other costs	4,243	1	5,745	1
Total	$309,763	100%	$821,010	100%

Present treatment cost at Dallas (averaging three plants) with sand-filtration and coagulation lime softening, and intermittent use of powdered activated carbon is:

Capital (based on original cost)	.034
O & M	.037
Total	.071 per 1,000 gal

Thus, a GAC requirement could easily double or triple the present treatment costs.*

*Ed. Note: The reader should be warned that because historic costs are used and capital costs written off over a bonding period rather than useful equipment life, these comparisons can be misleading--especially by understating the real capital costs facing the current system without new GAC units.

As stated previously, 7,500 gal of water per month cost the Dallas, Texas, homeowner 76¢ per 1,000 gal. An increase of 15¢ per 1,000 gal can be converted to $13.50 per year, or 4¢ per day, or 1¢ per capita per day in an average-sized family of 3.5 persons. This is cheap, of course, but the typical urban dweller who equates this to an overall increase in his water bill of 20 percent based on his 76¢ average rate is going to demand more than a "possible health effect" explanation. GAC's best selling point, insofar as the average citizen is concerned, is that it eliminates taste and odors arising from natural organic material and aquatic organisms in reservoirs and streams. Water which makes bad coffee is noticed by a great many people. Admittedly, if we can sell rate increases for growth, we can sell them for valid quality or health benefits. You might ask, Why not proceed then?

The Dallas example seeks to illustrate the financial consequences of just one facet of the Safe Drinking Water Act. There are even more capital-intensive areas which will need correction. There are many uncertainties inherent in the act that affect the ability of a local utility to finance needed system improvements and revisions. This law, as do many others today, reaffirms state's rights and local prerogatives. However, it also describes, in considerable detail, administrative procedures and local responsibilities. The federal bureaucracy is directed to write the rules and regulations "which in the opinion of the Administrator" are considered proper. The rights of states now have become the rights of a puppet to do as it is told "in its own fashion." Congress, having invaded the administrative area, and at the same time having given the bureaucracy license to cover anything it forgot or might think of later, then compounds and confuses the issue by granting open license to any special interest group or person to file suit in a federal court if it or he or she disagrees with the administrative interpretation of the law.

Any discussion of how the local utility can finance improvements required by the Safe Drinking Water Act simply has to begin with a decision as to what the future holds, at least as far as federal regulation and legislation is concerned. Under the circumstances this is hard to do. Right now consultants are pondering what provision to make for future process requirements. What began as a very simple law, supported by all, to establish drinking water standards for the nation is now well on its way to becoming enmeshed in a maze of legislation, regulation, and litigation. If the proposed standards for synthetic organic treatment are any example, investor-owned water companies who are required to install GAC may face financial ruin, and many publicly owned water systems will experience real financial difficulties if they must accomplish all the mandated improvements at one time. The result of this is predictable: the municipalities will cry for a federal grant program. The investor-owned system will possibly have to sell to a public agency, or public utility commissions must become much more generous in granting rate increases. I have witnessed abandonment of privately owned water systems when they become a financial liability. Unfortunately, the mayors of most of our major cities are so constrained by the public attitude that their first reaction is to cry for federal money.

There _is_ justification for acceptance of federal funding of local projects when financial burdens are of such consequence and are suddenly imposed upon a local entity. The burdens are all too often prompted by impatience on the part of the consumer advocates for immediate rectification of what they consider first-priority problems. It seems as though national publicity is necessary in order to obtain passage of legislation and funding of justifiable projects. For example, the Safe Drinking Water Act did not get off the ground until a great crisis was exploited. For a long time most scientists in the water-supply industry have been aware of the growing need for research on preventing toxic materials from entering drinking water sources. However, without a crisis, real or exaggerated, little progress was made.

I think we were all glad to see a reasonable piece of legislation evolve, establishing (1) meaningful national standards for drinking water; (2) funding of the federal agency for administration of the drinking water standards; (3) training of personnel, both in the regulatory and operational areas of the water industry; and above all (4) funding of research efforts to work hand in hand with other federal agencies regulating toxic materials in our environment. Welcomed by all (and I want to emphasize this) is the federal government's authority to declare an emergency in the instance of any municipal or investor-owned utility where an immediate threat to public health exists. Public notification by the regulatory authorities is a power largely available to the states now, but it is often influenced by local political consideration. The right of the federal government to declare such an emergency in the absence of such action by the state, or a default on their part to properly regulate, was a provision recommended by the water industry itself. However, failure of the law to allow the administrative arm of government to exercise discretion in the matter of public notification is both counterproductive to cooperative efforts and, notably in the instance of bacteriological standards, is making a joke of public notification. Advertisements in local papers, or radio announcements of a bad bacteriological sample appearing weeks after its collection destroys public confidence.

Then there is the matter of the citizen suit. The Safe Drinking Water Act is better than most legislation in that it discourages capricious suits, but it certainly does not eliminate them all. The fact that the Environmental Defense Fund filed suit under the Safe Drinking Water Act may have had some influence on what in my opinion is an unwise and premature decision to prescribe a treatment technique for any water source that is subject to contamination by synthetic organics. This, in essence, declares all surface water supplies in the country unsafe until proved otherwise. I believe that citizen suits should be further restrained in this and future legislation. If the administrative processes of government cannot or do not function, then it is the legislator's responsibility to supply the remedy. Failure to do so is an admission on the part of the legislative branch that it cannot properly respond to the public need and that it does not trust the administrative branch to properly implement the law. The Congress seemingly prefers to partially accomplish through legislation what should be done by administrative functions and to depend on the courts of the land to assume an administrative role. The unfortunate result of this dilemma is that no decision is made, while plans are drawn and redrawn. In many cases, project funding melts away in the face of inflation, indecision, and costly additions. Solutions are simply not accomplished. This has been proved time and again in the stream-pollution program under Public Law 92-500.

Joe D. Moore, formerly commissioner of the Federal Water Pollution Control Administration, recently made a statement before a public hearing of the Water Resources Policy Committee of the Water Resources Council in Texas, which I think summarizes succinctly the problem which we face with the public participation and citizen suits in general, in terms of accomplishing goals and objectives. He stated:

> Current preoccupation with public participation tends to leave policymakers with the conclusion that every decision must accommodate every viewpoint expressed or that a consensus must be achieved, thus decisions are often so long postponed until this ideal result can be achieved, or the conclusion that is impossible of achievement [sic], that no public participants are satisfied. Our whole government structure is founded on the premise that issues will be precipitated so as to produce a majority and a minority; it cannot function when it tends to act on the basis of three or more minority viewpoints.

The conclusion that must be reached is that in the absence of a decision, the "hows" of local financing can hardly be assured.

I realize these statements are in direct opposition to the viewpoint of the many who feel public participation actually avoids delays. I agree that an informed public constructively employed in the decision-making process is of value. The many citizen boards and commissions which have long functioned well in local government testify to this fact. These, however, are selected groups, representing a cross section of opinion and professional backgrounds. Properly selected, they represent and consider all points of view rather than advocating a particular and narrow point of view. This is, I believe, the "representative" philosophy under which our government is supposed to function. The problem with unrestrained public participation is just this: a multitude of narrow points of view are presented, all too often in direct opposition to one another. Hearings for federal regulations face this dilemma. At some point and at some place, however, the decision-making process must function. To the extent that the preamble to federal regulations discusses pros and cons of public comment and explains the agency's reason for acceptance or rejection of public recommendations, recognition is granted to the public participation process. The administrative arm of government often finds itself between divergent views and lawsuits from one or more sides of an issue. This is getting us nowhere, and somehow will have to be moderated if any real progress is to be made in implementing voluntarily and cooperatively legislation such as the Safe Drinking Water Act (PL 93-523) and the Federal Water Pollution Control Act Amendments of 1972 (PL 92-500).

Another impediment to local financing is the tendency of both federal legislation and regulation to inject social values into technical considerations. Increasingly, both laws and regulations prescribe the methods to be employed, including financing. An example here is the criteria for sewer rates prescribed under the industrial recovery provision of Public Law 92-500. If there is an area where the federal government has no real ability, it is in the area of rate making and enterprise financing of local functions such as the water and sewer utility. Those who print money have little understanding of return on investment and equity in rate making. Their preoccupation with social values generally conflicts with established and proved rate-making principles.

So, how can we finance the necessary improvements to the water supply industry? Let me repeat, I am by no means opposed to legislation which enables the regulatory arm of government to function quickly and effectively in those cases where there is a threat to the public health, happiness, and welfare. For the past forty years, I have been employed in the water supply and wastewater industry at local, state, and federal levels both as a regulator and as an operator. I think I have some understanding of the responsibilities and the problems of each. Twenty-five of those years were spent as director of a municipal water and wastewater utility. I am well aware of the kinds of pressures brought to bear upon local officials to defer needed improvement programs and future plans. Federal legislation and regulation can be helpful to local officials in overcoming the local political realities. This kind of help functions best, however, when it is supportive and not preemptive. I can truthfully say that most local officials consider the present attitudes of both the legislative arm of the federal government and its bureaucracy as largely preemptive of local initiative and prerogatives. In the final analysis, getting the job done remains with the localities. The means by which this can be accomplished, however, have been made more costly, difficult, and uninspiring. The federal grant is often more destructive than constructive insofar as the local enterprise operations are concerned, because it creates dependency. It would seem that after forty five years of grant programs, we should have found a way to prevent destruction of the initiative of the recipient and to avoid application of leverage through the grant process. The final result is somewhat less than satisfactory from the standpoint of all levels of government.

The enterprise functions of government lend themselves to user charges or they would not be so named. The most equitable and humane approach to service is to put the cards on the table and charge a fair price based on the cost of service, considering both how much service is used and how it is used by all concerned. Transfer of wealth through loss rates, either via "life line" or industrial inducement rates should be avoided. The demand charge or increasing unit price for seasonal peak use, which leaves idle system capacity, is a much needed reform in the construction of water rates.

Federal or state grants disguise the cost of service and destroy local resolve to be self-sustaining and to develop equitable rates. If one city accepts a grant, another city which does not will have commensurately higher charges and cannot compete. Soon, all must fall into line. In my experience, I have observed the sequence of (1) the initial grant or "seed" money; (2) the commitment to an obligation by local interest; and (3) withdrawal by the federal government of its financial support. I have seen federal "conditions" of a grant grow to a point where the financial advantage of a grant becomes minimal. In fact, the main inducement to accept federal funding is often only the inability to finance locally the federal "extras" imposed on the project. I have seen instances of water resources development where a reservoir project could be built cheaper and quicker with local funds. I have seen water pollution control projects, financed under Public Law 84-660 (An Act to Extend and Strengthen Water Pollution Control Act, 1956), built twice as fast and at half the cost of the same size and type of plant built under Public Law 92-500.

There are well intentioned people in all levels of government who simply wish to find a way to get on with the job. Somehow we are all caught up in a procedural web, woven with great skill, which is well on its way to dismantling the separation of powers upon which our form of government has been based. Again, I will say the real frustrations and the real barriers to accomplishment of our joint objectives are the frustrations and delays inherent to the decision-making process in government today. Our founding fathers clearly declared their intent to establish a government at all levels which was strong but responsive to the public need. Thomas Jefferson said it best, "I know no safe depository of the ultimate powers of society but the people themselves--and if we think them not enlightened enough to exercise control with a wholesome discretion, the remedy is not to take it from them--but to inform their discretion by education."

I understand him to say that government must have authority to act, but must use its authority discretely and sparingly, and above all, it should hasten to restore authority to the lowest level of government as soon as possible. Unfortunately, bureaucratic control once obtained and centralized is seldom restored to the lower echelons of government. Special interests find centralized authority easier to control, and organizations (not just bureaucracy) stoutly resist its decimation or elimination.

Where the improvements required exceed the ability to finance locally on the public market at reasonable and affordable water rates, I suggest, in lieu of the grant process, long-term loans, second-lien bonds, federal guarantees, or some similar approach to assisting the local water utility to meet drinking water standards for treatment and system reliability and adequacy. Farmers' Home Administration grants and loans are available now for small utilities, but no similar program is available for the larger community systems. Whatever remedy is adopted must offer equivalent assistance to investor-owned systems. The basic criteria should be self-reliance and respect through eventual repayment by user charges. Very small systems faced with exorbitant costs should be allowed alternatives, such as distribution of bottled water treated to drinking water standards. Of primary importance is an R&D grant for a full-scale application of prescribed treatment techniques in order to assure that pilot studies are applicable to actual situations.

I suggest patience. Of first priority are treatment plant and system improvements necessary to remove those threats to the public health of immediate significance. It does not seem a proper order of priority to defer correction of a health defect capable of transmitting hepatitis to thousands immediately, in preference to avoiding a possible long-term health effect with an estimated risk factor on the order of 200 to 262 excess fatalities a year nationwide (Tarcliff, 1976a and b; Federal Register, 1978), as a result of a lifetime exposure to low levels of chloroform. Every life is important. True, we want to attack cancer on all fronts, but the simple fact is we cannot afford to do so all at once. The proposed emphasis on education and publicity to reduce smoking would deserve our first attention. Newsweek (1978) reports that 220,000 deaths from heart attacks and 100,000 from cancer can be attributed to smoking. I believe that the average citizen, if he or she were confident in the odds, would agree that a tax increase equivalent to the annual costs for GAC, applied to better ambulance service with paramedics in attendance, would benefit him more than an equivalent increase in his water rate to extend one life to 1.3 lives in 1 million.

Are we so engrossed in a power struggle and publicity contest that the mundane but important tasks essential to premium water service are not getting the attention they need? I feel we should ask a few questions, such as:

1. Fundamental to implementing a program of this scope is to first find the problems, evaluate them as to significant values, seek viable solutions and then sell the solutions. Have we inspected all of these

water systems and classified problems in order of their importance so that we can attack and solve them in an orderly process?

2. How good are our laboratories? Are they adequate to the requirements of the law? Can their results be trusted? For example, there is considerable variation in the results of testing for organic compounds.

3. Cross-connection programs and plumbing code revisions are needed in a great many systems. Are we making progress here? Many older cities face major replacements in their water distribution systems. Often, distribution systems are inadequate in size or they experience constant service interruptions because of breaks in corroded mains.

4. Are trained, professionally competent personnel available to regional offices of federal and state field divisions who can work cooperatively with the managers and operators of small local utilities who are willing but technically inadequate to their tasks at this time? Permanent progress can be attained only through willing and enthusiastic efforts.

5. How best can we upgrade the financial and organizational structure of local utilities so as to protect water revenues from diversion to other government functions and provide qualified management capabilities? This is the key to local finance!

In conclusion, let me express appreciation of the accelerated program of research that has occurred over the past two years. This is where we started, that is, (1) a growing concern for the increasing presence of toxic organics and inorganics in our drinking water sources; (2) concern for the adequacy of water production and distribution facilities; (3) quality of personnel to manage and operate the water utilities of the nation; (4) assurances of adequate revenues through water user charges in order to support these needs and contribute to the search for, and application of, new and improved quality control measures and treatment processes. Better knowledge of the whole problem, based on the results of sanitary surveys coupled with accelerated research, animal testing, and epidemiological evidence, is of the first priority. First we must implement those programs necessary to protect water systems from contamination by bacterial, viral, or toxic concentrations of organic and inorganic material of immediate health significance. New systems or plant expansions should provide for potential process applications that current toxicological data indicate will likely be required in the future. For example, space should be provided for carbon contactors.

We can and should be responsible locally to pay for necessary improvements. We must, however, find a progressive plan of action within local financial capabilities.

References

Federal Register. 1978. vol. 43, no. 28 (February 9) pp. 51-69.

Newsweek. 1978. January 30.

Symons, James A. 1977. "Interim Treatment Guide for Controlling Organic Contaminants in Drinking Water Using Granular Activated Carbon" (Washington, EPA).

Tardiff, R. G. 1976a. Testimony before Safe Drinking Water Advisory Council (December 15) p. 122 of Hearing Transcript.

______. 1976b. "Health Effects of Organics: Risk and Hazard Assessment of Ingested Chloroform." The 96th Annual Conference of the American Water Works Association, New Orleans, Louisiana.

Discussion by Helen Ingram

Mr. Graeser has written a wide-ranging paper with, in my view, the intelligent bias that beneficiaries should pay. While the paper is entitled "Financing System Changes," Mr. Graeser touches upon many of the great issues of American government. For instance, Mr. Graeser has something to say about all of the following:

Intergovernmental Relations. Graeser sees and laments the tendency of the federal government to preempt local initiatives and prerogatives. He observes that grant programs end up being a tangled web of restrictions that outweigh whatever benefits local governments derive from them. The rights of the states, he says, are the rights of puppets.

Checks and Balances. Graeser is concerned about the role of citizen suits and the courts in circumscribing administrative action.

Majority Rule and Minority Rights. Graeser believes that government must be able to act, and recognizes that if public participation means everyone must reach a consensus before action, few government projects will be completed.

Proper Balance of Scientific and Technical Information in Decision Making. Charges, according to Graeser, should be tied to costs. When the distributive issue of helping the disadvantaged becomes involved in rate making, the process of economic choice is distorted.

It may seem curious that under the rubric of financing, Graeser should choose to cover so many subjects. Why, when the author espouses the economic concept that users should pay, does he write so much about politics? The answer, as I see it, is that the issue of who pays is at the heart of politics. For that reason, I have chosen in my brief remarks to explicitly address the issue of the politics of financing. I hope to place the financing of system changes for safer drinking water in a more general explanatory framework through which it can be better understood.

Patterns of Federal Funding in the Environment

Throughout Mr. Graeser's paper lurks the shadow of a less rational alternative to the financing of safe drinking water by user charges; that is, federal grants-in-aid. During the two and a half days of this conference we have similarly been preoccupied with the possibility of federal grants for this purpose, even though Philip Cummings of the Senate Committee on Environment and Public Works stated as strongly as he could that there would be no construction grant program for safe drinking water. All of us have anticipated the issue of grants because this means of financing has been so much a part of the past pattern of evolution of environmental policy.

Davies and Davies (1975) and other policy researchers (Ingram and Mann, forthcoming) have traced the evolution of financing arrangements for national environmental policies. Federal water pollution control policy fits the pattern exactly, but there are other examples. The development of federal aid to the states proceeds as follows:

1. The program begins as strictly research. Sponsors of federal legislation in Congress insist that control of the actual operating program rightfully belongs to state and local governments. These levels of government could do their jobs better, it is argued, if they had better information and technical assistance.

2. A federal, low-interest loan is introduced. Congress recognizes that responding to an environmental problem puts a financial strain on localities, and low-interest loans are suggested as relief.

3. A federal grant is initiated. Congress views the state and local response to the problem as inadequate, and grants-in-aid provide incentives for more action. Typically, grants are restricted to small localities least able to pay, and require localities to pay half or more of the cost.

4. Incrementally the restrictions on eligibility are broadened so that all localities qualify for grants. The federal share of the cost also moves up bit by bit so that federal funds come to constitute the lion's share of the financing.

Causes and Results of the Pattern

This pattern occurs for reasons that have powerful political force. Environmental legislation often contains regulatory provisions that engender strong political opposition. Grant programs act as a sweetener to elicit support in Congress in order to overcome that opposition. Grant programs are also helpful in buying the acquiesence of regions of the country that may not place a high priority upon the particular environmental goal in question. Impatience is also a factor: the seriousness of a particular environmental ill is sometimes oversold in building political support for regulation, and if, after the initial law is passed, conditions do not change promptly, supporters become frustrated and grants are supposed to quicken the pace of change. Another reason grants are employed to finance the costs of regulations is simply habit. Politicians tend to fall back upon the established mechanisms to incorporate into legislation. Grants-in-aid have become the usual way of accomplishing national goals in a federal system.

While federal grants-in-aid are a well worn mechanism, their record of achievement is not particularly strong. Many students of federal grants, especially economists, are sharply critical on the following grounds:[1]

1. Because grants deal only with capital costs they distort factor costs to localities and bias the choice of solutions toward those that are capital intensive.

2. Grants distort federal allocations of funds. Federal agencies seize upon grants as a means to build constituencies. Therefore, grants are widely distributed to applicants with little regard for efficient implementation of federal goals. The record of federal agencies in effectively steering a program by the strings attached to grants is very poor. Agencies are reluctant to deprive localities of grants even if they are not performing, for fear of an adverse congressional reaction.

3. Federal grants-in-aid enlarge the obligated portion of the federal budget. The burden to the federal government of the waste treatment grant program is large, for example (Kneese and Schultze, 1975).

---

[1]For a more complete discussion of the politics of grants, see Ingram (1977).

Future Financing of Safe Drinking Water

For all of these reasons, the past practice of throwing federal money at problems through grants is increasingly being questioned. Further, the important role of states and localities in implementing environmental legislation is increasingly being realized. The 1977 amendments to both the air and water acts recognize the critical factor of state and local commitment, without which environmental clean-up cannot occur. It is from the perspective of this past experience of financing environmental controls that we can now turn to a consideration of the Safe Drinking Water Act.

It seems to me there are three plausible futures for the program:

1. It will follow past patterns and in a few years grants will be initiated.
2. There will be a very low level of implementation of the act. Federal standard setting for which locals must pay flies in the face of experience. Cities and towns are likely to balk at installing the expensive equipment required by the EPA regulations.
3. A new, more enlightened pattern of federal-state-locality cooperation will emerge. Standards will be set with the participation of state and local officials and will be sensitive, in timing and in severity, to the limits on local ability to finance changes. The basic principle to be followed will be that users must pay. Whatever financial incentives emerge will reinforce rather than displace this principle.

It is difficult to say which of these three alternatives is most likely. Obviously, Mr. Graeser and I would prefer the third. Its chances for success depend on a number of factors beyond the scope of these brief comments. Most have been identified by others at this conference. To my mind we must at least find answers to some of these questions.

Can implementing safe drinking water standards become politically profitable to local officials? As it is now, local officials must increase bonded indebtedness and increase water rates to achieve something for which the EPA may take all the credit. It will take aggressive action on the part of local politicians to get action. What is their incentive? As Mayor Moody has said, nobody gets elected on a clean water campaign promise, but a number of incumbents are likely to disappear when the bill for the Safe Drinking Water Act comes in.

Can safe drinking water attract a public constituency that chooses to pay? This is an issue to which others, including particularly Professor Baumann and Dr. Rosenfield, have spoken usefully. Getting the attention of the public and getting people to believe professional judgments, especially when they are not unanimous, is very difficult.

Can user charges be made equitable? Mr. Graeser expressed a bias against life-line rates. Yet people differ in their ability to bear sharply increasing obligatory costs, and the burden placed on poor families may not be perceived as fair. The burden of the Safe Drinking Water Act upon small towns, which receive very little of the large federal assistance going to large cities through all kinds of federal programs, may seem disproportionate.

How can federal agencies provide the most helpful incentives for localities? Under what circumstances might demonstration grants and incentive grants be successfully employed without undercutting the user-should-pay principle? What role might the EPA play in technical assistance?

References

Davies, J. Clarence, 3d, and Barbara S. 1975. The Politics of Pollution (Indianapolis, Pegasus).

Ingram, Helen. 1977. "Policy Implementation through Bargaining: The Case of Federal Grants-in-Aid," Public Policy vol. 25, no. 4 (Fall).

______ and Dean Mann. Forthcoming. "Environmental Politics from Innovation to Implementation" in Theodore Lowi and Alan Sloan, eds., Public Policies in America (Beverly Hills: Sage Publications).

Kneese, Allen V., and Charles L. Schultze. 1975. Pollution, Prices and Public Policy (Washington, Brookings Institution).

Chapter 10

CONSUMER PERCEPTION AND CHOICE OF DRINKING WATER QUALITY

Duane D. Baumann and Daniel M. Dworkin

In planning for water supply, the influence of specialists, particularly engineers, is well established. Recently, however, the public's faith in professional capability to assure safe drinking water has begun to be tried. Public confidence was tested in 1970 when the U.S. Public Health Service noted that only 59 percent of the surveyed systems produced water that met drinking water standards, and it was shaken when the Environmental Defense Fund (1974) released its study claiming the occurrence of carcinogens in New Orleans drinking water.

We argue that it is time to actually involve the public more in the formulation of public policies concerning water quality standards and treatment. Too little attention has been given to consumer preference in the formulation of drinking water standards and the full range of alternatives to provide appropriate qualities of water have not been adequately assessed.

This chapter is concerned with those consumers who prefer a higher-quality drinking water than is currently available. The objective of this paper is to (1) identify consumer attitudes toward drinking water; (2) to identify the range of alternative point-of-use treatment technologies; and (3) to assess consumer preferences and behavior.

It simply is not known how many people desire a higher-quality drinking water than is currently available. Although a Gallup Poll (Gallup Organization, Inc., 1974) showed that 17 percent of the respondents had some type of treatment equipment, most often a water softener, it is not known what consumer preference would be if a full range of options were available that would accommodate demands for improvements in taste, and odor, and reduction in risks to health.

The drinking water for many Americans is simply not treated and consequently taste, odor, or risks to health are not controlled. A recent report (Temple, Barker and Sloane, Inc., 1977) for the U.S. Environmental Protection Agency (EPA) estimated that there were a total of 35,000 community water systems serving 192 million persons in 1976 (table 10-1). The report noted "that 45 percent of the systems serving under 10,000 people provide no treatment to the water which they distribute. Among the systems in the larger categories (10,000 to 1 million) approximately 9 percent do not treat water which they distribute." Before we review the options open to the consumer who desires higher-quality water, we shift our attention to research on public attitudes toward water quality.

Attitudes Toward Water Quality

We know little about public attitudes toward drinking water quality. Indeed, until recently it was probably safe

Table 10-1. National Estimates of Population Served and System Size, 1976

	Population category									
	25-99	100-499	500-999	1,000-2,499	2,500-4,999	5,000-9,999	10,000-99,999	100.000-999,999	1 mil-million	Total
Population served (millions)	0.4	2.8	3.4	7.6	8.7	10.4	73.8	58.9	26.2	192.2
Percentage	Less than 1	2	2	4	4	5	38	31	14	100
No. of systems	6,308	11,714	4,932	4,850	2,496	1,646	2,442	232	11	34,631
Percentage	18	34	14	14	7	5	7	1	Less than 1	100

Based on EPA Inventory (adjusted) and Community Water Systems Survey, TBS, 1976. Excludes systems which are federally owned and directly serve fewer than twenty-five permanent residents, or those which are located in the U.S. territories, Alaska, and Hawaii.

Source: Temple, Barker, and Sloane, Inc. (1977), p. II-4.

to say that except for occasional problems of taste and odor, the general public did not consider that there might be adverse effects on human health from its drinking water. There seemed no need to undertake research concerning consumers' attitudes toward the quality of their drinking water.

Threat to Human Health

From a general perspective, two Gallup Polls (Gallup Organization, Inc., 1973 and 1974) provide some insight into Americans' attitudes toward their drinking water. Approximately 70 percent of the respondents rated their drinking water as good (a nebulous adjective); and, the most common complaints among the 30 percent who were dissatisfied were not related to health, but to hardness, taste, and odor. Moreover, only 40 percent of the surveyed population perceived a relationship between their drinking water and health: a majority of the respondents (53 percent) did not consider their drinking water as having an effect on their health. We suspect, however, that such complacency will diminish as our knolwedge about health and drinking water increases. There is already some evidence to support this assertion. Among the respondents of the 1973 Gallup Poll it was the more educated and those from the professional and business group that were more likely to feel that water affects one's health (55 percent of the college educated as compared with only 22 percent with a grade-school education). It appears that as education and information increase, so does concern.

What is startling is that those persons from smaller communities are <u>less</u> likely to feel that pollution is a threat to their drinking water, even though the water produced in those communities is more likely to be threatening to human health. For example, over 80 percent of the respondents who live in communities greater than 500,000 viewed pollution as a threat to the safety of their drinking water; whereas only 40 percent of the respondents who reside in communities under 2,500 persons perceived such a threat (table 10-2). However, according to the 1970 U.S. Public Health Service report, it was the smaller systems that were less likely to deliver water of satisfactory quality (table 10-3).

Finally, the Gallup survey found that the most preferred sources of drinking water were springs and wells, not that currently available from municipal water systems. While only suggestive, this response points to the possibility that the consumers are not as pleased with the esthetic qualities of their tap water when the primary source is surface water instead of groundwater (approximately 20 percent of the respondents' tap water was from the ground), or possibly to the need for a better educated public, or to both.

Taste and Odor

Although little information exists about consumer perception of the relationship between water quality and health, considerable research has been directed toward the specific problems of taste and odor. One might speculate that consumers

Table 10-2. Percentage of Respondents Who View Pollution as a Threat to the Safety of Their Drinking Water by Size of Community

View pollution as	Size of community				
	Over 1 million	500,000-1 million	50,000-499,999	2,500-49,999	Less than 2,500
Great threat	47.2	43.2	35.9	25.7	16.6
Somewhat of a threat	33.8	38.3	41.3	44.1	22.6
No threat	11.3	11.8	17.2	23.6	53.4
Don't know	7.7	6.7	5.7	6.7	7.4

Source: Gallup Organization, Inc. (1973).

Table 10-3. Summary of Community Water Quality Evaluation by Size of Community

	Population group served			
	Less than 500	500-100,000	Greater than 100,000	All populations
No. of systems	446	501	22	969
	Percentage of systems			
Evaluation				
Met drinking water standards	50	67	73	59
Violated recommended limits	26	23	27	25
Violated mandatory limits	24	11	0	16

Source: U.S. Public Health Service (1970) p. ii.

consider water with unsatisfactory taste and odor to be unsafe for consumption; however, in a recent survey concerning consumer perception of taste and odor, only 6 percent of the interviewees perceived that drinking water influenced health (Bruvold and Mitchell, 1976, p. 31).

Bruvold (1975) published a significant review of consumer perception of taste and odor in drinking water. The major conclusions are summarized below.

First, as total dissolved solids (TDS) of the drinking water increases, consumer evaluation of the water quality becomes increasingly negative. Bruvold and Mitchell (1976) interviewed 1,500 respondents in fifteen California cities--the cities were ranked according to the amount of TDS in the drinking water (table 10-4). In Oakland, where average TDS was lowest, 44 percent of the sample rated their drinking water as good; however, 7 percent or less of the sample rated their water as good in San Diego, the city with the highest mean TDS (see table 10-4). Moreover, the consumers' general quality rating of drinking water, based upon a 5-point scale with 1 representing good, ranges from 1.97 in the sample with the lowest TDS to 3.51 in the sample with the highest TDS (see table 10-4).

Second, Bruvold and Mitchell (1976) concluded, in general, that although the presence of a clearly detectable mineral taste may not necessarily imply unacceptability in daily drinking, drinking water containing more than 500 mg per l of TDS

Table 10-4. Consumer Perception of General Water Quality

City and zone	Mean TDS	Affective evaluation of groundwater quality: Good	Neutral	Bad	No comment	Total	General quality rating[a]
Oakland-Northeast	31.07	44	38	10	8	100	1.97
Oakland-South	31.94	28	48	8	16	100	1.95
San Francisco-West	77.86	16	48	21	15	100	2.32
Sacramento-North	177.40	11	61	22	6	100	2.63
Los Angeles-San Fernando Valley	180.90	11	25	47	17	100	2.90
Los Angeles-Venice	181.52	14	45	30	11	100	2.80
Los Angeles-Westchester	193.57	15	51	22	12	100	2.53
Los Angeles-San Pedro	267.65	16	33	34	17	100	2.78
Los Angeles-Downtown	340.00	8	17	47	28	100	3.15
San Jose-North Central	350.30	9	53	23	15	100	2.90
Santa Ana-Central	469.85	10	32	32	26	100	3.04
Garden Grove-Central	534.25	10	22	39	29	100	3.13
San Diego-San Ysidio	713.30	7	23	53	17	100	3.37
San Diego-North	727.30	4	27	54	15	100	3.50
San Diego-Central	743.45	2	26	59	13	100	3.51

Source: Modified from Bruvold and Mitchell (1976) pp. 28, 31, and 33.

[a]Rating: (1) Very good, (2) Good, (3) Average, (4) Poor, and (5) Very poor.

is perceptually unacceptable. The chemical composition of the TDS influences consumer acceptance, with bicarbonate being rated least offensive, sulfate next least offensive, then chloride, and carbonate as most offensive or unacceptable.

Finally, Bruvold and Mitchell (1976) concluded that consumers are willing to pay more provided that there is a significant improvement in the quality of their drinking water as measured by taste and odor. More persons were willing to pay more for improvements in taste than for clarity or softness. Among the 1,500 respondents in the fifteen California cities, from 16 to 44 percent of the respondents were willing to pay more for improved water quality (Bruvold and Mitchell, 1976, p. 38).

Recycled Renovated Wastewater

Another area of research into consumer perception of drinking water problems is the potential use of recycled renovated wastewater. In a review of the literature, Baumann and Kasperson (1974) conclude that there is little evidence to support the widespread conviction among those charged with proposing solutions to the nation's water supply problems that public opposition constitutes the most important obstacle to the adoption of wastewater reuse systems. Indeed, there is evidence that the public will accept renovated wastewater for potable use provided that they are aware of the technological characteristics of water treatment. Based upon survey data in five communities, Sims and Baumann (1974) suggest that what the consumers know and feel about drinking renovated wastewater is related to the individual's general level of education and his knowledge about water treatment and is not related to unconscious threats or specific concerns such as fear of contamination or beliefs concerning nature, technology, esthetics, authority, progress, or destiny.

Bruvold's (1975) review of the literature on public acceptance of drinking renovated wastewater concluded that the _initial_ response was negative, that over 50 percent of the surveyed population responded negatively, that approximately 40 percent were favorable, and the remainder had no opinion. Also, previous research found that it was the young and those persons from higher socioeconomic levels who were more willing to drink reused water. Other factors that were related to individual acceptance were the extent of prior knowledge, and the individual's perception of the existing water quality and quantity.

These findings seem consistent with the favorable public response to questions about the acceptability for drinking of renovated wastewater in Denver. In a survey of 500 people, the _initial_ response to the concept of recycling renovated wastewater was primarily negative. However, as the respondents received additional information on the implications of water reuse planning, the rate of public acceptance increased until 85 percent of the respondents expressed a willingness to drink renovated wastewater (Heaton, 1978). In Windhoek, Namibia, where renovated wastewater has served as a source of potable supply for over a decade, "public acceptance has been very good." (U.S. National Water Commission, 1973, p. 313).

Aside from the considerations of health risks in drinking renovated wastewater, a common reason given by managers and professionals who opposed the concept was an anticipated rejection by the consumers. Johnson (1971, p. 92) found that "it would appear that water managers know very little of consumer responses concerning renovated wastewater, but generally consider the public would not accept it." Bruvold (1975) notes that although government officials' attitudes toward reuse are approximately the same as those of the general public, they perceive the public's as being less favorable.

An important question, then, is why managers and engineers perceive the public as unwilling to accept recycled, renovated wastewater when the available evidence suggests that the consumers would not be, in fact, an obstacle in community adoption of such a program. I hypothesize that the training of these officials, especially those trained in public health, predisposes them to see problems in particular ways and to be rather rigid about admitting the possibility of valid competing conceptions--whether held by other professionals or by the public at large. In one study, for example, ninety-eight consulting engineers (from thirty-three firms) and twenty-two state public health officials (from nine states) were interviewed on the practicability of a community program using renovated wastewater

(Sims and Baumann, 1976). This study concluded that public health officials "hold the more negative position--they begin by not liking the idea, then raise many and major objections to it, and in the end, find their reflection has strengthened their antagonism. Consulting engineers . . . begin with a far more favorable attitude, raise fewer objections, and conclude with a perfectly even split between endorsement and rejection" (p. 9).

This matter of professional and public perceptions is clearly related to the potential place of home treatment devices in drinking water policy. It is, for example, possible that professional biases toward central treatment may lead to the rejection of decentralized alternatives without adequate research into their cost, effectiveness, or acceptability to consumers.

A Need for Public Involvement

Because of very sketchy data about consumer preferences over the risks and benefits related to drinking water quality, a major conclusion is that there is a greater need to survey public preferences in the process of formulating policy. At the same time the need is equally great to provide the public with more information in order to produce a more sophisticated response. Recurrent criticisms are (1) that too often policy decisions are made by professionals who either ignore or are misinformed regarding consumer preferences; and (2) that public participation means participation of the ignorant. Therefore, we need programs to provide an interface between professional expertise and the public resulting in a better articulated and informed response in formulating policy. Recently, Okun (1977, p. 1041) has summarized the problem faced by environmental engineers: "If we do have the convictions that would lead us to opt for sound engineering decisions, we are fearful that society will not accept them, yet we have not sought to enlighten society as to the principles that should help guide their choices."

The recent declaration by the EPA concerning the reduction of trihalomethanes illustrates the likely absence of information on public preferences. Those systems serving fewer than 75,000 persons are not required to comply with the MCL or to conduct monitoring under this regulation. This means that approximately 48 percent of the population served by water systems is not protected. According to the Federal Register (1978, p. 5770), one of the primary reasons for excluding this group was "because GAC (Granular Activated Carbon) is a costly technology, it was not deemed economically feasible for the smaller systems to incur the relatively higher per capita costs of GAC treatment technology at the present time. Larger systems will be better able to keep the added cost burden on the individual consumer to a reasonable level."

The question immediately arises as to the empirical basis of this decision especially for the use of the word "reasonable." Were consumers presented with the risks to health and the costs of treatment utilizing GAC? Contrariwise, does the EPA have evidence that consumers in the largest cities are willing to pay more for a reduction in the risk to health resulting from trihalomethanes? What alternative treatment technologies, such as home treatment, were considered for the consumers in small communities?

While information concerning the public's perception of water quality is meager, at best, it is important to know at least what the current choices are for those consumers who <u>are</u> dissatisfied with the quality of their present source of drinking water.

Consumer Alternatives for Improving Water Quality

As we have observed, most consumers appear to be satisfied with the quality of their present drinking water. However, those persons who are dissatisfied with their present supply have three choices: (1) to continue using their supply; (2) to modify the quality of their present supply; or (3) to obtain a new source (table 10-5).

Modifying the Present Water Supply

There exists a wide choice of technologies that can be used to modify the quality of present supplies of water at the household level. These range from such simple methods as boiling or cooling to such sophisticated techniques as reverse

Table 10-5. Alternatives for Consumers Dissatisfied with Drinking Water Quality

Do nothing	Modify present supply	Seek alternative source

Use present supply	Filtration
	Adsorption
	Deionization
	Distillation
	Disinfection
	  Chlorination
	  Ozonization
	  Bromination
	  Iodination
	Reverse osmosis
	Oxidation
	Silver impregnated
	Activated carbon
	Neutralization
	Ion exchange
	Boiling
	Cooling
		Wells
		Springs
		Vended Water
		Bottled Water
		Rainwater

osmosis. Commercial water treatment units are available that use combinations of technologies designed to eliminate all unwanted components of the water. Again, there is a paucity of information: little is known concerning the effectiveness of the available home treatment devices but there exists a wide range of prices.

The water treatment units available for household use cost from less than $10 to over $1,000 and can be used to eliminate, to decrease, or to modify the level of any of the pollutants in the original water source. Some of the devices are countertop units which process a batch of water at each filling; others continually process the incoming supply for either the whole house, selected faucets, or a single faucet.

(Estimates on the prices of the technologies were supplied to the authors by the major manufacturers.)

Filtration. Filtration is a mechanical process for removing solids from the incoming water. Filter devices are rated for the size of the smallest particles removed. Filters can be used to process all incoming water. The cost for a household unit is from $250 to $400, or units may be rented for between $5 to $15 monthly. Smaller filters are available to filter water for a single faucet. These devices cost from less than $10 to $50.

Adsorption. Adsorption removes dissolved organic and inorganic materials and suspended solids, usually by means of activated carbon. This results in taste and odor control as well as some removal of toxic organic compounds dissolved in the water. When new, activated carbon filters are very effective in reducing the concentration of organic chemicals including pesticides, trihalomethanes, and other products of chlorination, pesticides, heavy metals, and chlorine. There are, however, two problems implied by their use which have not yet, in our opinion, been satisfactorily dealt with.

First, there is the troublesome question of judging the effective life of the carbon unit, for the ability of the carbon to adsorb impurities deteriorates as it does its job. There is no simple way for the consumer to tell how far this process of deterioration has gone at any particular time, nor is there any universally applicable rule of thumb for judging the length of time between filter replacements. One such unit comes with six disposable chlorine test tubes with instructions for their use. Replacement of the cartridge is indicated when less than 25 percent of the original residual chlorine is removed (Vaughan, 1978). If chlorine removal is indeed an accurate measure of the performance level of the carbon filter, then this test indicates replacement only when the unit has lost more than 75 percent of its effectiveness.

The Environmental Defense Fund, an organization which has been active in the drinking water-quality field, has published a do-it-yourself manual for constructing an activated carbon filter for home use (Environmental Defense Fund, n.d.). The organization has also filed a petition with the Federal Trade Commission concerning the advertising claims of manufacturers of home water filters (Environmental Defense Fund, 1978).

While waterborne pathogens are not affected by the filter, the residual chlorine which controls their growth is very largely adsorbed. Thus, there exists the possibility that these organisms will multiply in the unit when water is not running over the surface. Such a buildup could then be contained in the next water drawn. One method of controlling the growth of pathogens is to impregnate the activated charcoal with silver. This inhibits the growth of organisms, but the inhibiting mechanism is not completely understood. There are also the problems of loss of effectiveness over time and of the presence of trace amounts of silver in the filtered water.

The Environmental Protection Agency, Office of Pesticides, tests and registers units of this type because of the claim that they control pathogens (U.S. EPA, 1976). As of 1978, twenty-eight units and replacement filters have been registered (U.S. EPA, 1978).

Adsorption units are available for treating all the water in a house or for a single faucet. The price range is between \$200 and \$300 for the full house unit and from \$9 to \$50 for treating water from a single faucet. Silver-impregnated units are only available for single-faucet installation and cost between \$9 and \$80.

Deionization. Deionization provides almost complete removal of inorganic elements (minerals) and compounds. Installed for single faucets, the equipment costs from \$20 to \$225, or it can be rented for between \$10 and \$15 monthly.

Distillation. Distillation can remove 90 to 99 percent of organic and inorganic materials and completely destroy all pathogens. The operation of the process is critical, since too high temperatures will allow undesired solids and organics to break over and pass through to the treated water. Home distillation units require substantial inputs of energy and have a capacity of between 1 and 10 gal daily. The operation is not continuous; a single batch is processed at each filling. These units cost from \$70 to \$800.

Reverse Osmosis. Reverse osmosis removes 80 to 95 percent of dissolved solids, both organics (except those with low molecular weights) and inorganics. It also removes all waterborne bacteria. In the process of reverse osmosis, pressure applied to a raw water source will force the water through a semipermeable membrane, but most solids will be blocked. It is available for a single outlet. Reverse osmosis units have only recently become available for household use, although they have been widely used for commercial and industrial applications. The unit costs per faucet may be $300--with a maintenance cost of $20 to $30 per year.

Disinfection. Processes to destroy waterborne organisms range from relatively simple chlorination to more sophisticated processes using ultraviolet light, ozone, iodine, or bromine. All processes are available to treat the water for the entire household. Ozonators are also available for a single faucet or can be used to treat single batches of water. While each is of proved efficiency in large-scale installations, their use, with the exception of chlorination and bromination, has been limited in small-scale applications. And the application of chlorination in the home raises the same possibilities for the production of carcinogens as does central disinfection. The cost of equipment ranges from $150 for chlorination to $2,200 for bromination equipment.

Oxidation. Oxidation is used to prevent the staining of clothes and plumbing fixtures by iron and manganese and to eliminate the odor caused by hydrogen sulfide. The process is accomplished by the use of an oxidizing filter, such as those containing manganese treated zeolite. Units are available to treat water for the entire household. The cost is between $150 and $300. Rental units are available for between $10 and $15 monthly.

Neutralization. Neutralization is used to alter the pH of acid water to neutral or slightly alkaline. The most common method is by adding soda ash. It reduces the tendency of acid water to corrode plumbing and thus add lead and cadmium to the water. These metals when ingested represent a health problem.

Softening by Cation Exchange. Softening water by cation exchange involves the displacement of calcium and magnesium ions with sodium ions, thus reducing the hardness of the water. Sodium ions are present in all water supplies, and in hard water areas softening can add appreciably to the natural sodium content. Two health problems may result. First, there is the problem of increased sodium in the water that might contribute to a daily intake of more than 2,000 milligrams (mg). (It is estimated that 40 percent of the population would benefit by a sodium ion intake of no more than 2,000 mg per day. A smaller group, estimated at 3 percent of the population, are already on diets which call for restricting their sodium intake to less than 2,000 mg daily, and, finally, some persons must restrict their daily intake to a total of only 500 mg (National Academy of Sciences, 1977). The second problem is the possible effect the decreased levels of magnesium and calcium will have on cardiovascular disease. Epidemiological evidence indicates that cardiovascular disease rates are high in areas which have lower levels of magnesium and calcium in the water supply (National Academy of Sciences, 1977). It appears that cardiovascular disease might be reduced by 15 percent in the United States with optimal levels of hardness. Hence, if further research were to confirm these tentative findings, then consumer preferences for soft water would be in opposition to the demand for a lower risk of cardiovascular disease.

Softening units are available to treat the entire household or selected water lines. These cost between $200 and $600. Units can be rented for $5 to $11 per month.

Combined Treatment Methods. Finally, in many cases, combinations of treatment methods are used to provide the desired water quality. For example, chlorination may be used for disinfection and for the precipitation of iron and hydrogen sulfide, which are then removed by filtration. Organic material low in molecular weight can be removed by activated carbon in combination with reverse osmosis aimed at organics with higher molecular weights. Such combined treatment units retail for approximately $300.

As a general matter of concern there is at present no sure method of evaluation by the untrained consumer of the efficiency of most home treatment devices. While these devices at best are effective methods of treating unsatisfactory sources of supply, they are at worse not only ineffective, but indeed may be hazardous. The untrained consumer is not in a position to judge unless detailed instructions are provided and appropriate tests of efficiency are available. However, the National Sanitation Foundation has established a committee to study and develop performance standards for point-of-use treatment devices (private communication from T. Gable, Feb. 16, 1978).

Seeking Alternative Water Sources

Another set of choices available for those who are dissatisfied with their present source of water is the provision of alternative sources for some or all household uses. The most widely used alternative source is bottled water. Other alternatives available are water collected by the user from wells, springs, or other sources; water from vending machines; and rain water from roofs or other collecting areas. Some persons with a public supply choose to develop private supplies of groundwater or, in some instances, surface water, while others with private supplies can opt to convert to public sources.

Bottled Water. Two basic types of bottled water are available:

1. Drinking water prepared by treatment through distillation, deionization, or reverse osmosis, from public or private sources. Minerals are sometimes added to provide a distinctive taste. Fluorine may be added to retard tooth decay.

2. Spring water or well water which is disinfected water collected from a groundwater source.

All commercially available bottled water in the United States is disinfected by ozone which leaves no residual aftertaste. The Food and Drug Administration (FDA) sets quality standards and regulates manufacturing practices for bottled water. The standards provide for a maximum of four coliform organisms per 100 mm, with an average not exceeding one coliform organism per 100 mm.[1] Since there is no residual chlorine or other disinfecting material in the bottled water any organisms present can multiply after bottling. There are also, at present, no standards for chloroform, a recognized carcinogen, even though the Delaney clause would seem to indicate that it should be banned. Turbidity, color, odor, chemical composition, and radiological quality are regulated (FDA, 1973). Manufacturing practices include requirements for testing and approval of the product water source and for coding and monitoring the distribution of the bottled water (FDA, 1975).

Bottled water produced in the United States costs between 50¢ and $1.00 a gallon. Imported bottled water can cost up to $10.00 a gallon.

Well or Spring Water. Groundwater sources can provide an alternative to public supplies. These sources, however, are not inherently safe from pathogens. Illness from untreated

---

[1]The American Bottled Water Association, an industry organization whose members produce over 90 percent of the water sold in the United States, has set a standard for water bottled by its members which allows no coliform organisms. Water produced by the membership is not always identified on the label. Industry spokespersons state that there is now a program to identify water produced by its members.

groundwater increased in the United States in the years 1961-70 when compared to 1946-60 (Environmental Defense Fund, 1974).

Vended Treated Water. One source of water available in a limited number of communities is the vending machine. Such machines treat a public supply by reverse osmosis, filtration, or ultraviolet radiation. The consumer has a choice of water which is mineral-free or with selected minerals added. Water is supplied to the consumer's container, and savings of up to 60 percent over the cost of bottled water are claimed. The vending of water from coin-operated treatment machines started in California in 1964. The distributor now claims that units are available throughout California, as well as in Arizona, Florida, Louisiana, and Texas.

Rainwater. Rainwater can be collected from roofs and other surfaces. Such water is of variable quality, depending on the contaminants in the atmosphere or on the collecting surface. In addition, if the water is stored without some method of control, algae and bacteria will multiply.

Consumer Behavior

As we have just seen, the alternatives available to consumers include (1) doing nothing; (2) modifying the existing supply; or (3) providing a new supply. We may ask what their responses to these alternatives have been. While the data are incomplete, some evidence on consumer behavior is available by examining sales of bottled water, water softeners, and the relatively new treatment devices which are available for attachment to a single faucet. In a national survey (Gallup Organization, Inc., 1974), 17 percent of the sample had some water treatment equipment (softener, ice maker, filter, iron remover, or cooler). This figure varied from 13 percent in soft water areas to 21 percent in hard water areas.

Cost was one factor mentioned by nonusers for not purchasing equipment. Twenty-seven percent of the national sample who do not have water treatment equipment would use it if price were not a problem.

Aggregate Indicators

Bottled Water. Bottled water sales in 1977 were estimated at 400 million gal (*Chicago Tribune*, 1978) at a cost of $160 million (F. H. Jones, private conversation, February 17, 1978). For the past five years sales have been rising between 10 and 12 percent each year. California accounts for more than 50 percent of all sales, while Florida, Illinois, New York, and Texas combined represent an additional 37 percent (American Bottled Water Association, 1977). Route delivery is the predominant means of distribution.

Home Treatment Devices

An industry survey of the total sales for water treatment devices for home and industry indicate total retail sales in 1975 for products and services of over $70 million. Over 500,000 water-softening units were sold in the period, nearly

1 million were on rental, and over 2 million units were exchanged (Water Quality Association, 1977). While most of this volume represented water-softening units, the industry includes the full range of home treatment devices.

The market for individual low-cost, single-faucet, silver-impregnated, activated carbon devices has been growing rapidly. Water Pik had estimated sales of 600,000 units the first year and 800,000 the second year. The advertising budget was $5 million. Competitors are planning to introduce a similar product, also providing $5 million in advertising, and expect similar results. The price of several systems ranges from approximately $30 to $300. A division of the Coca-Cola Company has also announced a unit at the same price level (Vaughan, 1978).

In summary, the sales of bottled water and home treatment devices are increasing rapidly. This trend suggests consumer dissatisfaction with the quality of available drinking water.

The Duluth Experience

The recent controversy concerning the presence of asbestos in Duluth's drinking water allows the unique opportunity to observe consumer perception and response to a specific threat of unsafe drinking water.

Duluth, Minnesota, located on the shores of Lake Superior, has used the lake water for municipal supply with only minimal treatment. In 1973, asbestos fibers were found in the water supply. These fibers, which have been identified as possible carcinogens, result from the operation of the Reserve Mining

Company, which processes taconite ore. The object of the processing is to enrich the low-grade ore, separating out the wastes which are then disposed of by dumping in the lake. The plant began operations in 1956, and by 1976 had disposed of 200 million tons of waste or tailings. At present 67,000 tons are being discharged daily.

A series of court actions to prevent discharge of additional wastes has been underway since 1970. The presence of asbestos in Duluth's water was suspected in late 1972, confirmed in early 1973, and publicly revealed in June 1973. There was concern on the part of public officials that there might be panic when the news was released. The response, however, with few exceptions, ranged from acceptance to anger, not at Reserve Mining, but at the officials responsible for releasing the information. Since that time there has been extensive publicity given to the problem in most of the area.

A mail survey was conducted in Duluth, in the summer of 1975, in order to identify consumer use of alternative sources of water. At the time of the study, individuals, public and private agencies, and organizations had taken the following actions to make available alternate sources of water or to process the public supply to eliminate the asbestos: individuals purchased home filters, bought imported water, or drew water from private wells. The city government provided filtered water at locations throughout the city and subsidized with federal funds the purchase of bottled water for low-income families with children under the age of six. WDIO, a local television station, provided and promoted the use of its well. Hence, individuals could use either the unfiltered public water supply or an alternative source.

The central research question in that particular study was to determine the numbers of persons who used the unfiltered tap water and who chose an alternative source. The study also sought to characterize the two groups on the basis of differences in personality, information, socioeconomic situation, and perception of the risk involved.

A random sample of 500 households of Duluth was selected from a city directory. Two follow-up letters were sent to the nonrespondents, and eventually 400 usable responses were received. Selected preliminary results of the study are summarized.

Less than half, 47 percent, of the families in Duluth used the public tap water without further treatment; 53 percent used an alternative source of water for potable use. Of those who responded by using alternative sources, most (51 percent) used specially filtered water available at public buildings throughout the city (figure 10-1). The next largest fraction, 29 percent, obtained water from wells. Bottled water and, more often, water packaged in cardboard containers by a local dairy, was used by 6 percent of the families, and 6 percent of the respondents reported using water filtered by devices in the home. The remaining 8 percent of the respondents reported obtaining water from other sources, such as from local

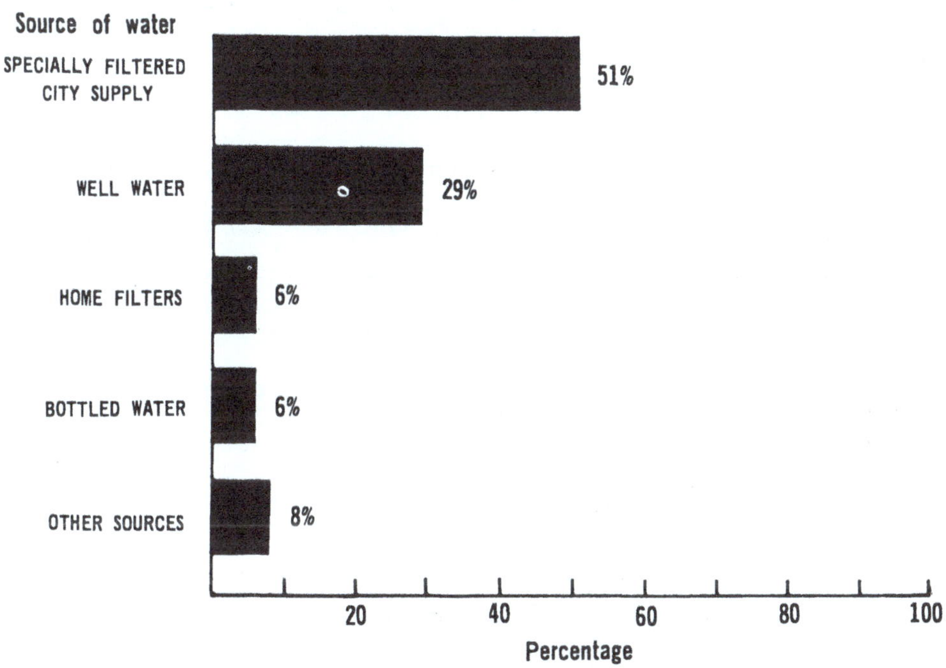

Figure 10-1. Choice of Alternative Sources of Supply in Duluth.

farmers, rain barrels, and water from an adjoining city--Superior, Wisconsin--which had no asbestos problem.

There were no differences between those persons who obtained an alternative source of water and those who continued to drink the water from the tap with respect to socioeconomic characteristics, sex, or personality. However, there were important differences between the two groups in their willingness to pay for the elimination of environmental pollution in general and in perception of the judgment of water system professionals. The higher the person's willingness to pay increased taxes to alleviate pollution, the more likely the person was to use alternate sources of water. And those persons who perceived ambiguity among the professionals were more likely to continue to drink tap water than those persons who perceived professional judgment to indicate asbestos as threatening to health.

Summary and Recommendations

Our major conclusion is that we do not have sufficient knowledge of the risks and benefits of point-of-use treatment, commonly referred to as home filters. The data are simply not there. Considering the rapid increase in the use of such devices, there is an urgent need to evaluate the available point-of-use treatment technologies. Until this is done, one cannot formulate an informed policy. We shall be left with the unsatisfactory assertions of the parties at interest: for example, the EPA's argument that until such information is available it is "probably" safer to rely on public water supply; and the word of the promoters of point-of-use treatment that these devices really do live up to their sales claims. Answers to the questions concerning point-of-use treatment must await the results of further research, such as the study directed by the Water Supply Office of the U.S. Environmental Protection Agency, which is to be completed in March 1979.

A second recommendation concerns the role of public preferences in formulating the standards for drinking water quality. Incorporating public preferences requires not only additional research to measure consumer perception of risks and benefits, but will necessarily involve programs to better educate the public about the state of knowledge concerning the relationship between water quality, health, esthetic characteristics, and the associated costs over the full range of practicable alternatives. What are the risks associated with various qualities of water, and what are the associated costs? No one body can answer except on the basis of an informed public.

In our judgment, the public has exhibited confidence in the capability of water management to provide safe drinking water; however, as the complexities concerning health and drinking water continue to grow, the public must not only be kept abreast of the risks and benefits but also must be brought into the formulation of the policy. The current situation is that consumers are poorly informed, their preferences for improved water quality are inadequately understood, if recognized at all, and the efficacy of the full range of alternatives in providing a higher-quality drinking water are simply not known.

References

American Bottled Water Association. 1977. Fact Sheet: The Bottled Water Industry (Los Angeles, Calif., The American Bottled Water Association).

Baumann, Duane D., and Roger E. Kasperson. 1974. "Public Acceptance of Renovated Waste Water: Myth and Reality," Water Resources Research vol. 10, pp. 667-674.

Bruvold, William. 1975. "Human Perception and Evaluation of Water Quality," Critical Reviews in Environmental Control vol. 5, pp. 153-231.

______ and Raymond N. Mitchell. 1976. Consumer Evaluation of the Quality and Cost of Domestic Water, No. 15 (Davis, Calif., Water Resources Center).

Chicago Tribune. 1978. "Bottled Water Sales Soar." January 26.

Environmental Defense Fund. 1974. A Summary and Evaluation of Selected State Drinking Water Systems (Washington, D.C., The Environmental Defense Fund).

______. 1978. "Petition for Investigation of Advertising Claims of Home Water Filters," filed with the Federal Trade Commission on March 28.

______. n.d. "Construction of an Activated Carbon Filter for Home Use" (Washington, D.C., The Environmental Defense Fund).

Food and Drug Administration. 1973. "FDA Quality Standards," Federal Register vol. 38 (November 26).

______. 1975. "Bottled Water," Federal Register vol. 40, no. 49 (March 12).

Gallup Organization, Inc. 1974. Public Attitudes Toward the Quality of Drinking Water, (Princeton, N.J., The Gallup Organization).

______. 1973. National Survey Conducted for the American Water Works Association (Princeton, N.J., The Gallup Organization).

Heaton, Richard. 1978. "Planning for Water Reuse in Denver, Colorado," in Planning for Water Reuse (Chicago, Maroufa Press).

Johnson, James F. 1971. Renovated Waste Water: An Alternative Supply of Municipal Water Supply in the United States, Research Paper No. 135 (Chicago, University of Chicago, Department of Geography).

National Academy of Sciences and National Academy of Engineering. 1972. Water Quality Criteria (Washington, D.C., EPA).

______. 1977. Drinking Water and Health (Washington, D.C., NAS).

Okun, Daniel A. 1977. "Principles for Water Quality Management," Journal of the Environmental Engineering Division, ASCE, vol. 103, no. EE6, Proc. Paper 13434, pp. 1039-1055.

Page, Talbot; Robert H. Harris; and Samuel S. Epstein. 1976. "Drinking Water and Cancer Mortality" Science vol. 193 (July 2), pp. 55-57.

Sims, John H., and Duane D. Baumann. 1974. "Renovated Waste Water: The Question of Public Acceptance," Water Resources Research vol. 10, pp. 659-665.

______. 1976. "Professional Bias and Water Reuse," Economic Geography vol. 52, pp. 1-10.

Temple, Barker and Sloane, Inc. 1977. Survey of Operating and Financial Characteristics of Community Water Systems (Washington, D.C., EPA).

U.S. Environmental Protection Agency Office of Pesticide Programs. 1976. "Interim Requirements for Registration of Bacteriostatic Water Treatment Units for Home Use," Federal Register vol. 41, no. 152 (August 5) pp. 32778-32782.

U.S. Environmental Protection Agency. 1977. "Drinking Water and Health," Federal Register, vol. 42, no. 132 (July 11) pp. 35764-375779.

______. 1978a. "Control of Organic Chemical Contaminants in Drinking Water," Federal Register vol. 43 (February 9) pp. 5756-5780.

______. 1978b. Registered Water Treatment Products. Mimeographed list.

U.S. National Water Commission. 1973. Water Policies for the Future. (Washington, D.C., GPO).

U.S. Public Health Service. 1970. Community Water Supply Study. (Washington, D.C., U.S. PHS).

Vaughan, J. 1978. "Internal Memorandum on 1978 Housewares Show." Culligan International.

Water Quality Association. 1977. Industry Profile Study. (Chicago, Ernst and Ernst).

Discussion by Alan H. Molof

Let us take a moment to evaluate what the consumer will now see as his options in getting drinking water of higher quality on both a health and esthetic basis.

The Safe Drinking Water Act has two levels of standards. The first level, concerned with health, is termed primary standards. The second level concerned with esthetic factors is termed secondary standards. It should be noted that the secondary standards are the ones that the consumer can directly experience such as taste, color, foaming, and odor.

Upgrading of the primary or health standards in the Safe Drinking Water Act was proposed in the Federal Register (1978) and includes both the regulation of trihalomethanes and the use of a treatment involving granular activated carbon (GAC) for removal of organic substances. Over three years have elapsed since the Environmental Defense Fund and the EPA released information involving the problems of New Orleans drinking water due to organic substances. Furthermore, these regulations are such that one and one-half years will elapse after the regulations become final before the regulation of trihalomethanes is in force even for those communities with populations over 75,000. There will be a five-year lapse after regulations become final before the use of granular activated carbon will become operational for communities with populations over 75,000. It should be noted that these proposals are now in the public comment stage, and neither their final adoption date nor their final form, if adopted, is known at this time.

Thus, we can see two patterns emerging from the consumer's point of view. One, an excessive time delay and, two, an omission of a large segment of the population. The consumer response to this action could be to do nothing or it could be positive, and involve one of two available actions: to immediately start buying bottled water for drinking and cooking, or to immediately install a point-of-use water-treatment device for drinking and cooking water. The person who decides to wait, and is in a community with a population greater than 75,000, knows roughly how long the wait will be for collective action. If one lives in a community having a population less than 75,000, the time of waiting is undefined.

One cannot, of course, confidently predict which response consumers will adopt. But it can be stated with great certainty that the bottled water option, and the point-of-use water treatment device are immediately available. If this immediacy is attractive to the consumer, then sales of bottled water and point-of-use water treatment devices will increase.

It was an observed increase in the sale of unregulated point-of-use or home water treatment devices that prompted me to appear before the National Drinking Water Advisory Council, in April 1976, in order to suggest that the EPA Office of Water Supply was the only group qualified to take control of this area. Although not taking control of the area, the EPA Office of Water Supply asked in early 1977 for proposals to study the ability of point-of-use water treatment devices to remove organic substances. The contract was to be awarded in early 1978. In addition to this direct activity, the EPA, in July 1976, encouraged the formation of the Water Quality Improvement Standards and Certification Council for industry self-regulation of point-of-use devices.

When considering long-term, in contrast to short-term or immediate solutions to drinking water quality problems, other methods became relevant. For upgrading water quality, one suggested approach is a dual-distribution system involving two pipes, each carrying different quality water. One variation of this alternative involves a single raw water source, two treatment plant processes (each treating the water to a different degree--one for drinking and the other for all other uses); two distribution pipe systems from the water treatment plant to the home; and two piping systems in the home, separating the two grades of water. A second variant is similar to the first, except that there are two separate raw water supplies, one of high-quality drinking water and the other for all other uses of water. This results in two completely separate systems.

A single-pipe distribution system can also be altered to make it roughly equivalent to a dual or two-pipe distribution system. The water delivered by the existing system could be used for bathing, cleaning, irrigation, and so forth; with bottled water used for drinking and cooking. Alternately, a point-of-use water treatment device could be used to treat a small quantity of water from the single-pipe system in order to upgrade it to drinking water standards.

Short- and long-term policy decisions for drinking water policy must consider all possible methods which can be used to solve the perceived problem. Only in this way will the consumer be able to intelligently evaluate his individual response on both a short- and long-term basis. A drinking water policy that ignores either aspect will result in a confused consumer response and a generally confused policy. Dual-pipe water systems with a single or separate water source are more of a long-term than short-term solution except in new communities. The single-pipe water systems, with bottled drinking water and point-of-use water treatment devices for drinking water, can constitute both a short- and long-term solution.

References

Federal Register. 1978. vol. 43, no. 28, p. 5756.

Discussion by Donald L. Porth

As president of both the Water Quality Research Council, a nonprofit group devoted to research and education in water quality, and the Culligan International Company, a manufacturer of point-of-use water-treatment equipment, I have long been concerned with research, products, and services that fulfill consumer needs for improved water quality.

The Water Quality Research Council has sponsored seven international water quality symposiums dedicated to examining the very problems we are discussing here today. The point-of-use water treatment industry, of which Culligan is a member, has, can, and intends to serve public and consumer needs.

I appreciate the opportunity to comment on Dr. Baumann's excellent and informative paper. I agree with most, if not all, of what he has said, but I would like to add some salient facts.

First, I would like to point out that the water used in drinking and cooking amounts to only one-half of 1 percent of the total used in a community, and that it does not appear to make economic sense to treat 100 percent of all water to drinking water quality, when only a small percentage requires such treatment.

Second, I would like to touch on the plight of small community systems and of those forgotten Americans--the people on private water supplies. They are not covered by either interim or proposed Safe Drinking Water Act regulations.

Third, I want to say something about the omitted alternatives--point-of-use water treatment and bottled water. Nobody seems to know they exist.

Finally, I want to mention barriers that stand in the way of progress.

Table 10-1 shows the number of community water supplies and the populations that they serve. From those figures, we can see that only 45 percent of the population served by community water supplies lives in cities with a population of 100,000 or more. Fifty-five percent of the population lives in smaller communities, where central treatment becomes a serious burden in terms of investment, maintenance, and operating costs, and the technical capability available for providing adequate treatment.

Perhaps the most startling fact which affects the water supply problem is that more than one-half of all community water systems serve fewer than 500 persons, and over 6,000 community systems serve fewer than 100 people. Additional treatment to bring these 6,000 systems to present standards could cost as much as $21 million to $95 million per year, or $52 to $240 per person per year. This does not include treatment to remove trihalomethanes, which could cost another $4 to $4.5 million per year. And most of this additional treatment will be needless, because out of the present total community water use of 175 gallons per capita per day, less than 1 gal per day (see table 10-6) is used for drinking and cooking. The other 174 gallons are used for such purposes as sprinkling lawns, flushing toilets, fighting fires, cleaning streets, or for laundry, bathing, and industrial uses.

This means that to meet the EPA Drinking Water Regulations, 99.5 percent of water supplied by communities will be purified needlessly. Does this make economic sense?

What is going to be the per capita cost of improvements needed in small communities to comply with federal regulations? Might bottled water or some alternative source be more economical than a central system? The EPA is aware of the potential cost impact on users in small community water systems. At the

Table 10-6. Community Water Use

Type of use	Gal per capita per day	Percentage
Municipal	24.5	14.0
Industrial	80.5	46.0
Household[a]	70.0	40.0
Total	175.0	100.0

Source: AWWA USGS Circular 765 (1977).

[a]Of this, drinking and cooking utilize 0.875 gallons per capita per day, or 0.5 percent.

community level, one Illinois city has already proposed giving its citizens coupons for bottled water, rather than building a costly central plant to remove nitrates.

Consider also the small systems that are totally outside the Safe Drinking Water Act authorities or concern. Most important, there are about 25 million people not served by community systems with twenty-five or more customers. There are also 200,000 noncommunity systems that principally serve transients in such places as resorts, hotels, restaurants, commercial establishments, institutions, and factories. These people, too, do not have the protection of the Safe Drinking

Water Act because of the minimal standards that have been established or proposed.*

In sum, the Congress should be aware that a large number of Americans are currently forgotten, or only partially protected by the Safe Drinking Water Act.

But they are not alone; ignored to an even greater extent are the potential contributions of the bottled water industry and the point-of-use water treatment industry. Their alternatives have not really been recognized, considered, evaluated, or even acknowledged by the EPA. There is at present, no representative of these industries on the National Drinking Water Advisory Council, nor has there been since the council was formed.

And look at what the point-of-use water treatment or bottled water, or both, have to offer. This is the present state of the art in these industries--without bond issues, without huge construction projects, without delay, without obsolescence--available to consumers today.

Bottled water and point-of-use treatment have the potential for providing or treating water to meet all requirements of both the interim, primary, and proposed secondary regulations, implementing the Safe Drinking Water Act. These alternatives

---

*Ed. Note! Because the most controversial features of the Safe Drinking Water Act and implementing regulations are concerned with long-term exposure to trace amounts of carcinogenic chemicals, the protection of supplies for transients is not especially germane here.

can provide water for private systems, small community supplies, noncommunity supplies, and for those persons needing special types of water in times of illness or during different periods of life, such as nitrate-free water in infancy. Most important, only that water which needs to be of drinking water quality is treated to that quality.

If consumers in Boston want lead removed from their water, it can be done now. If consumers in Nevada want arsenic removed from their water, it can be done now. If consumers in Illinois or Iowa want radioactivity removed from their water, it can be done now.

Of great concern too, is the degradation between the water produced by the municipal plant and that from the point-of-use tap or faucet. Water passing through antiquated water mains, water meters, and home plumbing systems--systems that can contain a variety of metals such as brass, copper, black iron, or galvanized steel--can pick up undesirable pollutants. Of special concern in many places are the acidic waters, low in total dissolved solids, that can pick up lead and cadmium from the piping, and can be deleterious to health.

The point-of-use industry is a significant one. It is an industry consisting of over 5,000 businesses with nearly $1 billion in annual sales. Employing 50,000 men and women, it serves nearly 11 million families and over 25,000 commercial-industrial and institutional establishments.

Add to this the bottled water industry, comprised of 500 bottlers with total annual revenues of about $200 million.

If these are the capabilities of the point-of-use and bottled water business, why are they not more widely used? Why are we not moving ahead? What is stopping us in our efforts to supply better water to consumers?

Let us identify some of the barriers to progress. First, there are the legislative barriers. There is confusion and uncertainty as to what legislation applies, and what regulations control point-of-use water treatment. There are at least five congressional mandates that would appear to apply: the Safe Drinking Water Act; the Federal Insecticide, Fungicide, and Rodenticide Act; the Consumer Product Safety Act; the Food, Drug and Cosmetic Act; and, finally, the Toxic Substances Control Act.

Most legislative confusion creates another barrier--a barrier to investment. In the present situation, even if an industrial company goes ahead with the research, does the design work, and builds the equipment, it has no assurance that the government rules are not going to change in such a way as to put him out of business. Until the situation is clarified, the private sector, depending as it must on investor input, cannot do the job of which it is capable.

There is one more area of regulatory clarification that is needed. In 1976 the EPA suggested that the point-of-use industry write standards for self-regulation rather than undergo government regulation. The industry subsequently contracted with the National Sanitation Foundation, an independent research group, and together they drafted these standards. There is need now for government to reaffirm this self-regulation by industry.

A final and most important barrier to progress is that there is no federal recognition that point-of-use water treatment is a viable, workable, existing, and often less costly alternative. As a result, public attention and public policy are focused entirely--and in many cases to the consumer's disadvantage--on municipal water treatment as the only way to go.

Dr. Baumann said that we need consumer input to balance professional judgment, and I agree. If we listen only to professionals, we might think that community water systems are the only choice.

Dr. Baumann said that more research is needed, and I agree. But our industry cannot do it alone. We need the assistance of academic and federal resources. We pledge to make every contribution we can to this coordinated effort, and we ask that we be permitted to participate.

The facts are simple and I would like to restate these points:

1. Water for drinking and cooking amounts to only one-half of 1 percent of the water used in a community. The most economical approach is to treat this amount only to drinking water standards, and point-of-use equipment can do it, or bottled water can provide it.

2. Forty million people are supplied by private wells; 30 million people are served by noncommunity systems, and 32 million people are served by small community systems. All are outside the Safe Drinking Water Act. They should not be forgotten.

3. There are alternative systems available, and they should be explored.

4. We need consumer involvement in the study of public policy, and it should be encouraged.

5. There needs to be better dialogue within the water treatment sector; perhaps the National Drinking Water Advisory Council is an appropriate forum for this.

I would like to propose that we mobilize and focus all of our resources in the most effective, efficient manner possible. Let us centralize our drinking water problems; let us centralize the planning, the research, the regulations; let us eliminate the legislative uncertainty and the regulatory confusion; and let us avoid the delays and high costs associated with the uncertainties of our present multiagency approach.

No one can disagree with the basic congressional intent and focus of the Safe Drinking Water Act. Its goal to provide safe and esthetically pleasing drinking water for every American is certainly commendable, but as the regulations unfold, and complexities mount, we strongly believe that it is a time for complete review, reassessment, and reappraisal. Such reappraisal could bring to light and put into focus problems that now appear to be virtually unsolvable.

I believe that this is the primary purpose of this National Conference on Drinking Water Policy Problems, and we commend Resources for the Future, and this conference's supporters.

References

Nebraska Water Resources Center. 1976. Water Problems in the Rural Environment, Proceedings of a Conference, November 4-5 (Lincoln, Neb., Nebraska Water Resources Research Center).

"Percolation and Runoff." 1978. Journal of the American Water Works Association (January) p. 26.

Temple, Barker, and Sloane, Inc. 1977. EPA Survey of Operating and Financial Characteristics of Community Water Systems (April), (Washington, EPA).

U.S. Environmental Protection Agency. 1975. "Interim Primary Drinking Water Regulations," Federal Register vol. 40, no. 248 (December 24).

______. 1976. "Interim Primary Drinking Water Regulations, Promulgation of Regulations on Radionuclides," Federal Register vol. 41, no. 133 (July 9).

______. 1977. "Proposed National Secondary Drinking Water Regulations," Federal Register vol. 42, no. 62 (March 31).

U.S. Geological Survey. 1977. Estimated Use of Water in the United States in 1975, Circular 765 (Washington, USGS).

Discussion by Patricia L. Rosenfield

Many questions raised throughout the conference discussion have involved the degree of public confidence in the judgment of professionals of all types--managers, policymakers, and scientists. Public understanding of problems in drinking water quality and public participation in policymaking have also been discussed. In their thoughtful paper, Baumann and Dworkin have emphasized the role of the individual consumer in policymaking. The technical aspects of improving drinking water quality at home are also discussed thoroughly. The thorny issues in consumer choice, however, are given less attention. These issues--ones of consumer attitudes, consumer behavior, and public involvement--are worthy of more detailed consideration. In fact, greater understanding of consumer preferences and responses could help both businessmen and policymakers estimate the demand for improved water supplies.

In their discussion, pertinent questions are asked, but few suggestions for action are given. Especially important, the equity concerns are only implied. Ambiguity surrounds the nature of health hazards related to drinking water. What constitutes appropriate policies and choices under such uncertain, unclear conditions requires more careful analysis. The range of uncertainty is certainly wider than the general public might realize. The authors stress the need to educate the public and water-system managers so that more appropriate decisions may be made. Yet, scientific evidence is still not firm, and how to inform the nonscientists about uncertainty in a credible way is of critical concern. More detailed studies are needed on how consumer perception of health risks and consequent behavior are shaped by the controversy surrounding the health issues. Moreover, how can scientists and policymakers clarify the controversy and perhaps make broader statements to a public that is now aware of health concerns? What is scientific and political responsibility for determining the most appropriate way in which to make the public aware of the lack of certainty in some of these issues? The results from these studies would make possible a more informed weighting of costs, risks, and benefits by individuals.

The authors show, however, that when the public is informed, they are willing and able to make rational choices concerning water supplies. The example of using recycled, renovated waste water effectively makes this point. The example from Duluth concerning asbestos demonstrates that when uncertainty prevails and the public is not well informed, the path of least exertion--no change in supply or somewhat passive acceptance of city-subsidized filtered water--if often taken. In Duluth, 47 percent of the population chose to continue to use water coming out of the tap without any additional treatment. Of those who chose alternative sources for drinking, 51 percent used the city-provided water. That leaves a small percentage (about 20 percent) that actually made a substantial effort to improve their water supplies.

Consumer preferences in drinking water, as the authors discuss, seem to center on issues of taste, odor, and hardness. These are only esthetic concerns, yet surprisingly they lead to massive purchasing of home devices that "improve" such conditions. Taking advantage of incipient public awareness of possible carcinogenic hazards in drinking water, many companies are now claiming that their products, registered and approved by the EPA, also protect against this new problem. The validity of such claims needs to be looked into since it appears that consumers, uncertain of risks, scared about cancer, and affluent enough to purchase home devices, will be highly susceptible to such advertising.

The authors, as well as the Environmental Defense Fund and EPA staff, make clear that home devices now available are not easily maintained by the purchasers, possibly leading to serious, immediate health hazards through increased bacteria concentrations and even increased silver concentrations in the water being consumed. Purchasers are not capable of analyzing the quality of their home-treated tap water. Because of professional arguments surrounding long-term serious health hazards, purchasers are not in a position to evaluate advertising claims and make a rational choice between use of home devices and no modification of supply.

It is also of interest to ask, Who can take advantage of such products? Given the high cost of the products that seem to be the most reliable, it is those who are concerned enough and knowledgeable enough to seek out the products and affluent enough to afford them. Those unable to take advantage of these products are the poor and the uninformed. The results of a Gallup Poll, cited by Baumann and Dworkin, indicated that cost was a major factor in nonuse of home devices. Even devices at the low end of the price range might be too costly for low-income persons to purchase and maintain correctly. Reliance on home devices by middle-class families might well make those families unwilling to pay higher rates for improved municipal water treatment facilities or at least uninterested in agressively advocating such improvements. The EPA has estimated that the installation of granulated active carbon treatment in municipal treatment plants would increase an average annual water bill for a family of three by from \$6 to \$10 a year (U.S. Environmental Protection Agency, 1978).* It should be noted that this number is only an estimate and might not apply to all users of the system. In my opinion, it is likely to be both more equitable and more efficient to approach improvement of drinking-water quality through systemwide changes designed and operated by professionals.

Experience in developing countries, especially in rural areas, where consumers without adequate training or economic incentives are expected to maintain their own water supply systems shows too clearly the unfortunate results of such

---

*Ed. Note: But see estimates and comments in the final chapter of this volume, "Lessons from the Drinking Water Conference."

expectations. It was estimated in 1970 by the World Health Organization (WHO) that average annual investment for construction of rural community water supplies in developing countries amounted to $932 million. Despite this, in 1970, it was also estimated that only 22 percent of the rural populations of developing countries had access to working community water supply systems. It was estimated by WHO that over $6 billion would be required to serve another 36 percent of the global rural population with water by 1980. Many of these earlier systems were built with the idea that the consumer would maintain them (World Health Organization, 1976); but no matter how much money is spent constructing new systems, unless greater concern is given to system maintenance, these systems, too, will fail. Without adequate training, information, and incentive, users of water supply systems cannot be expected to maintain their systems efficiently.

Questions of consumer willingness to pay for home devices were briefly raised in the paper. Consumer willingness to pay may be thought of as a function of "consideration of external benefits, extent of consumer knowledge, and the ability to pay" (Saunders and Warford, 1976). A clear description of this concept in relation to consumer behavior regarding modifying existing supplies or providing new supplies is critical if consumer preferences are to be taken into account. Questionnaire results in the Duluth example may accurately reveal consumer preferences, but the provision of guidance for policymakers in making municipal water system investments and in setting standards will require more detailed expressions of consumer willingness to pay for improved water quality. The authors also present preliminary data on the cost of home drinking water devices. Such data could supplement what we know from questionnaires and help round out our picture of willingness to pay.

The need for increased public participation in decision making and in standard setting appears in almost every aspect of regulatory activities. Yet in contemporary water and health issues, public preferences seem to tend toward complacent acceptance of official judgment unless controversy arises. Moreover, the public becomes confused when controversy does arise. This is hardly surprising given the amount of confusion existing on all sides in the area of drinking water. Nonetheless, the authors' implication that decision makers and businessmen should suspend judgment and activities until all knowledge is in and all consumers are educated could lead to serious, unhealthy consequences. Consumers, policymakers, and investors are making choices now. These are based on dubious knowledge and may result in increased health hazards and inequitable public investments. What is needed in this area is an appropriate device for interpreting scientific findings for the public and for determining what policies should result from these findings that will limit in the most equitable fashion human exposure to carcinogens.

References

Saunders, R. J., and J. J. Warford. 1976. Village Water Supply (Baltimore, Md., World Bank, Johns Hopkins University Press).

U.S. Environmental Protection Agency. 1978. Fact Sheet on Granulated Activated Carbon (Washington, D.C., EPA Office of Water Supply).

World Health Organization. 1976. Community Water Supply and Wastewater Disposal: Mid-Decade Report (Geneva, WHO).

Part III

Problems with the Safe Drinking Water Act

Chapter 11

OPTIONS FOR COPING WITH THE SAFE DRINKING WATER ACT

David J. Eaton*

This chapter explores the problems of moving from the passage of a law to its implementation. The focus on the regulation of contaminant levels in drinking water as proposed in the Safe Drinking Water Act (PL 93-523). A dichotomy exists between those who decide and those who carry out the act. Those who decide include the U.S. Congress, which passed the law, the U.S. Environmental Protection Agency (EPA), which has promulgated standards, and the states, which in most cases will accept a primary role in implementing the act. However, ultimate responsibility for improving drinking water rests upon local water utilities and their customers, who pay the bills.

There are four types of implementation problems:

1. The roles of different agencies and the rules they will follow are not yet certain.

2. The act may impose harsh consequences on small water suppliers not in current compliance with its requirements.

---

*Although Professor Eaton wrote this paper, the ideas were developed in a graduate seminar, the Safe Drinking Water Policy Research Project of the Lyndon B. Johnson School of Public Affairs, The University of Texas at Austin. Professors Gerard A. Rohlich, R. Barry Lovelace, and David J. Eaton have directed both years of the project.

3. The meaning of safety and the balance between benefits and costs complicate any effort to regulate contaminants in drinking water.

4. The act does not address the fate of water suppliers too small to be subject to its jurisdiction.

Cooperation, co-optation, and conversion are three parts of a strategy to administer the act under existing conditions of uncertainty. The implementation process should provide the time, technical assistance, and, on occasion, the financial leverage to allow these forces to work.

Strategies for Implementation

Through the Safe Drinking Water Act, the Congress has directed the EPA and the states to require water suppliers to limit the contaminants in drinking water in order to reduce health risks. This task of persuading people of their own best interest has a venerable and honorable history in the waterworks profession. It may be useful to review how two civil servants involved with water supply in earlier times approached implementation.

Vitruvius

In the halcyon era of Augustus Caesar, a bureaucrat named Vitruvius stunned the waterworks community by suggesting the elimination of lead pipes because "Water . . . conducted through lead pipes . . . is found to be harmful for the reason that white lead is derived from it, and this is said to be hurtful to the human system" (Vitruvius, book 10). Although the journals of the period are not available, we can speculate that some Etruscan engineers objected to Vitruvius's recommendation on three grounds. First, the aqueducts had been used "safely" for centuries and water was already extensively treated, prior to entering the distribution system, in settling basins at the terminus of each aqueduct (deCamp, 1963, p. 211). Second, Vitruvius did not show a cause-and-effect relationship between lead pipes and health. Rather, he extrapolated from epidemiological (community use of lead paint) and occupational health (lead fumes) experiences (Vitruvius, books 10 and 11). Third, Vitruvius did not demonstrate that health benefits would exceed the costs of replacing lead with earthenware pipes.

Vitruvius's suggestions were never implemented. His failure might have reflected a lack of consensus among water supply professionals that there existed a problem, or that his solution would be best. Another reason for inaction could have been that the magistrates, rather than the Roman Senate, held the purse strings for public works (deCamp, 1963, p. 212). Imagine how the magistrates, who used their personal fortunes to build civil works, must have felt to be accused of endangering the health of Rome!

It is possible to draw certain apparent parallels between Vitruvius's times and our own. The EPA has promulgated regulations limiting water contaminants. Water-utility managers are not yet fully convinced that the health risks of certain contaminants justify the substantial costs to treat water or develop new sources. The notion that drinking water may not be "safe enough" may be perceived by suppliers as an accusation. While federal or state governments may sing for safe water, they do not pay the piper. The incremental costs will be paid by the persons who consume the water, whether or not they perceive that their health is endangered.

Vic Ehlers

The story of Vic Ehlers, Texas' first State Sanitary Engineer, provides an alternate paradigm. When Ehlers began work in 1915, major epidemics of communicable diseases were commonplace. For example, from January, 1915 to September, 1917, 1,682 Texans died of typhoid fever. Less than thirty towns in Texas had sanitary sewage collection systems, and only seven had facilities for even partial treatment of drinking water (Dabney, 1974, p. 20).

Vic Ehlers recognized that small Texas communities could not easily finance water treatment plants. He encouraged compliance through the persuasion of public officials and the mobilization of public opinion.[1] He also fostered a sense of pride and developed a locus for peer group pressure among water-treatment operators by organizing professional organizations, such as the Southwest Water Works Association

---

[1]Based on interviews with Clayton Billings, Vic Ehlers, Jr. and David Smallhorst, conducted by Mary Jo Seemann in 1977.

in 1920. These groups in turn spread the gospel of safe water, along with technological information, even to isolated water and sewer districts. Despite the fact that Texas had no law permitting the state to supervise or regulate suppliers (LBJ, 1978a, p. 117), drinking water quality improved measurably. By 1933, there were forty towns that could boast a perfect bacteriological record for twelve consecutive months (Texas Department of Health, 1934). By the late 1930s, Texas had more sewage treatment plants than any other state in the union (Dabney, 1977).

One key to Ehlers' success was that he understood the water supply industry. This industry is composed of a large number of independent providers operating under a variety of legal charters. For example, in Texas there are approximately 2,000 water supply systems, whose ranks include municipalities, special water districts, water supply corporations, and private providers, such as mobile home parks. The populations served by these systems range from more than a million to less than twenty-five persons.

One parallel between Vic Ehlers' times and our own is that voluntary cooperation of water suppliers is a necessary condition for regulation. Most water suppliers are public or nonprofit entities that neither pay income taxes nor receive federal or state financial largesse. Thus, they are unaffected by either income tax incentives or traditional grand-in-aid programs. A state could threaten to revoke a supplier's certification of water as "acceptable for human consumption." However, unless water users are convinced that their water is unsafe, they may continue to use and pay for the water. The problem of implementation is not changed much since Vic Ehlers' day--rarely have so few had to regulate so many with so little leverage.

Administrative Strategies

A comparison of Vitruvius and Vic Ehlers leads to alternate strategies for implementing the Safe Drinking Water Act. Vitruvius, of the righteous ends, did not blanch at the implications for existing water providers. Vic Ehlers, of the reasonable means, was as concerned with building momentum for incremental improvements over time as with immediate results. Although the states are beginning to implement the act, it is uncertain whether the EPA and the states will emulate Vitruvius or Vic Ehlers. The remainder of this chapter considers some administrative problems that state or federal regulators could face. In developing recommendations, this chapter will let the spirit of Vic Ehlers motivate its methods. The theme is how to resolve the uncertainties that may inhibit the achievement of safe drinking water.

There are four types of problems that complicate the administration of the act--procedural ambiguities, unintended harsh consequences for suppliers, inherent enigmas of the regulatory process, and drinking water risks not addressed by the act. Each of these complications will be considered in turn.

Roles and Procedures

There are four role problems related to ambiguities in the act. First, federal agency mandates appear to overlap. Second, the federal concept of primacy may conflict with state expectations. Third, the strength of consumer action to push water system compliance is uncertain. Finally, variance and exemption procedures have yet to be fully defined.

Federal Agency Mandates

The Safe Drinking Water Act clearly indentifies the EPA as the leading federal agency in administering it. However, the EPA interests overlap the mandates of the Food and Drug Administration (FDA) and the Farmers Home Administration (FmHA).

The FDA is responsible for assuring safe preparation and processing of foods and drugs. When water is used in processing, does the EPA or the FDA have the responsibility for enforcement of limits on contaminant levels in process water? Since this subject is the topic of other discussions in this proceedings volume, no more will be said other than to recommend that these two agencies establish a common basis for limiting contaminants in water used in drug or food processing.

The FmHA administers a grant and loan program to assist in the construction of rural water supply systems. The FmHA loan program is predicated upon the ability of the water supply system to repay the note over its normal term. The enforcement of the act may require some rural systems to install costly treatment equipment and may jeopardize the ability of the system to repay the FmHA note. Who is responsible if there is a conflict between meeting the requirements of the act and maintaining the economic viability of rural water systems?

A good example of this unintended overlap of EPA and FmHA roles is the case of the Grassland Water Supply Corporation (LBJ, 1978a, p. 86). Until the FmHA intervened, the seventy-five individuals in Grassland, Lynn County, Texas, drew their drinking water from private wells. The FmHA, following federal policies to develop rural water systems, encouraged these families to join together to form the Grassland Water Supply Corporation. The FmHA provided a low-interest loan to construct a community water system. Returns from the completed operation barely meet the FmHA repayment terms. There are no salaried officials responsible for the system; all labor is provided on a "good neighbor" basis.

With the advent of the EPA drinking water standards for fluorides, Grassland finds itself in conflict with one set of federal policies after having closely followed another set. Funds are not available to build a plant to remove fluorides. Should the community return to private wells and abandon a community water system so recently constructed? The seventy-five individuals of Grassland really want little to do with government at any level. Now, with a community water system in place, the FmHA, the EPA, the Texas Public Utilities Commission, the the Texas Department of Health have entered their lives. A Grassland spokesperson, Gene Lawson, summarized their frustrations during an interview in 1977: "We come under a federal agency ourselves (the FmHA) and we think that the rules should come through the FmHA. . . . There has to be some form of reconsideration for these small systems. You (the interviewer) should not even be here. You don't have any business coming down here to this system duplicating the efforts of the FmHA. . . . This overlap is in direct violation of President Carter's program to streamline the government. Leave it all up to the FmHA since they are the ones we owe the money to."

On Primacy

State health departments may find the procedures surrounding the granting of state primacy, in connection with administration of the act, to be puzzling. Prior to 1974, the states alone were responsible for regulating public drinking water supplies within their jurisdictions. In 1974, Congress took away their monopoly through passage of the act, but gave the states the option to reclaim control under EPA supervision through primacy. As an incentive, Congress authorized renewable federal grants to support up to 75 percent of a state's water system-supervision program. However, the act clearly authorizes such aid on a year by year basis (PL 93-523, sec. 1443). It will be interesting to see how the EPA administers the act in ambivalent states such as Pennsylvania, which has declined to seek primacy. Even in states with primacy, the fate of Public Law 93-523 administration in the absence of administrative grants is uncertain. What would keep an administrator from firing the staff that federal funds had supported? In the next round of Safe Drinking Water Act amendments, Congress may wish to clarify its intentions. If Congress intended to extend the scope and quality of state public water supply supervision, why did it not authorize long-term financing for state administration? If the goal is to wean states from federal aid, then Congress may have to consider fiscal disincentives to termination of primacy.

On Consumer Action

One pathbreaking provision of the act requires water systems to notify users and the news media of violations of EPA regulations (*Federal Register*, 1975, sec. 141.32). Once alerted, water consumers can pressure the supplier to comply with the act. One problem is that some suppliers are still not familiar with the act and the notification regulations. Last spring, a group of graduate students and faculty at the Lyndon B. Johnson School of Public Affairs at The University of Texas at Austin conducted a survey of sixty-three water suppliers in Texas whose water exceeds the maximum contaminant levels (MCLs) for at least one contaminant.[2] Only one-third of the suppliers surveyed could report receipt of the Texas Health Department compilation of information about the act.[3] Of those surveyed, less than half the municipal water systems and only one in six private suppliers reported receipt of these books. Only half of the respondents reported contact with the Texas Department of Health regarding water hygiene; only one in twenty reported that they had conversations with the EPA. States and the EPA should continue their efforts to instruct suppliers regarding notification responsibilities. Participation of water consumers would be expedited if the state would distribute sample forms to be used by noncompliant systems in their required communications with users and the media.

Variance and Exemption Procedures

For Texas, where there are 573 suppliers with drinking water that exceeds at least one MCL (LBJ, 1978a, p. 88; and

---

[2]The survey was part of a study of the impact of the Safe Drinking Water Act on Texas, which was made under contract to the Texas Department of Health (LBJ, 1978a).

[3]Project participants interviewed 63 of 573 water suppliers not in current compliance with the MCLs. It is difficult, if not impossible, to make sweeping generalizations from such a small survey of Texas systems. Water systems were selected on the basis of the size of population served, type of ownership, financial status, geographic location, raw water source, and type and severity of contaminants. The communities were not selected to be a random or even a representative sample. Therefore, all results should be viewed as illustrative, and should be qualified by the phrase "for the water suppliers surveyed."

Texas Department of Health Resources, 1967a and 1967b), the variance/exemption process could be complicated. For a state to grant a variance or an exemption, at least one public hearing is required. However, a "worst case" could arise if (1) each system would request a variance from an MCL issued by Texas; (2) each community would wish to be involved, via a hearing, in the decision process affecting both the variance per se and the compliance schedule; and (3) EPA would resolve that the compliance schedules did not meet with its expectations. Under such conditions, Texas could face as many as 1,700 separate hearings to administer variances and exemptions. Such hearings could be costly for a state. In Texas, the possibility exists that two Department of Health attorneys may have to attend a single hearing, one to act as a hearing examiner and fact finder, and one to represent the Office of Water Hygiene as a party.

An alternative to hearings for individual systems would be hearings by classes of systems--but how to define a class of systems? Water systems grouped as to geographical propinquity might not share identical contaminant problems. Classes grouped on a "severity of contaminant" basis might differ substantially in terms of alternate sources of water or the prospects for combination with nearby suppliers with similar problems. Neither of these schemes takes account of system size, even though size is a prime determinant of the ability of a system to finance improvements.

Unintended Consequences of the Act

The sponsors of the act intended it to be a reasonable vehicle to achieve safe drinking water. Opponents feared that it could have unintended fiscal consequences for rural systems (Congressional Quarterly, Inc., 1974, p. 425). This section explores small-system costs to meet the requirements of the act.

Private Systems

Privately owned water systems such as trailer parks or cotton gins (Texas Department of Health Resources, 1976a) are as liable as other systems to the provisions of the act. In theory, municipal water systems can spread cost liability among consumers via water rates and federal or state subsidies. Private water supply corporations can spread the burden of improved treatment among the corporation owners, who in many cases are the system users. Owners of mobile home parks or cotton gins, however, bear sole liability for costs. The owner may attempt to pass on a portion of these costs to those who use the facility, but consumers can refuse to pay by moving elsewhere (as could consumers in municipalities). Such costs put a private park owner at a disadvantage versus competitors. One owner put his dilemma succinctly, "People would move from (the) park with increased rates. . . ." Another states that "this will put all independent people out of business." Because the trailer parks are small,[4] even if a few families move in response to higher rents due to increased water costs, the business impact could be substantial. The act's impact upon private systems will be aggravated by their relative ignorance of its contents. Of twenty-four private systems responding, only four indicated that they had received the Texas Health Department compilation of information on the act (LBJ, 1978a, p. 95).

Small Public Systems

Of about 550 community water systems with drinking water that exceeds at least one MCL, 191 supply fewer than 100 persons. Another 177 systems serve 100 to 500 persons (LBJ, 1978b). How can these systems reduce contaminants and what will it cost to do so?

The principal contaminant problems in Texas are fluorides and nitrates that exist naturally in groundwater. Numerous water treatment processes and nontreatment alternatives can reduce the level of these inorganics. Table 11-1 lists a few of these options (Sorg, 1978). The choice among these or other options depends upon many factors, such as the availability of alternate water sources; the volume of system flow;

[4]Of the twenty-four private systems responding to a University of Texas Project question regarding system population, seven (29.2 percent) had less than 99 residents and another eleven (45.8 percent) had populations between 100 and 400--and most were closer to the lower end of the scale (LBJ, 1978a, p. 95).

Table 11-1. Options for Reducing Fluorides and Nitrates

Water Treatments	Nontreatment Alternatives
Nitrate removal	Switch to surface waters
Demineralization (ion exchange)	
Selective ion exchange	Change groundwater sources
Reverse osmosis	
Electrodialysis	
	Blend raw waters
Fluoride removal	
Demineralization (ion exchange)	Purchase of finished or bottled waters
Reverse osmosis	
Lime softening	
Tricalcium phosphate adsorption	

and the contaminant form (anion or cation), concentration, chemical oxidation state (valence), and solubility.

Inorganics removal and water source development could be expensive. According to one recent engineering study (Bernard Johnson, Inc., 1977, pp. I-1 through I-3), the total cost of bringing Texas suppliers into fluoride and nitrate compliance could exceed $73 million in capital investment and $15.5 million of annually recurring operating and maintenance costs. Several LBJ School graduate students developed a simple simulation to calculate the water rate impacts of reverse osmosis treatment for several hypothetical but not unrealistic Texas communities (LBJ, 1978a, pp. 142-178). This analysis found that the expenses of fluoride or nitrate removal could result in doubling of water rates for a very small community. Such result would be consistent with the evidence presented by Clark, Stevie, and Sinnot (see Chapter 8).

How May Small Systems Respond?

Are small suppliers in Texas willing to remove fluorides or nitrates from drinking water? The LBJ Project members, who interviewed water system personnel, found resistance to voluntary compliance. If system managers recognize that their water supply does exceed the MCLs, they do not always view the water as a health risk. Forty percent of the systems studied had excess nitrates but their managers expressed the feeling that the current standards are too strict. Nearly two-thirds of the operators of systems with excess fluorides felt that the MCL is too low (LBJ, 1978a, p. 90). Feelings about the safety of current water supplies often surfaced during the questionnaire interviews. Some managers would refer to elderly town residents who had consumed that water over a lifetime with no apparent detrimental health effects. One water system owner stated that "other things will kill you graveyard dead"--but not fluorides. Some were simply skeptical of the MCLs: "I don't know where these magic numbers come from. I've never heard of any blue babies."

The federal attitude, according to Douglas Costle, the EPA administrator, is that Congress intended local users to bear the costs of compliance (Costle, 1977). This attitude is consistent with the notion that drinking water systems should be self-supporting entities. The LBJ School survey made a special effort to elicit community expectations regarding the appropriate roles of federal and state governments and the local suppliers. The most frequent response was stated bluntly by a Seagraves, Texas, official, "If they're gonna make us do it, they ought to pay for it . . ."

When asked about state or federal assistance, fifty-five of sixty-three system representatives interviewed (87 percent) felt that the federal government should financially aid local water suppliers. Forty-six systems (73 percent) responded that the state should share the financial burden. These attitudes were consistent for all types of suppliers--municipalities, water districts, and private water systems (LBJ, 1978a, p. 91). A follow-up question asked which level of government should bear the financial burden of system compliance. The most frequent response (twenty-one responses or 33 percent) was "mainly federal." Seventy-one percent of system representatives interviewed (forty-five systems) indicated some expectation of federal responsibility. Those who did not favor federal involvement were the private water suppliers (LBJ, 1978a, p. 92).

Constraints to Small-System Improvement

Even if a supplier accepted (1) that fluorides or nitrates are a risk to health; (2) that the benefits of compliance exceed the incremental costs; and (3) that such improvements are a local responsibility, are small water utilities capable of improving their drinking water? Unless financing and appropriate personnel are available, small suppliers may be unable to deliver acceptable water. This section considers financial impediments to system construction, and personnel problems for operation.

Even for a system willing to bear the financial burden, the first question prior to building capital improvements is, "Where will the money come from?" Sinking or operating funds may be insufficient to finance development of new water sources. What about grants or loans to assist construction?

There are numerous federal and state programs that exist to provide aid to communities for water supply improvements. These programs, listed in table 11-2, are concerned either with drinking water, community development, or water development (LBJ, 1978b). Although the following discussion will focus on Texas, it may be applicable to many other states.

Two agencies, the EPA and the Texas Department of Health, regulate the safety of drinking water. These agencies also provide, upon request, limited technical assistance to communities not in compliance with with act.

A second group of agencies is concerned with community development. At the federal level, the Economic Development Administration (EDA) and the Department of Housing and Urban Development (HUD) give grants and loans for infrastructural improvements. Although HUD and EDA funds could be used to finance water treatment or source development, safe drinking water is not a high priority with either of these agencies. In addition, most of their aid is targeted at communities larger than those Texas towns with fluoride or nitrate

TABLE 11-2. Sources of Assistance to Small Public Water Systems in Texas

Agency	Program	Type of Aid	FY 1977 aid ($ millions)		FY 1977 aid for drinking water quality ($ millions)	Clients
Farmer's Home Administration	Rural water and waste disposal systems	Loans	$31.46	(Texas)	Not available	Political subdivisions, private or public non-profit corporations
		Grants	$17.347	(Texas)	Not available	
Economic Development Administration	Public works facilities	Grants	$26.822	(Southwest)	Not available	Associations representing an EDA-designated area or EDC, such as: political subdivisions; Indian tribes; private or public nonprofit corporations
	Supplemental 304 projects	Grants and loans	$ 0.694	Texas allocations	Not available	
Department of Housing and Urban Development	Community development	Grants	$21.073	(Texas)	Not available	SMSA or non-SMSA local governments
Environmental Protection Agency	Safe Drinking Water Act	Loans	$0.0	(U.S.)	$0.0	Public Water systems
Texas Water Development Board	Water development	Loans and research	$19.339	(Texas)	$0.067 in research	Political subdivisions
Texas Department of Health	Surveillance and technical assistance	Laboratory testing	$ 4.300 (est.) in services	(Texas)	$4.3 (est.) in services	Licensed water systems

problems. In Texas, the regional Councils of Governments (COGs) try to foster regional development. The COGs provide neither grants nor loans, and are not experienced providers of technical assistance on problems of drinking water.

A third group of agencies view development of water systems as their business. As previously mentioned, the federal FmHA provides grants and loans for construction of rural public water supplies. It is uncertain whether the FmHA will place a high priority on water treatment applications. The Texas Water Development Board (TWDB) purchases bonds of local government entities as a means of loaning funds. The TWDB is a possible source of funds, but their financial resources are limited by legal and historical constraints. By law, the TWDB can provide loans but not grants to political subdivisions. Water Board funds are loaned to a jurisdiction only if it can convince the loan officer that it can repay them. In passing on an application, the board is required to consider Texas constitutional limitations on the bonded indebtedness of municipalities (25 percent of assessed property values) and board guidelines for loans to special water districts (no more than 10 percent of assessed property value). In the past, drinking water treatment may have been a lower priority for the board than water acquisition, storage, transmission, and distribution.

In short, small public water suppliers in Texas may not be able to afford water treatment to remove inorganics, even if they are inclined to comply with the act. If funds to build treatment works were to materialize, it is uncertain whether users could afford the water rate increases needed to retire bonds and pay for system operation and maintenance.

Assume for a moment that money problems can be ignored. Do these small public systems have the operator talent necessary to manage sophisticated treatment plants? The LBJ survey disclosed that many system operators work part-time, without pay, and lack experience with even conventional treatment systems. Chlorination is the only treatment of 87.3 percent of the systems responding in the survey (LBJ, 1978a, p. 90). Thirty-nine of sixty-three systems reported that they pay operators (LBJ, 1978a, p. 95). Part-time operators, certified only for chlorination, would be ill equipped to operate and maintain sophisticated fluoride or nitrate-removal plants.

On Small-System Compliance

There is an expectation gap between the federal government and small public systems with respect to compliance. The local systems do not feel willing or able to finance by themselves the necessary improvements. The federal government stands on the tradition of water supply as a distinctly local public utility and has few fiscal inducements to encourage water system compliance.

If there is no basis to induce, does there exist a way to compel local compliance? Federal and state governments have remarkably little leverage over local water suppliers.

As discussed in the previous section, there are few federal subsidies that can be held hostage to enforce community compliance with the act. Investment tax credits are incentives only to the private owners of public water systems. The branding by a state of local water as "unfit for human consumption" may have little effect if consumers do not share federal definitions of their own safety. Going to court to force either compliance or system shutdown would be a difficult decision for state water hygiene officials, particularly if they sympathize with people who argue that they have used the water for generations without apparent ill effects. Compulsion is not an easy route to compliance.

An alternate strategy would be to work with affected communities over a reasonable time period in order to achieve compliance, in the manner of Vic Ehlers. Such an approach would recognize that there are alternate routes to deliver acceptable water to consumers--via treatment, water-source development, point-of-use purification, or purchase of bottled water. It could be combined with appeals to professional pride and pressures by water consumers to facilitate compliance.

In order to resolve technical uncertainties, the EPA and the state could advise system operators and consumers on questions concerning the validity of the standards, the range of technical options, and associated costs. The EPA has begun this process, through (1) development of nontechnical discussions of the relation between human health and inorganic contaminants in water, and (2) its research program on removal of inorganic contaminants from water (Sorg, 1978). The EPA and the states should cooperate with manufacturers of treatment equipment and vendors of bottled water in order to estimate the costs associated with treatment and nontreatment approaches. One emphasis of all these programs should be upon the problem of small water suppliers.[5]

To resolve doubts regarding system financing, the EPA and the states could devise new fund sources to assist systems in investment or operator training required to implement the act. The EPA should collaborate with Congress to develop amendments to the Safe Drinking Water Act to provide for federal grants or loans for public and private nonprofit suppliers who must make major capital improvements to a system in order to comply with the provisions of the act. Such aid programs could be targeted toward small water suppliers who have no access to other federal financial assistance.

Two Enigmas and an Omission

The two preceding sections have discussed problems that arise because of the act. This section focuses on issues that exist despite the act. Two subsections discuss the questions, Are there risks to health?, and, Are the benefits

---

[5]Ed. note: Such a study is underway, as noted by Bauman and Dworkin in chapter 10.

worth the costs? A third subsection investigates water-related health problems that do not fall under the jurisdiction of any regulatory program.

Are There Risks to Health?

Lest we leave the reader with a bad impression of Vitruvius, let us give credit where credit is due. Vitruvius was the first to recognize the difference between safety and risk. This distinction was rediscovered by Snow in London in 1854, by the Treasury Department's 1914 drinking water standards committee, by the current generation of regulators, and will continue to be rediscovered as long as societies regulate impurities in drinking water. The distinction is that safety is an enigma and risk is a question.

Safety is an enigma because scientists cannot conclusively answer the question, Is water with contaminant X safe? To answer, "Yes, the water is safe," requires knowledge of the behavior of chemicals and living things far beyond the level mankind is likely to possess. To answer, "No, the water is not safe," implies the design and implementation of a controlled experiment capable of demonstrating an unambiguous cause-and-effect relation between the contaminant in drinking water and ill health. Such experiments are rare.

Scientists can make declarative statements on risk, by alluding to different levels of association between a water contaminant and ill health. Vitruvius could reason from epidemiological and occupational health experience to the assertion that lead in water is a risk to health. Snow could reason from the coincidence of cholera and water supply in areas of London to a relation between the two. Likewise, our contemporaries observe statistical associations between cancer events and chlorinated organics in drinking water, and conclude that a relation between the two exists (Federal Register, 1978).

To those in the water supply business, this distinction between safety and risk is a fundamental part of the professional consciousness. It is useful to stress this distinction to those outside the profession, because it underlies expert disagreements relating to removal of contaminant X from drinking water. The answer to the question, To remove or not to remove? does not spring from ultimate truth, but is rather a reflection of a dominant professional and public consensus on acceptable risks at some point in time. Vitruvius was right, but in his time he was a lone voice. As Doll (1977) put the issues, "Scientists may be able to measure risks to health, but the real question of safety is not the effect per se, but the value judgment to determine a socially acceptable level of risk. . . . Judging safety is a normative, political activity."

An underlying dilemma for any drinking water-regulation program is the need to build a professional and public consensus that will legitimize appropriate water improvements. The EPA has taken the first step--to declare contaminants.

However, disputes still exist among professionals over the legitimacy of removing at least two contaminants, fluorides and chlor-organics, from drinking water. These disputes pose an implementation problem for the act and deserve prompt attention.

Questions have been raised by knowledgeable water researchers and dentists regarding the labeling of fluorides as contaminants. Fluorides at low level in water are effective in the reduction of the incidence of tooth decay. At increased concentrations, fluoride consumption produces tooth mottling. At much higher levels, fluorides have a tendency to corrugate tooth enamel. Fluoride intake in excess of 20 mg per day for twenty or more years has been observed to cause a crippling skeletal fluorosis (Safe Drinking Water Committee, 1977, pp. 377-380). What are the health risks and health benefits of fluoride at differing levels of intake? Some persons question whether the risks, as opposed to esthetic sensitivities, are sufficiently recognized in the regulations implementing the act.

Another question is the degree of risk associated with chlorinated organic by-products from the use of chlorine as a disinfectant in drinking water treatment. Several studies have shown apparent associations between the levels of chlor-organics in drinking water and the number of carcinomas in the associated human populations (_Federal Register_, 1978). No cause-and-effect relation of particular cancers in humans have been demonstrated; indeed, we are ignorant of any potential effects of some of the chlor-organic species identified in drinking water.

Although apparent risks of fluorides and chlor-organics at some levels are sufficient to warrant reductions of these contaminants, the EPA and the states should continue their current research efforts to quantify the relation between dose and effects. However, an administrative strategy should establish a clear standard based upon the best available health data. In this perspective, some persons question whether recent proposals to establish different classes of MCL violators based upon population size alone are wise. These proposals, which have surfaced both for chlor-organics and fluorides, may involve the granting of essentially unlimited time periods for compliance to systems with water that exceeds MCLs if the system (1) serves fewer than a designated number of users and (2) is contaminanted below some upper bound. One example would be the granting of an open-ended compliance schedule for suppliers with excess fluorides if the fluoride concentration in water is below 0.4 mg per liter and if the system serves fewer than 1,000 persons. The advantage of this approach is that it reduces the number of noncompliant communities for enforcement purposes. One disadvantage of violator classes is that the scheme is not easy to explain to noncompliant systems. How can one justify removal of fluorides from a water supply serving 1,010 persons, while a neighboring town of 990 need not comply? What happens if fifteen persons move from one town to the other to escape increased water rates? What is the answer if a resident of the smaller town questions whether the water is safe? Rather than change the standard, why not adopt the principle that while the MCL is fixed, the tempo of compliance will reflect the circumstances of the technology and the water supplier?

Are the Benefits Worth the Costs?

When conventional benefit-cost analysis was in vogue, analysts would approach the issue of choice between capital investments and human health by attempting to portray existence as a stream of discounted benefits. Since the advent of multiobjective analysis, the dominant procedure for trading off incommensurates is to array and choose among them, rather than hide the value judgment in a maze of methodology. At what point do potential health benefits outweigh the costs of compliance?

Under any set of standards, there would be disagreement whether the costs of compliance are balanced by reductions of health risks. Under the present maximum contaminant levels, some observers have questioned whether the costs of activated carbon treatment are balanced by the possible reductions in human cancer rates. Such a value question has no ultimate "scientific" answer; the solution at any point in time will reflect the consensus of professional and, ultimately, popular opinion.

As a society, our choice to reduce chlor-organics is reflected in passage of the act and in the regulations (Congressional Quarterly, Inc., 1974, p. 423). However, many professionals are still uncertain whether the decision was correct. The EPA should recognize the legitimacy of such skepticism in the administration of its regulations and should further explore the relation between chlor-organics and cancer to provide a stronger base of information. All involved should view implementation as a conversion process; a long-run change in the climate of professional opinion is just as important as a shift in the treatment practices of the moment.

An Omission from the Act

There is at least one health issue related to drinking water that is not covered by the act. Public Law 93-523 requires that all piped water systems with at least fifteen service connections or twenty-five customers comply with federal drinking water standards (PL 93-523, Sec. 1401 (4)). What about those water providers with no piped distribution systems, such as the colonias of the lower Rio Grande Valley in Texas?

The colonias are poor, rural unincorporated communities that exist in Cameron, Hidalgo, and Willacy Counties in Texas. These collections of dwelling units have no formal ties with the governments of cities or towns. Therefore, the primarily Mexican-American residents do not pay property taxes, nor do they receive the typical public works amenities of piped water, treated sewerage, or street maintenance.

In 1976 an LBJ School Project studied these colonias. The project participants conducted a 1 percent, spatially stratified random sample of colonia households in sixty-five colonias in Cameron and Hidalgo Counties. The project found that although forty-five of the sixty-five colonias had access to treated drinking water, an estimated 57 percent of all colonia households do not receive treated water. Of the households surveyed, 45 percent obtained water from a public water supply stream, 40 percent from wells, 6 percent from irrigation ditches, and 7.5 percent from other sources. Of the households surveyed, about half disposed of sewage by cesspool or septic tank and about half by outhouse. None of the colonias has access to a sewage treatment facility (LBJ, 1977, p. 7).

One consequence of the primitive water and waste disposal systems is that the rate of diseases which are spread by waterborne fecal contamination (viral hepatitis, typhoid, bacillary and amoebic dysentery) are higher in the lower Rio Grande Valley than in almost any other part of the United States (LBJ, 1977, p. 10). These diseases could be eliminated, for all practical purposes, with the introduction of disinfected drinking water and sanitary sewage treatment. Should not these conditions be addressed by a national safe drinking water law?

Final Comments

A lesson of the stories of Vitruvius and Vic Ehlers is that the influence of a sensitive and patient civil servant may exceed the power of an idea, such as national drinking water standards, whose time has come. Those administering the act should recognize and work to resolve producer ambivalence about contaminant health risks. The EPA and the states should listen when water suppliers question whether health benefits exceed the costs of compliance. An implementation strategy should appeal to the professional pride and peer group pressures within the water works community. It should further provide the time, technical assistance, and on occasion the financial leverage to allow these forces to work. The interest groups involved should move beyond rhetoric that appears to be as sensitive to each new trace contaminant discovered in water as our detection devices.

All involved should recognize that the process of regulation of contaminants in water did not begin, nor will it end, with the Safe Drinking Water Act. This is but another step to achieve the goals set by the first commission on federal drinking water standards in 1914, namely, to provide water that is "free from injurious effects . . . and . . . offensiveness"; that is obtained "without prohibitive expense";

and that can meet water examinations that are "as few as simple as consistent with the end in view" (Public Health Reports, 1914).

References

Bernard Johnson, Inc. 1977. Economic Impact of the Safe Drinking Water Act on Texas (Austin, The Texas Water Development Board).

Congressional Quarterly, Inc. 1974. 1974 CQ Almanac (Washington, D.C.).

Costle, Douglas. 1977. Congressional Record (July 12) p. 6858.

Dabney, Henry L. 1974. "TSDH's Role in Water Pollution Control: Past and Present," Southwest Water Works Journal vol. 56, no. 4, pp. 20-27.

deCamp, L. Sprague. 1963. The Ancient Engineers (New York, Ballantine Books).

Doll, Laura. 1977. Unpublished graduate course paper on risk assessment.

Federal Register. 1975. vol. 40, no. 51 (March 14).

______. 1978. vol. 43, no. 28, part II (February 9).

LBJ Lower Rio Grande Valley Policy Research Project. 1977. Colonias in the Lower Rio Grande Valley of South Texas: A Summary Report (Austin, Lyndon B. Johnson School of Public Affairs, The University of Texas at Austin).

LBJ Safe Drinking Water Policy Research Project (1976-77). 1978a. Impact of the Safe Drinking Water Act on Texas (Austin, Lyndon B. Johnson School of Public Affairs, The University of Texas at Austin).

______ (1977-78). 1978b. Options for Coping with the Safe Drinking Water Act (Austin, Lyndon B. Johnson School of Public Affairs, The University of Texas at Austin).

Public Health Reports. 1914. vol. 29, pp. 2959-2967.

Safe Drinking Water Committee. 1977. Drinking Water and Health (Washington, D.C., National Academy of Sciences).

Sorg, Thomas J. 1978. "Treatment Technology to Meet the Interim Primary Drinking Water Regulations for Inorganics," Journal of the American Water Works Association vol. 70, no. 2, pp. 105-112.

Texas State Department of Health. 1934. Bulletin of the Texas State Department of Health (Austin, Texas).

Texas Department of Health Resources. 1976a. "List of Systems in Excess of the Fluoride Maximum Contaminant Level" (Austin, Texas) mimeo.

______. 1976b. "List of Systems in Excess of the Nitrate Maximum Contaminant Level" (Austin, Texas) mimeo.

U.S. Public Law 93-523. 1974. 42 USC 300 (December 16).

Vitruvius, Marcus Pollio. 1960. The Ten Books on Architecture, book VIII, chapter VI. A translation by Morris Hicky Morgan (New York, Dover).

Discussion by Christopher Bellavita

I use the word "administration" to mean "carrying out the purposes of the state." If one cuts through the mountains of words that have been spoken and written about the Safe Drinking Water Act (PL 93-523), it is evident that the primary purpose of the act is to make sure that water is safe to drink. But how do we determine what is safe? That has been one of the major questions posed at this conference. David Eaton, in chapter 11, argues that safety is an enigma that cannot be resolved, at least not by science. I agree, in part. Safety _can_ be defined, but its definition is the product of political and administrative activity, not a life science experiment. I will return to this theme at the end of this paper. First, however, I would like to describe briefly California's experience with the Safe Drinking Water Act.

In most respects, California's experience has been similar to that of Texas: cooperation, co-optation, and conversion have been the keys to what up to now can be described as the successful implementation and administration of Public Law 93-523. Three types of actors have been involved in the administration of the act. There are the rule-makers--Congress, the EPA, state legislatures, and state regulatory agencies. These actors are responsible for providing much of the formal structure and guidelines for administration. Next there are the professionals in water utilities and water works. These are the people who actually provide drinking water, the people who are being directed by the rule makers to provide their product in a particular way. Third are the consumers, in many respects the first cause of Public Law 93-523 and certainly the ones who experience the ultimate effects--financially and physiologically--of the act. Each of these actors has a part to play of the purpose of the act is to be achieved. How have they done so far?

The Rulemakers

The role of Congress and the EPA has been discussed at length. Here I will focus on the California rulemakers--EPA Region IX and the state's regulatory agency.

Region IX EPA has not been what one would call an activist administrative agency when it comes to the Safe Drinking Water Act. There are several reasons for this. First, the act was designated so that states would be encouraged to assume primary responsibility for enforcement, with the federal government playing essentially a back-up or support position. In California, there has been little need to date for EPA involvement. Second, prior to the passage of the Safe Drinking Water Act, California had one of the most advanced water quality programs in the country. Some states and localities in Region IX--the Pacific Trust Territories for example--have little or no drinking water-regulatory program at all; consequently Region IX resources are used more effectively in those areas than in California. Third, drinking water is not a number one priority issue in the region.

Other issues, such as water pollution control, receive more attention and resources. In sum, California is left largely on its own to administer the act.

The California Department of Health's Sanitary Engineering Section (SES) is the agency responsible for the bulk of regulatory and administrative activities under the act. Henry Ongerth (chapter 12) deals specifically with the SES perception of the act, so here I will make only a few observations.

Administrative agencies tend to execute a new law in terms of what is legally required, what is politically feasible, and what is organizationally possible. So far the Safe Drinking Water Act has not presented any major problems to the SES. The act has brought with it a dramatic increase in the SES budget (approximately 30 percent). The organization has translated this increase into more staff, more equipment, and more knowledge about the quality of the state's providers of drinking water. Indirectly, the federal funds have also helped to foster a closer and more cooperative relationship between consumers, utilities, and regulators.

The act has also made the SES dependent upon the federal government. Federal funds can become addictive, and when one actor is dependent upon another to satisfy a need, the dependent has lost autonomy and power. There is little to stop the EPA from attaching conditions to subsequent grants not directly related to drinking water quality. Who will pick up the costs of personnel and procedures required under Public Law 93-523 if, in a few years, the Congress decides to terminate administrative grants? The loss of state autonomy (some would call this co-optation) combined with uncertainty about the extent of the federal resolve to continue to assume administrative costs are two reasons why several states have not rushed--and in some cases even refused--to seek primacy. But this is not the case in California. The carrot has been accepted. The full consequences of that decision remain to be discovered.

The Water Utilities

The water utilities are the second member of the administrative triumvirate. They are responsible for translating the words of the Safe Drinking Water Act into the final product--safe drinking water.

The majority of the men and women in the waterworks field are professionals in the finest sense of the word. Some of them may view their job as a business, while others see it as a public service, but there are few villains in the industry. Go out into the field and you discover people who really care and who feel a sense of responsibility about the quality of the water that is being delivered to consumers.

But quality is not all they care about. When faced with a decision about water quality, they act according to a criterion of "reasonableness," defined according to their professional experiences and training, the facts at issue, and economic and political feasibility. No one advocates "unsafe" water, particularly not water professionals. They will, however, debate the meaning of safety, and for the most part they do not view zero risk as a reasonable goal. Their arguments concern the <u>acceptability</u> of risks when compared with the costs of reducing those risks. In this issue their motto might be "A dollar's worth of increased safety is not worth a fifty-dollar increase in treatment costs."

The water professionals have made important contributions to the development of the Safe Drinking Water Act. It would be a major error for policymakers to dismiss their views as being too parochial to warrant serious attention. As David Eaton has persuasively argued, no drinking water-quality law can work without the cooperation of water professionals. And the best way I know to encourage their cooperation is to include them as <u>active</u> participants in the policymaking process.

The act is still too new to assess the impacts it has had on utilities. But some preliminary judgments can be made. In 1974, when the act was signed into law, many water professionals cried loudly and ferociously about the public-notification provision. "It will only scare people," some said. "It's foolish," said others. "It will encourage people to go to unsafe sources," argued still others. Few people in the industry liked the idea that they would have to tell customers when they failed the federal standards.

Now it is 1978 and some evidence is coming in. The sound and fury of the profession signified very little. Public notification has been tried, and not only has it so far failed to generate many of the problems feared; it has actually done much of what its designers intended. Consumers are beginning to discover that there is something called drinking water quality. Not only that, the mere threat of notification has caused many utilities to literally clean up their act. In California, the number of bacteriological failures by utilities declined steadily as the date for the public notification provision to go into effect came closer.

One difficulty with the provision is what one might call "retronotification," notifying consumers after the fact when there has been a sampling or bacteriological failure. Utilities argue that there is little point notifying customers after the problem has been cleared up, particularly when the failure (as is true with numerous bacteriological failures) did not pose a public health threat. Some utilities have tried to subvert this system by, for example, oversampling so they could report less than 10 percent of their samples were positive, hence avoiding the public notification requirement. The utilities maintain that requiring notification in situations not related to public health considerations is a form of regulatory overkill that should be eliminated.

Consumers, and some regulatory people, argue that a well-run utility should never have any sampling or bacteriological problems. Perhaps, but the state of the art does not permit an infallible testing system. Some failures will occur that are a result of sampling error, not a water quality problem. But if a utility fails too often (three times in three months, for example) there is probably something more than a sampling error involved. In sum, policymakers should consider making the public notification provision more flexible. One suggestion is to permit the state regulators to grant a utility a notification variance for some monitoring and sampling failures that, in the regulator's judgment, are not related to public health; such variances to be limited to, say, no more than three nonconsecutive months in a twelve-month period.

So far, the largest water utilities in the state (typically those serving urban areas) have not been greatly affected by the Safe Drinking Water Act. Most have been able to meet the requirements of the act with little difficulty. (Most were meeting the relatively strict state standards in force before PL 93-523 was passed). Some systems have not been able to meet some standards--Los Angeles Power and Water is not meeting the turbidity standard, for example--but they are taking action to meet these standards. This behavior is not necessarily caused by the fact that they agree that, say, a 0.5-turbidity water is significantly safer than a 1-turbidity water, and thus worth achieving; more likely it reflects the fact that the issue is marginal and probably not worth the time or expense arguing about. This "let well enough alone" attitude will likely not carry over into areas that large utilities believe constitute "unreasonable" demands. Policymakers can expect considerable reluctance and political action by large utilities if some of the proposed organic standards are not modified, or at least justified by more conclusive data.

Small water systems are another story. As has been well documented and discussed, many small systems are faced with a considerable economic burden as they try to satisfy the requirements of the Safe Drinking Water Act. Some people have called this a major unintended consequence of the act. I doubt however that anyone who reads the testimony attendant to or the House Report accompanying Public Law 93-523 will conclude that this consequence was unintended. People knew that the small systems would be hurt economically. In my judgment, one of the aims of the act was to put small systems out of business. The economies-of-scale shibboleth was used to justify this action. In a poorly thought out scenario, the small systems were expected to be unable to bear the costs of the act on their own; hence they would be forced either to merge with other systems or to go out of business. Well, that has not been what has happened.

Small systems were responsible for much of the problem that gave rise to the Safe Drinking Water Act: inadequate monitoring, unqualified operators, deteriorating distribution systems, and so on. If one examines the character of the small water systems, one will discover why the act has treated them so harshly.

Small water systems have little or no political voice, and there are few who will speak for their cause. They are unorganized--there is no Small Systems Association--so they have no muscle. They have hardly enough money to run their own systems, let alone organize. If they are to meet the provisions of the act, they need help. Not only help in the form of money, but help in the form of new technologies that are more appropriate to the problems and economics of small systems; and help in the form of new management strategies and institutional arrangements that are designed for the specific administrative problems of small water systems.

Small systems will not go away. The barriers to regionalization (a favored alternative of the framers of PL 93-523) are formidable. Further, it is not likely that the small systems will be shut down simply because they cannot meet certain standards; if they are closed down, where will their customers get water? I look forward with anticipation to the small water-system study called for in the recent amendments to the Safe Drinking Water Act. At this point, however, I doubt that much of significance will come out of that study. We know what the problems are. If those problems are to be resolved, we need creative attention by the rulemakers in the form of money, technology, and managerial assistance.

But where is the political pressure to come from to encourage such action? Without a realignment of political groups, it will be easier and less expensive (in the short-run anyway) to do another study. That is what I expect will happen.

Consumers

Consumers are the final element in the three-part administration of Public Law 93-523. In most cases they are silent partners. Yet the potential power of consumers with respect to water utilities and regulatory agencies is considerable. Many utilities in California and elsewhere are public bodies, governed by elected officials. And though water district elections are notorious for their low voter turnout, that may not be critical to effective consumer action. There is considerable evidence that a few dedicated people working together in the right forums are more than equal to the power of the rulemakers and the utilities.

This is a time of consumer awareness in many arenas. There are cases in California in which consumer involvement in drinking water has significantly affected the legislative process, as well as the operations of utilities and regulatory bodies. The big problem, however, is to get consumers to even bother about drinking water, to get them to complain when they are dissatisfied. On the surface, consumers have other concerns which exert a stronger claim on their attention--property taxes, schools, the cost of living and so on. Yet half of all the bottled water sold in this country is sold in California. Now this may mean that those who sell bottled water are outstanding salesmen, or that more persons than first meet the eye are dissatisfied with the quality of the water they are provided by utilities. In my judgment, the latter is more likely the case since I know of few cases in which a product can be sold to people consistently when they have no real need for it. Bottled water sales are increasing from 10 to 15 percent each year. Something is going on.

Consumer quiescence may quickly become a thing of the past, however. The populist-oriented public notification provision of the act brings drinking water to the attention of consumers precisely at the time when consumers have a reason to do something about it. Most of the time in most places the quality of the drinking water is satisfactory. There is little reason for consumers to be involved--indeed their involvement is most unlikely--when business is as usual. But when the status quo changes, when the quality of the water deteriorates, the utilities must notify their customers; this is the best time for consumers to enter the picture. This is when the limited amount of attention that consumers can give to the various issues that affect them can receive the greatest returns. In one California community, a utility took several high-nitrate wells out of service <u>rather</u> than go to public notification, even though the utility did not believe the nitrate level posed a health hazard. "We weren't sure how consumers would react to the notices," one utility executive explained, "so rather than take chances we just took the wells out of service."

In other instances, when utilities have notified the public, consumers have not ignored the notices. They have called or written the utilities, said one utility representative, to find out "what the notices really mean," and to press for further information about what the problem is and why it exists. Drinking water consumers--and that includes all of us--are not dumb and can tell a snow job from an honest explanation. If explanations are not satisfactory, consumer awareness can easily become citizen action.

In my opinion, there is one major flaw in the populist approach of the Safe Drinking Water Act. The act permits consumers to choose a higher-quality water, but not a <u>lower</u> quality. Consumers can sue the EPA or the state if consumers believe those bodies are lax in carrying out the provisions of Public Law 93-523--in effect saying, "We want a higher quality of water than you are willing to see us provided with." But consumers can do nothing if they believe the act is being enforced <u>too strictly</u>.

The act states that the EPA will set specific standards for drinking water "contaminants," that is, anything in water other that two atoms of hydrogen to each atom of oxygen. These standards then become "the will of the people." How? By an exercise of logic which argues that (1) since the Safe

Drinking Water Act was passed by the Congress, and (2) Congress represents the people, then (3) the Safe Drinking Water Act represents the will of the people. In addition, (4) anything that the act requires the EPA, for example, to do also touches the mantle of the "will of the people."

The belief that the Safe Drinking Water Act represents the will of the people is an example of federal arrogance. In fact, the act represents the will of a few congressmen, some consumer advocates, regulatory and utility interests, and a few other special interest groups. Most people in this country did not know about the Safe Drinking Water Act when it was passed, and they remain ignorant of it today. Obviously the notion of the "people's will" is a political fiction that is probably justified under numerous conditions. This just happens not to be one of them.

Some communities may desire lower standards than the federally imposed levels. Some communities would rather keep their high-fluoride water and risk the consequences than pay for improved treatment. Other localities would rather _give_ bottled water away to at-risk population groups, than to issue bonds to pay for new treatment plants. These communities, under present law, must follow the federal "will"; they are not permitted to exercise their own.

The purpose of the Safe Drinking Water Act is to make sure that the water is safe to drink. Although the word "safe" may not have a scientific definition, it does have a political one. A substance is safe if the risks associated with its use are judged acceptable. The life scientist assesses risk, and we have their most recent findings in the National Academy of Sciences report on drinking water. It turns out that there is still much we do not know about the health effects of different levels of drinking water "contaminants." I suggest that where there is significant professional disagreement about the public health effects of certain _levels_ of substances found in drinking water (such as fluoride at concentrations of 1 mg per liter), the _users_ of the water should be given the _opportunity_ to decide whether they wish to accept the risks associated with that or some other level. Is water with a turbidity value of 1 unit safer than water that is 0.5 unit of turbidity? Is the difference in safety worth the costs of a new filtration plant? Let the people who pay the bills and drink the water decide.

There is a provision in the California drinking water statute that, under certain specified conditions, permits consumers served by a utility to petition for mandatory secondary treatment of drinking water. The consumers served by that utility will of course have to pay the bill for the improved treatment, but they also get to make the decision. They can if they wish be content with a lower water quality, or they can choose to pay more for a higher quality. In the words of one member of the drinking water arena, "Consumers should be able to decide if they want a Cadillac or a Chevrolet water supply." The California statute can serve as a national model for the consumer choice approach that I am advocating. It allows, indeed, insists upon, both consumer and health department involvement in the decision-making process. While the outcomes of these local decisions may not reflect the will of all the people in the community, I suspect that it may come closer than having a federal agency develop and then promulgate a national will. This local option mechanism would also provide a fertile field for some needed epidemiological research into the effect of drinking water constituents.

In essence I am maintaining that the life science experts should make available to consumers and public health officials the best information they have available. Where there is agreement about the public health effects of substances, incorporate that agreement into the drinking water standards; that is, specify a "safe" level for that substance. But when agreement is something <u>less</u> than considerable, give what information is available to consumers, let health and utility interests make the best case they can to try to <u>convert</u> consumers to their side, and then let the consumers decide the level of water quality that is acceptable to them. Consumers are quite capable of deciding issues that affect their health and finances. They should be given the opportunity. One of the legacies of democracy is the right to make mistakes; it is from mistakes that we learn what to do differently the next time.

Recommendations

The Congress is scheduled to hold oversight hearings on the Safe Drinking Water Act later this year. Based on my judgments of the California experience with the act, I suggest the following:

1. Water professionals should continue to play an active, as opposed to a symbolic, role in the process used to change the current act. The law will not be successful without the cooperation of water professionals, and participation facilitates cooperation.

2. The public notification provision of the act should be reviewed to eliminate "frivolous" notices; that is, notices that do not contribute to ensuring the public health is protected. State regulatory agencies or the EPA should be permitted to grant notification variances under certain specified conditions.

3. More creative attention to the problems of small systems is needed. This should be in the form of more money, new technology, and new institutional and management arrangements.

4. Consumers should be given the option to be the ones to decide, under certain conditions, the acceptability of the risks they face drinking water containing levels of contaminants in excess of the federal standards. They should be the final arbitrators of the safety of drinking water.

Discussion by Jacqueline M. Warren

The Safe Drinking Water Act was passed amid a national controversy over the presence of chemical contaminants, especially carcinogens, in drinking water. The statute mandates a preventive approach to the problem: The administrator of the EPA is required to regulate contaminants that may (rather than will) have any adverse effect on human health. In accordance with this requirement, the EPA promulgated national drinking water standards in December 1975, and, more recently, proposed a regulation to control organic chemicals in drinking water.

While the EPA has the responsibility for standard-setting under the act, the burden of compliance is borne by the states and, ultimately, by local consumers of the water. Because the public will have to pay for improvements in drinking water quality, the dissemination of information explaining the basis for the regulations is critically important. The process of consensus-building through education of public water suppliers and their customers must be given a high priority by the EPA if the goals of the Safe Drinking Water Act are to be realized.

In the past, the EPA has not done an effective job of explaining the basis of its drinking water regulations and thereby gaining public support for them. At the same time, the water supply profession, characterized by some as "conservative and cautious," has been outspoken in its criticism of the EPA's efforts to improve the quality of drinking water. In particular, water supply professionals have been loudly opposed to the EPA's proposal to address the problem of organic contaminants in drinking water. The EPA's failure to publicize the evidence of health risks, which led to the recent proposal to control organics, has left the public with only one source of information--the water-supply industry--which is telling only one side of the story. As a result, the public is hearing only that regulation of organics is unwarranted by the evidence of risk, that no proof of causation between contaminated water and cancer rates has been shown, that the proposed regulation is too stringent and too expensive, and that there is no need to regulate organics at the present time. To the EPA, the water-supply industry adds that the public is unwilling to pay the cost of the treatment technology required to reduce the concentration of organics in drinking water.

The Safe Drinking Water Act established a structure for improving the quality of drinking water which depends upon a high level of public awareness of local water conditions. The statute requires water suppliers to notify their customers whenever the local water supply violates existing standards for drinking water contamination. In theory, this requirement will alert the public to the existence of a problem, increase demands for correction of the problem, and thereby ensure that a continuous level of water quality will be maintained. Many water suppliers opposed this requirement, but Congress felt it was necessary to provide incentives from above as well as below to make the system work. It cannot work, however, if the only voice the water consumer hears is that of the supplier saying that the regulation is unnecessary, for whatever reason.

Perhaps realizing that public support on the issue of safer drinking water should be sought and can manifest itself by demands upon local suppliers for improved water quality, the EPA has undertaken a new public education program. Hearings on the proposed organics regulation will be held in seven cities around the country. In addition, the education program will include seminars, radio and television segments, pamphlets and other materials for distribution, and a variety of other techniques for reaching the public. By presenting the proposed organics regulation for what it is--a relatively inexpensive public health measure designed to prevent or reduce exposures to carcinogens in drinking water--the EPA will be educating the public and building a consensus that the proposal is both necessary and desirable.

One of the major criticisms leveled against EPA's drinking water program is that it is too costly to implement. Viewed objectively, however, none of the requirements promulgated or proposed to date by the EPA is prohibitive for existing water systems, except for the very smallest whose particular needs may require special attention. Water costs are very low by comparison to other utility costs such as electricity. The public is currently paying steadily increasing costs for the same service by electric utilities. By contrast, the modestly increased water bills projected by the EPA will be purchasing significantly improved drinking water quality.

Although there is no federal money currently available under the Safe Drinking Water Act for capital construction, there is funding available for technical assistance and system supervision. In fiscal year 1977, the total was $26 million. This was increased to $35 million in fiscal year 1978, and the proposed total for fiscal year 1979 is $47 million. Funds are also available through HUD, Farmers Home Administration, and a number of state and local agencies. In the absence of a new federal program to provide sufficient capital for construction to meet the technology requirements for compliance with EPA's drinking water standards, however, it is clear that the consumers of drinking water will have to pay the costs.

Despite the protestations of the water-supply industry that the public is unwilling to pay for improved drinking water quality, estimated consumer expenditures for bottled water and home water filters indicate that the public is concerned and is already turning to alternative sources in large numbers. According to rough estimates by both the EPA and the bottled-water/home filter industry, $400 million were spent on alternative drinking water sources in 1977--and predictions of continued increased growth in the industry are being made. If these estimates are accurate, the water supply industry's assessment of public attitudes about the costs and benefits of improving drinking water quality are incorrect.

Pursuant to its statutory mandate to protect human health to the maximum feasible extent, the EPA has made a judgment that the health benefits of reducing the levels of organic contaminants in drinking water are worth the costs. The risks of continued exposure to carcinogens in drinking water are not as uncertain as some opponents of the EPA's proposed organics regulation have suggested. Nevertheless, the public retains the option of pressing for a change in the law if, after hearing both sides of the issue, there is strong public sentiment against paying the costs of reducing organics in drinking water. Before an informed decision can be made on this issue, however, an active effort to educate water consumers must be made by the EPA. At the same time, water suppliers must retreat from the rhetoric of opposition to a fair determination of what public health protection requires. Finally, both sides must work together to develop constructive proposals for assisting small communities to meet existing and future drinking water standards so that the size of a community will not be determinative of the quality of its drinking water.

Chapter 12

JURISDICTIONAL ISSUES WITH THE SAFE DRINKING WATER ACT

Henry J. Ongerth

Although the Safe Drinking Water Act was passed over three years ago, and the Interim Primary Drinking Water Regulations (1975) became effective last June, only a year has elapsed since the first state qualified for primary enforcement responsibility. As a consequence, experiences with administration of Public Law 93-523 are limited. Part of this assessment of jurisdictional problems is based upon experience in development of the Interim Primary Drinking Water Regulations, and part upon a review of the potential for problems arising from the provisions of the act. Thus, this discussion in part is based on speculation.

The provisions of the act create a major jurisdictional problem because the EPA has the responsibility for enacting drinking water regulations and the state and local regulatory agencies have the responsibility for enforcing those regulations. The developing controversy over the organics standards stems partly from the strict charge which Congress has placed upon the EPA, allowing the administrator only limited latitude in setting the regulations. The House committee report on the act (U.S. Congress, 1974a, p. 10), as indicating legislative intent, is even more restrictive.

> Primary regulations must specify contaminants which in the judgment of the Administrator may have an adverse effect on the health of persons when found in drinking water. The words used by the Committee were carefully chosen. Because of the essentially preventive purpose of the legislation, the vast number of contaminants which may need to be regulated, and the limited amount of knowledge presently available on the health effects of various contaminants in drinking water, the Committee did not intend to require conclusive proof that any contaminant <u>will</u> cause adverse health effects as a condition for regulation of a suspect contaminant. Rather, all that is required is that the administrator make a reasoned and plausible judgment that a contaminant <u>may</u> have such an effect.

Section 1412(a)(2) of the act states that the interim primary drinking water regulations "shall protect health to the extent feasible, using technology, treatment techniques, and other means, which the Administrator determines are generally available (taking cost into consideration). . . ." Section 1412(b)(1)(B), referring to revised primary standards, in part states that the administrator "shall . . . established <u>recommended</u> maximum contaminant levels for each contaminant which, in his judgment based on the report [of the National Academy of Sciences] . . . may have any adverse effect on the health of persons . . . set at a level at which, in the Administrator's judgment, . . . no known or anticipated adverse effects on the health of persons occur and which allows an adequate margin of safety. . . ." Further, Section 1412(b)(3) in part states that revised national drinking water regulations shall specify a maximum contaminant level which "shall be as close to the recommended maximum contaminant level established under paragraph (1)(B) for such contaminant as is feasible. . . .

The term 'feasible' means feasible with the use of the best technology, treatment techniques, and other means, which the Administrator finds are generally available (taking cost into consideration)."

Section 1412(a)(2) read by itself seems to be extremely comprehensive. However, when read in the context of Section 1412(b)(1), which refers to "each contaminant," some room can be seen to have been left for administrative discretion. Further, Section 1412(b)(1)(B) calls for establishing recommended contaminant levels on the basis of the National Academy of Sciences report. When read in conjunction with 1412(b)(3), these "recommended" maximum contaminant levels (MCLs) can be seen to represent goals and the administrator is to establish revised primary regulations "as close to these goals as is feasible, using the best technology, taking cost into consideration." Here again, the Committee Report (H.R. Report No. 93-1185, p. 19) indicates a very strict intent as follows:

> The Committee wishes to ensure that the NAS report is based solely on considerations of public health and is not influenced by political, budgetary or other considerations....
>
> The NAS is further directed to develop its proposals for recommended maximum contaminant levels so as to protect susceptible groups in the population; so as to take account of long-term exposures, to contaminants in other media, and synergistic effects of multiple contaminants; so as to prevent body changes which are reasonably suspect of increasing the risk of severity of illness; and so as to incorporate an adequate margin of safety.
>
> In recommending an adequate margin of safety, the National Academy of Sciences is directed to consider, among other factors, the margins of safety

> which are currently used in regulating foods and drugs, pesticides, radiation, air and water pollution, occupational exposures to contaminants, and other relevant regulatory systems. However, as in the rest of the study, economic or technological feasibility is not to be weighed in deciding how much margin of safety is necessary to give reasonable assurance that health of persons will be protected. Economic and technological feasibility are to be considered by EPA (and under section 1416 by the States) and then only for the purpose of determining how soon it is possible to reach recommended maximum contaminant levels and how much protection of the public health is feasible until then.

The phrase, "taking cost into consideration," permits some application of judgment and contains the seeds for controversy. Congress has provided little guidance as to what constitutes a reasonable cost in meeting the standards. The Safe Drinking Water Amendment of 1977 (PL 95-190), the revision to PL 93-523, takes some cognizance of this issue, giving the EPA eighteen months (until May 1979) to submit a report to Congress which identifies and analyzes the following:

1. The anticipated costs of compliance with interim and revised national primary drinking water regulations and the anticipated costs to states and units of local governments in implementing such regulations.

2. Alternative methods of (including alternative treatment techniques for) compliance with such regulations.

3. Methods of paying the costs of compliance by public water systems with national primary drinking water regulations, including user charges, state or local taxes or subsidies, federal grants (including planning or construction grants, or both), loans, and loan guarantees, and other methods of assistance in paying the costs of such compliance.

4. The advantages and disadvantages of each of the methods referred to in clauses 2 and 3.

5. The source of revenue presently available (and projected to be available) to public water systems to meet current and future expenses.

6. The costs of drinking water paid by residential and industrial consumers in a sample of large, medium, and small public water systems and of individually owned wells, and the reasons for any differences in such costs.

While this amendment does not provide criteria for making the judgment on what is a reasonable cost, it at least provides for the development of data which will aid in making the necessary value judgments. The EPA has obtained a brief consultant report discussing economic criteria for granting exemptions. To the extent this is relevant, it provides some indication of a possible approach to the "reasonable cost" issue. The consultant report concludes that a reasonable cost for treatment might be 2 percent of the community's median family-income level. On the basis of 1975 dollars, the national median family-income level is approximately $12,000, so that a reasonable annual cost would be approximately $240.

The central issue in this matter is how much risk is acceptable in water delivered to the people of this country. Lowrance (1976) provides an excellent discussion of this subject. He points out that in determining "safety" two activities are involved: first, the measurement of risk--measuring the probability and severity of harm; and the judging of safety--judging the acceptability of the measured risks. Lowrance points out that between the two activities--measuring risk and judging safety--there is a no-man's land, "Scientists on the fringe of the political arena, attempting to avoid charges of elitism, are looking for more objective ways to appraise society's willingness to accept various risks. At the same time, political officials confronted by scientifically controversial facts that never seem to gain the clarity promised by textbooks are exploring the possibilities of advisory assistance, factfinding hearings, and formal technology assessments."

As has been pointed out, Congress has given only some very general guidelines as to acceptability of risk. These are utilization of the best technology, taking costs into consideration. There is scant guidance about how costs should be taken into consideration, what constitutes a reasonable cost, or an unreasonable cost. The committee report, on the subject of reasonable cost, but with a focus not directly applicable to this discussion, says:

> It is evident that what is a reasonable cost for a large metropolitan (or regional) public water system may not be reasonable for a small system which serves relatively few users. The Committee believes, however, that the quality of the Nation's drinking water can only be upgraded if the systems which provide water to the public are organized so as to be most cost-effective. In general, this means larger systems are to be encouraged and smaller systems discouraged. For this reason, the Committee intends that the Administrator's determination of what methods are generally available (taking cost into account) is to be based on what may reasonably be afforded by large metropolitan or regional public water systems.
>
> This, of course, means that some small water systems which cannot afford the methods determined by the Administrator to be "generally available" will be

> unable promptly to comply with all primary regulations. For this reason, authority to grant exemption from the effective date of the primary regulations and thus to delay the date for compliance by public water systems has been provided in section 1416. However, this period should be used to develop a regional water system which can afford to purchase and use such methods, to seek additional sources of funding such as State aid, or to develop a plan for otherwise serving the affected population after any existing inadequate system is closed.

The emphasis on consolidation and regionalization will be applauded in some circles, but it is quite unreasonable and unrealistic to apply the large-system cost yardstick to all small systems, since many are isolated, have no aid funds available, and cannot be closed as the committee report suggests. More importantly, this committee comment does not provide any real clue as to what a reasonable cost may be.

It is clear that with these national standards, local communities are given no opportunity to make their own value judgments as to what degree of risk they are willing to accept and what price they are willing to pay.

Congress has dictated a very strict standard but has not faced the difficult problem of how the necessary improvements are to be paid for. Neither has Congress taken any cognizance of the fact that water utilities throughout the nation have major unmet needs, for example filtration of surface-water sources, or upgrading of old and inadequate distribution systems, which require major financing. Ideally, each water utility should arrange its needs in a priority order and set about making the necessary improvements. In this ideal arrangement, organic quality issues would take their logical place on the list. State regulatory agencies are accustomed to working with water utilities along these lines. Parenthetically, the interference with local opportunity to set priorities in part accounts for the current state agency reactions to the new EPA proposals on standards for organics in water.

Some advocates for organics standards point out that the costs for these improvements in water quality are not excessive, ranging from a few dollars per capita per year up to levels approaching $10 or $20 or $30 per capita per year, perhaps doubling the water bill--a small price, they say, for insurance against the possible hazards. Considering other costs, these may not appear excessive. Unfortunately, small increments of costs have a way of mounting. California is now facing a taxpayer's revolt with relation to local taxes. The June 6, 1978, ballot will contain Proposition 13, the so-called Jarvis-Gann Property Tax Initiative, which would limit property taxes to one percent of market value, a reduction of nearly two-thirds or about $7 billion statewide in local tax revenue.[1] In face of such movements, it is easy to understand why there may be difficulty in imposing new local charges to correct what may be low-order risks.

---

[1]Ed. note: As the reader is no doubt aware, Jarvis-Gann passed with a sizable majority. It is interesting in this connection to note that conference participants frequently cited the independence of water systems for local general revenues. To the extent that customers perceive that they get what they pay for from these apolitical enterprises, the Jarvis-Gann movement may not be relevant.

Congress has preempted the states as well as local communities with respect to these water quality decisions. The final decision is made by the EPA with whatever "public" input it receives through the hearing process. State regulatory agencies and local communities have a major stake in the regulations to be promulgated. It is clear that EPA has one perception of what is reasonable and feasible, and the states and local agencies have other perceptions of this. As to interpretation of legislative intent, the final decision lies with the courts. Witness the February 10, 1978, court decision on the Environmental Defense Fund (EDF) suit in the U.S. Court of Appeals for the District of Columbia Circuit against the EPA (1978).[2]

In the development of the Interim Primary Drinking Water Regulations, the EPA and the states, working through the Conference of State Sanitary Engineers (CSSE), carried on negotiations over the terms of the proposed regulations. The differences which surfaced were not principally over constituents to be regulated nor maximum contaminant levels proposed, but rather, were largely concerns about the reasonableness of the monitoring requirements, among other matters. The Environmental Defense Fund is said to have strongly criticized the extent

---

[2]This decision involved a challenge by EDF of several features of the interim regulations issued by the EPA on 10 Dec. 1975. By the time the decision was issued the agency had issued proposed revised regulations for organics, thus removing part of the basis of the complaint. As Ongerth suggests, the opinion in large part rests on interpretation of congressional intent with respect to the act. It also stresses, however, agency discretion in the "drawing of lines"--the choice of particular MCLs, sampling frequencies, and so forth.

of state agency input into this process as representing undue influence by the states. From the states, point of view, since they have the responsibility for enforcement of the regulations, this was only an appropriate input to the process. A major point to be made here is that in this situation there was an opportunity for <u>negotiations</u>, working sessions where differences were minimized. It is strongly urged that provision be made for such negotiations in the future, preceding the formal hearing process, and perhaps following, with appropriate recognition of the public and water supplier interest in the matter. There is important support for this concept in the House committee report, with the statement that the committee is hopeful that the states will take the lead in adopting standards (House Report No. 93-1185, p. 21).

Section 1449 of the act provides that any person may commence a civil action against any governmental agency which is alleged to be in violation of any requirement prescribed under the act. Therefore, a serious jurisdictional problem may confront the states in the event that in the administration of the act, the EPA imposes too detailed requirements upon the states. There seems to be no doubt that the states do not now have the resources to carry out the total requirements of the act. There are two inescapable ingredients: time and money. In the development of the EPA regulations under the act, CSSE successfully negotiated for time, particularly with relation to application to noncommunity systems.

Until better inventory information is available, no one knows how large this regulatory task is. The best estimate that has been made up to now is that the entire regulatory activity may cost about $80 million annually. The present federal input in the state program grants is about $20 million. Although the state and local governments are also funding current activities, the total is nowhere near $80 million.

To some extent the state regulatory agencies are faced with an impossible situation--strict standards on one hand and insufficient means to enforce the standards on the other. Comparison may be made with Public Law 92-500, the Federal Water Pollution Control Act. Here also, Congress has enacted strict water quality standards. This act provides for federal construction grants, with financing of up to 75 percent of local costs to meet the federal requirements. Many states add to this; for example, in California an additional 12.5 percent is added. Thus, local communities must pay only 12.5 percent of the cost of correction. In contrast, Public Law 93-523 has no provision for construction grants and only a minor provision for a loan-guarantee program, which was never funded, and the authority for which has now expired.

To recapitulate, the jurisdictional problem concerning the drinking water regulations involves the EPA's authority to adopt the regulations and state responsibility to enforce the regulations if the state accepts primacy. This jurisdictional problem can be minimized to the extent that the states are able to have an input into the standard-setting process and to the degree to which the regulations are reasonable. The subject is complicated, however, because the EPA is bound by strict and quite rigid statutory requirements. Some narrow latitude for application of reason in this issue exists with relation to judgments about <u>adverse health effects</u> and judgments about cost. It is imperative that the most reasonable judgment possible be made with a close state-federal cooperation in arriving at ultimate decisions.

The potential for many other jurisdiction problems between the EPA and the states is minimized because of the congressional focus on state primacy. In essence, Congress seems to have established each state as the first among equals in their dealings with the EPA. The committee report on this is quite explicit: "While federal standard setting and back-up enforcement is authorized, the Committee is hopeful that state and federal cooperation will be the rule and that the States will take the lead in adopting standards, reviewing compliance strategies, and where necessary bringing enforcement actions" (House Report No. 93-1185, p. 21). This establishment of state primacy provides an important counterbalance to the otherwise overpowering federal authority.

In order for a state to qualify for primacy, Section 1413(a) of the act provides that a state must (1) adopt water-quality regulations no less stringent than the federal regulations; (2) adopt and implement adequate procedures for enforcement of the state regulations, including monitoring and inspections; (3) keep records and make reports as required by the EPA; (4) issue variances and exemptions which meet the provisions of the act; and (5) adopt an adequate plan for provision of safe drinking water under emergency circumstances. The committee report states that it is the committee's intent that the states and public water systems take the primary responsibility for assuring the safety of the nation's drinking water supplies. The report also states that the committee intends the EPA to exercise utmost care in passing upon applications for primacy and to deny any such application only upon a clear failure of the state to meet the requirements.

Here again, there is opportunity for differences of opinion and for jurisdictional problems. The federal program should be careful but general, rather than overly detailed and strict.

The EPA regulations (EPA, 1976) require that each state program contain ten program elements: (1) administrative and program development; (2) surveillance and technical assistance; (3) plan review; (4) laboratory certification; (5) laboratory capability; (6) training and certification; (7) enforcement; (8) data management; (9) disease surveillance and investigation; and (10) public participation.

Some current indication of the nature of the federal approach is contained in recent EPA-state communications concerning program grants. For example, the EPA has written to

California commenting on the California draft grant application for fiscal year 1978. As an example of a reasonable requirement, they state:

> In FY78 we must conduct a thorough review of the California Water Supply supervision program priorities, goals, objectives, procedures, schedules and relationships to other programs to insure that the SDWA is carried out and available resources are used effectively and efficiently. This process will be sequential and will have two focal points: first, the development of a detailed program plan for FY78 which defines the review process and which contains very explicit objectives, schedules and outputs. Secondly, preparation of a management plan which is based upon the results of this review. I believe that you will agree with the need for such a review since the FY77 and 78 programs both contained a similar objective. This process is necessary to insure that program grant funds are expended for the accomplishment of mutually acceptable goals. The program plan, once approved by EPA, constitutes a mutual reference point when measuring the adequacy of program accomplishments during the midyear program review.

Further EPA comments on the California fiscal year 1978 grant application have dealt in considerable detail with each of the ten program elements. Examples of overly detailed requirements from these comments are as follows:

> Surveillance and Technical Assistance: The major task under this element is the conduct of sanitary surveys. The program plan should provide numerical objectives for sanitary surveys of large, small and non-community water supplies by quarter. The plan should indicate who will conduct these surveys, what criteria will be utilized, the nature of reports and follow-up actions and quality control activities. If one does not already exist, a sanitary survey procedures and criteria manual should be prepared utilizing the soon to be completed EPA Sanitary Survey Manual.

> The nature, timing and resource requirements of any technical assistance programs should be discussed. Particular emphasis should be placed upon technology transfer to the small water supply systems. This may take the form of mass distribution of water treatment plant design and operating criteria to consulting engineers or the water supplies, response to individual written and verbal assistance requests, and informal assistance as part of surveillance activities. Consideration should be given to the expansion of the existing publications distribution activities.

These comments indicate a degree of detailed overview of state programs which probably will lead to significant jurisdictional problems if pursued further by the EPA. Present indications are that the regional office may not aggressively pursue these details.

Prevention is a basic principle applicable to all public health programs. As applied to community water supply, this requires the selection of the best available water sources, the development of good water treatment facilities and good water quality-management practices, and the creation of competent operating organizations. Collectively these constitute a water supply enterprise with characteristics which assure reliable production and delivery of safe water. Public Law 93-523 is based on a premise that reliable delivery of a good water supply can be successfully accomplished by a program focused on enforcement of a performance standard. Were all public health efforts so directed, the ultimate goal would be long delayed. Thus, the states must split their efforts. First, to satisfy the immediate requirements of Public Law 93-523, and second, to have enough resources available to carry out the indispensable preventive program which focuses on correcting defects in each of the basic elements of a water utility enterprise. The heart of this jurisdictional problem is that the EPA may interfere with these efforts.

The sections on issuance of variances and exemptions also provide opportunity for jurisdictional problems. In particular, the act requires an EPA review of variances and exemptions issued. Further, the act provides that if the administrator finds that a state has in a substantial number of instances abused its discretion in granting exemptions, the administrator shall notify the state of its finding and, with procedural limitations, shall propose revocation of specific variances and exemptions or propose revised schedules or other requirements for specific public water systems. The committee report states that it contemplates that the EPA will carefully review variances and exemptions which are granted by states to assure that the state has not abused its discretion and has not failed to impose reasonable control measures. Further, it states:

> This system of EPA oversight is intended by the Committee to confer maximum responsibility on States which make appropriate efforts to effectuate the purposes of the Act. While some EPA review of State [actions] was deemed necessary by the Committee to assure the effectuation of the national policy, it is not intended that EPA engage in a case-by-case review or substitute its judgment for the well-exercised judgment of a State. EPA notice to a State is warranted only when a significant number of cases can be shown of State action inconsistent with the intent of this bill.

Once again, it is too early in the administration of the act to have sufficient experience with variances and exemptions to know the extent to which jurisdictional problems may arise. Here, too, though the potential exists for interagency problems, the expression of legislative intent will tend to minimize the development of such problems.

Section 1413 provides emergency powers to EPA: "The Administrator upon receipt of information that a contaminant which is present in or is likely to enter a public water system may present an imminent and substantial endangerment to the health of persons, and that appropriate State and local authorities have not acted to protect the health of such persons, may take such actions as he may deem necessary in order to protect the health of such persons. . . ." This section might be put to use under two sets of circumstances. One would be occurrence of an emergency caused by discharge of a potent toxicant upstream from a water supply intake or by some sort of natural disaster. In either case, a state might choose to seek application of the EPA emergency powers in order to adequately deal with the situation. In such circumstances, with the state and federal agencies working cooperatively, there would seem to be little likelihood of jurisdictional problems. The second set of circumstances would be when a charge is made by a citizen or organization that a state or local authority has not acted to protect the health of the public where the drinking water standards are not being met, for example, under rather ordinary circumstances when bacteriological standards are not being met--but no major defects are known to exist. It is possible that in such circumstances the state and local authorities might consider they had acted adequately to protect the community. A federal finding to the contrary would certainly give rise to a significant jurisdictional problem. Once again, the language of the committee report (U.S. Congress, 1974a, p. 35) provides some insights to congressional intent as follows:

> In using the words "that appropriate State or local authorities have not acted to protect the health of persons," the Committee intends to direct the Administrator to refrain from precipitous preemption of effective State or local emergency abatement efforts. However, if State or local efforts are not forthcoming in timely fashion or are not effective to prevent or treat the hazardous condition, this provision should not bar prompt enforcement by the Administrator.
>
> In using the words "imminent and substantial endangerment to the health of persons," the Committee intends that this broad administrative authority not be used when the system of regulatory authorities provided elsewhere in the bill could be used adequately to protect the public health. Nor is the emergency authority to be used in cases where the risk of harm is remote in time, completely speculative in nature, or <u>de minimis</u> in degree.

An important part of the Public Law 93-523 process is based on laboratory analysis of water. As a result, the EPA has laid down requirements concerning certification of laboratories and requirements for specific analytical procedures. Experience to date with these programs is that the EPA has paid too great attention to detail and has placed quite specific requirements controlling laboratory methodology, details of laboratory operation, and laboratory certification. In developing this material, the EPA has been careful to involve states in discussions, but the end product shows very little evidence of attention to the state recommendations. In addition to this major problem, a few overlaps in federal-state jurisdiction appear. For example, the EPA has indicated that it intends to carry on its own program of certification of laboratories within a state when EPA plans to utilize such laboratories for its own analytical work. Also, when a state has requirements which go beyond the guidelines set forth by the EPA, some federal facility laboratories have chosen to seek the EPA rather than state certification. Although there are no problems at this point in the development of the program, such options for federal laboratories can lead to confrontations should such federal facilities be in a state of noncompliance with the regulations. Another jurisdictional problem involves the interstate use of laboratories. The EPA has recognized this problem and has attempted to work out some system of uniform approval of interstate laboratories. This remains, however, a problem not completely resolved.

Another problem lies with the matter of data management. The EPA has developed a Model State Information System (MSIS) and has given the states the option of participating in this system or in developing their own data management system. To the extent the EPA requirements for data stick closely to the specific requirements of the act, there is not much problem.

There have been instances, however, where in the interest of having nationwide statistics on water supply, the EPA has sought additional information which states are being asked to work into their data banks. For example, states have been asked to provide information on capacity of water treatment facilities and other water production information. Another example is the request to the states to provide latitude and longitude data on each water supply source. Development of such information is exceedingly time-consuming and most certainly is not necessary to carry out the requirements of Public Law 93-523.

The process of piecemeal modification of the drinking water regulations presents a serious problem to the states. State attention has been drawn to this issue since the EPA has already considered though not carried through on three or four regulation modifications. The federal procedure for regulation promulgation is relatively simple. On the other hand, states have various administrative procedures which make it quite difficult for some states to adopt new regulations. Some states require legislative review. Others require action by a legislative committee and other reviews of proposed regulations. In Virginia, the Administrative Process Act requires the legislative committee be informed by December 1 for each year of proposed regulations. Any regulation costing a governmental entity over $1,000 must be sent to all governing bodies. The regulation does not become effective until the following February.

The remaining comments deal with state-local jurisdictional problems. These most certainly will differ from state to state. In some states the entire water program is handled by the state and no problems are anticipated. In other states, including California, New York, and others, well-organized and strong local health departments carry on effective water supply programs. In California local health departments have the authority for public health regulation of community water systems with fewer than two hundred service connections and over all noncommunity water supplies. Thus, the satisfactory conduct of a domestic water supply program in California requires cooperation from fifty-eight local health jurisdictions. Except for a relatively small state subvention to local health departments, these departments are independent of state control. On the other hand, the regional office of the EPA has taken the position that it will deal only with the state health department with relation to meeting the primacy requirements. This is as the state would have it, but creates the possibility for serious state-local problems. The basis for this problem, once again, is money. All California local health departments are carrying on locally financed water supply programs. The provisions of Public Law 93-523, particularly when the noncommunity water supply requirements become effective in 1979, call for greatly increased efforts on the part of local health department water supplies. The state is sharing program grant assistance with local health departments through personnel hired with program grant funds in an arrangement mutually agreed to but containing the seeds for disagreements over what share each party is to receive. The state, in order to meet the EPA primacy requirements, must look to local health departments to provide inventory information, conduct sanitary surveys, public notification, and to police monitoring and standards compliance requirements. Here also, it is too early in the administration of the act to know how local health departments will perform. It is not hard to imagine there may be some difficulties in obtaining the required efforts and information from local health departments. When the non-community water supply provisions become effective and the local departments do not have resources to carry out all requirements under the act, problems of deciding on work priorities will arise. There well may be disputes between a local agency and the state and between the state and EPA on the choices made on priorities. In addition, the state and local agencies may be put on the defensive for failures to enforce requirements, though resources to do so may be inadequate.

In conclusion, it is important that Public Law 93-523 calls for a cooperative federal-state-local water supply program. An overriding consideration is that Congress intended that the states and public utilities take the primary responsibility for assuring the safety of the nation's drinking water supplies. Many states have good-quality water supply programs administered by highly qualified, experienced personnel. In these states there can be little justification for minute examination and policing of state administration of this program. The EPA must use the very best judgment it can muster in meeting its obligation under the act, but should not unnecessarily overregulate the state agencies. The states will continue to stand for a reasonable, minimum federal overview limited to taking action called for within the explicit provisions of the statute.

One state sanitary engineer expresses thoughts which speak for the majority (Vogt, 1978):

> If there are to be any positive benefits from the Safe Drinking Water Act, the enforcement must be a cooperative national effort of federal, state and those local agencies which are suppliers of public water supply. If the Safe Drinking Water Act is to provide the hoped for protection of the public health, a reasonable and practical approach must be used in developing regulations, guidelines, strategies, etc. If there is understanding and cooperation between the states and Federal Government, no force will be necessary to enlist the support of the states in the efficient implementation of the Safe Drinking Water Act and attendant regulations. Full and continuing involvement in the national effort may be more difficult for some states than others due to problems of obtaining all necessary approvals for changes in legislation and administrative rules. States cannot be expected to operate a program if someone in Washington is continually changing regulations, guidelines or the basic thrust of the program; particularly where there is not good cause to support the change.
>
> A question is raised as to state and local cooperation in working on the Safe Drinking Water Act. We look at this potential problem in the same light as we do the relationship between state and federal agencies. Local health departments will cooperate with the state only if the state operates as a reasonable, responsible and cooperative partner. This is the only possible approach for either the state-local relationship or the state-federal

> relationship. States must be asked to participate more actively in developing regulations, guidelines and other portions of the program. In this way, a more realistic and practical program can be developed, and the states will be much more likely to participate actively in the enforcement and support of the program with all the attendant benefits to the health of our nation's citizens.
>
> States will be responsive to EPA and its requests if EPA is responsible and cooperative in developing the program, regulations, etc. All of the theory and speculation in the world is of no benefit if the resulting program is not practical and if it cannot be demonstrated to be beneficial. Understanding and genuine cooperation between all participants in the effort is the only means of obtaining increased protection to the public health. (From a letter to H. J. Ongerth.)

Congress has written strict requirements for controlling water quality. The question has been asked, does any detectable quality of a substance that is considered harmful under differing testing conditions unrelated to human exposure dictate the setting of maximum contaminant limits for water supply? It is clear that water contains minute quantities of many substances. Moreover, as laboratory techniques improve, even more constituents will be identified. Congress should take a very careful look at this and seriously consider giving the administrator greater latitude in applying judgment in determining what constituents must be monitored and controlled.

In addition to the above, two further recommendations are made: first, that some mechanism be established for informed and responsive decision making at the community level; second, that a real partnership be created between the EPA and the states in establishing standards, policies, and strategies in carrying out the provisions of the act.

References

Lowrance, W. W. 1976. Of Acceptable Risk (Science and the Determination of Safety) (Los Altos, Calif., William Kaufmann).

National Academy of Sciences. 1977. Drinking Water and Health, A Report of the Safe Drinking Water Committee, Advisory Center on Toxicology, Assembly of Life Sciences, National Research Council (Washington, D.C., NAS).

U.S. Congress. 1974a. Safe Drinking Water Act, House Interstate and Foreign Commerce Committee Reports HR Report 93-1185 (Washington, D.C., GPO).

______. 1974b. Safe Drinking Water Act, Public Law 93-523 (Washington, D.C., GPO).

U.S. Court of Appeals for the District of Columbia Circuit. 1978. No. 75-2224. Environmental Defense Fund, Inc. v. Douglas M. Costle, Administrator, Environmental Protection Agency (Petition for Review of an Order of the Environmental Protection Agency).

U.S. Environmental Protection Agency. 1975. "National Interim Primary Drinking Water Regulations," Federal Register vol. 40, no. 248 (December 24).

______. 1976. EPA Regulations: 40 CFR 35.600-630, January 20 (Washington, D.C., GPO) p. 2914.

Discussion by Leo M. Eisel

Mr. Ongerth does a very creditable job in detailing the various problems and potential problems with the Safe Drinking Water Act. He correctly states that one of the major problems with the act is that the U.S. Environmental Protection Agency (EPA) has the responsibility for imposing drinking-water regulations while the state and local regulatory agencies have the responsibility for enforcing these regulations. He further attributes many of the problems of the Safe Drinking Water Act to the fact that Congress was very specific in the act and gave the EPA only limited latitude in setting the regulations. Mr. Ongerth also points out that Congress provided little or no guidance as to what constitutes a reasonable cost in meeting the standards. Consequently, the EPA was left to its own resources concerning the establishment of a reasonable balance between cost and achieving a standard.

The Safe Drinking Water Act requires important decisions concerning the acceptability of risk. As we all know, the reduction of risk in drinking water, as well as in many other aspects of our lives, requires the expenditure of funds. The reduction of risk to zero or near zero may cost very substantial sums. It may indeed cost more than the public is willing to pay.

Prior to my present position as director of the U.S. Water Resources Council, I was for a period of time the director of the Illinois Environmental Protection Agency. In this capacity, I was very aware that Congress had dictated a set of very strict standards in the Safe Drinking Water Act but had made no funds available to local communities for purposes of carrying out this act. This is in contrast to the very sizable grants program made available by Congress to local communities for purposes of building sewage treatment plants. Needless to say, state and local governments react negatively to spending funds to meet standards dictated by the federal government when they feel they have had no voice in setting these standards.

I would like to disagree with Mr. Ongerth's point that refusing primacy is not a real option for the state and would claim instead that this option cannot be easily dismissed. In some states, such as Illinois, present standards are very close to those required by the federal government. For all practical purposes there will be little change in the quality of water supplied to the large majority of Illinois residents as a result of the Safe Drinking Water Act. Therefore, in situations such as this, refusal of primacy is a real option for the state.

The author correctly points out the many potential opportunities for jurisdictional problems. He notes the potential conflict between state and federal governments with regard to the detailed overview by the federal government of the state problems. He also correctly indicates the many unresolved problems involving lab certification and data management.

Finally, he indicates the potential jurisdictional problems between state and local governments in those states, such as California, where local districts of governments carry major responsibility for providing the safe drinking water. Mr. Ongerth also points out the potential problems resulting from piecemeal changes of the drinking-water regulations. He refers to the fact that many state governments require approval of rules and regulations by the state legislature. I believe that the problem here is really with the state, and that this perhaps should force various states into providing more latitude to their agencies in promulgating rules and regulations. In this modern day of complex rules and regulations involving many aspects of our lives, I believe it is anachronistic to expect the state legislature to approve every change in rules and regulations.

In general, Mr. Ongerth has done an excellent job in detailing many of the problems with the Safe Drinking Water Act. As I neared the end, I expected to find a section concerning proposed remedies to some of the problems. Unfortunately, this section is not present in the paper and its absence leaves one with a rather unsatisfied feeling.

In some ways, the Safe Drinking Water Act may be a product of the late 1960s and early 1970s when the emphasis was on strong federal standards. In the late 1970s we are experiencing a swing away from this approach, and this may be causing some of the problems with the act. Nevertheless, if there is not some emphasis on providing common federal standards, then the states can expect significant pressure from industry and other entities to reduce state regulations in order not to have stricter regulations than surrounding states.

Discussion by Timothy L. Harker

In the law, jurisdiction means the power to decide. In this sense, there are two levels of jurisdiction under the Safe Drinking Water Act, that is, federal and state. There are also two kinds of jurisdictional conflicts presented by the Safe Drinking Water Act. One, which Mr. Ongerth addresses, is the political conflict arising because the act reserves most of the decision-making power for the federal government while imposing most of the financial and administrative burdens on state and local governments. I wish to address a few comments to this jurisdictional conflict. My comments pertain to problems which the Congress created when it decreed that "he who must pay shall have no say." But first I would like to discuss the second kind of jurisdictional conflict, that is, statutory overlap potentially resulting in confusion in safe drinking water law and policy.

Statutory Overlap of the Safe Drinking Water Act with Other Federal Statutes

Potential jurisdictional conflicts exist among federal agencies with respect to drinking water. Problems may occur because of separate authority over drinking water under the

Safe Drinking Water Act, the federal Food, Drug, and Cosmetic Act, the federal Insecticide, Fungicide, and Rodenticide Act, and the Toxic Substances Control Act.

Drinking water under the Food, Drug, and Cosmetic Act is classified as a food, subject to standards for food additives and unavoidable contaminants if sold in interstate commerce. But it is more than a food. It is an environmental medium through which the public may be exposed to contaminants, and as such drinking water is subject to the maximum contaminant levels of the Safe Drinking Water Act.

Additional statutory authority over drinking water may be indirectly exercised. Certain chemicals used in drinking water treatment, such as chlorine, are pesticides within the scope of the Insecticide, Fungicide, and Rodenticide Act. Pesticides must be registered and pesticide use must comply with label restrictions established by the Environmental Protection Agency (EPA), which also has the authority to prohibit the use of a pesticide that results in unreasonable adverse effects on the environment. The EPA may also establish tolerances for pesticide chemicals which may be added to raw water. Clearly, the use of pesticide chemicals in the production and treatment of drinking water may be regulated under this act.

Finally, indirect statutory authority over drinking water _may_ be achieved under the Toxic Substances Control Act. While this act exempts food additives, at least one theory holds that certain chemicals in drinking water are not "food additives" but rather, are chemicals used in the production of potable water. Pursuant to such an interpretation, existing and future chemicals added to drinking water may be subject to testing requirements and regulation of use under the Toxic Substances Control Act.

The statutory overlay with respect to drinking water is important for all forms of drinking water, that is, municipal water supplied by public water systems, bottled water, and drinking water used in the processing or packaging of other foods. While it may be true that the Food and Drug Administration (FDA) has not yet exercised its authority over public water supplies, an important distinction exists between having the authority to regulate (which FDA has) and choosing to exercise that authority. Regulatory priorities change from one administration to the next and may even change more frequently in response to public attention or the pressure of public or special interest groups. Reliable long-range business and public policy planning cannot be made on the basis of an assumption that because FDA has not yet exercised substantial authority over public drinking water, it will not do so in the future.

With respect to the other forms of drinking water--bottled water and drinking water used in the packaging or processing of other foods--the uncertainty caused by the statutory overlay is even more significant than that for municipal water supplies.

For example, a food under the Food, Drug, and Cosmetic Act is subject to the provisions of the Delaney Clause and to the FDA's aggressive posture toward the addition of carcinogenic substances to the food supply. Bottled water is a food, subject to the Food, Drug, and Cosmetic Act. It is not clear whether the FDA would view the <u>bottling</u> of municipal water which contains trihalomethanes within the EPA's standards (proposed at 100 parts per billion) as prohibited under the Delaney Clause. Such an interpretation would appear to be legally impermissible, but confusion will persist in the industry until Congress or the FDA provides a clear direction.

Even if the Delaney Clause is inapplicable, the potential for more stringent standards for bottled water than for municipal water creates an anomaly, that is, what is lawful when it comes out of the tap may become unlawful if put in a bottle. This may result in public confusion and clearly injects unnecessary uncertainty into business planning.

Another example of the potential conflict caused by overlapping jurisdiction pertains to drinking water used in food packaging or processing. The FDA has proposed to ban the addition of chloroform to food under section 409 of the Food, Drug, and Cosmetic Act. Yet the EPA will apparently permit 100 parts per billion of chloroform in drinking water. It is uncertain at this time whether the addition of such drinking water to other foods for processing or packaging would be construed by the FDA to constitute a violation of the Delaney

Clause. Again, the issue of Delaney aside, even the potential for requiring significantly tighter standards for a food because drinking water (itself a food) is added to it is anomalous and creates unnecessary uncertainty.

Similar problems of jurisdictional overlap exist with respect to food packaging, including water conduits. Drinking water at the point of treatment, which is in compliance with EPA standards under the Safe Drinking Water Act may, by the time it reaches the consumer, have picked up additional contaminants from metal cans used for food packaging, plastic containers used for water, or water conduits, including asbestos-containing pipe. If, for example, drinking water standards for such materials as lead, chloroform, or asbestos are established without taking account of the possible additional contribution of such contaminants from the drinking water package (that is, container, can, or pipe), then drinking water which complies with the EPA maximum contaminant levels (MCLs) at the point of treatment may exceed those standards by the time it reaches the consumer. In such a case, drinking water as a "food" may be in violation of the FDCA even though it complied with EPA standards at the municipal treatment plant.[3]

An even greater conflict could exist in the case of such leaching or migration. The EPA may set MCLs (in excess of zero) for carcinogenic substances in the municipal water supply

---

[3]Ed. note: The EPA standards, except for turbidity, are to apply at the tap, not at the treatment plant. This example, therefore, does not really illustrate the point at issue.

under the Safe Drinking Water Act. The FDA, on the other hand, may view the potential migration of minute quantities (even if unmeasurable) of such compounds from the "package" into drinking water as indirect additives, prohibited by the Delaney Clause. Again, drinking water "safely" consumed under the Safe Drinking Water Act, if consumed as a "food" under the Food, Drug, and Cosmetic Act, could become unlawful as well as unsafe.

Arguably there is a potential value to jurisdictional overlap. It can fill any gaps which may exist in the regulatory authority to protect public health. The EPA may not have all the authority under the Safe Drinking Water Act which it needs in order to deal with a particular problem. An example of such a gap may be the EPA's authority under the act enabling it to _directly_ prohibit or restrict the amount of certain water-treatment chemicals rather than setting an MCL for a particular chemical. Should the courts hold that the EPA lacks the authority under the act directly or prohibit the use of treatment chemicals, the FDA under the Food, Drug, and Cosmetic Act could regulate the addition of such chemicals to drinking water, even in municipal water supplies, on the theory that they constitute food additives.

Indeed, it appears, at least in part, that a concern for such regulatory gaps is causing the difficulty between the EPA and FDA in resolving their jurisdictional differences over drinking water. (See Commissioner Kennedy's letter to EPA Administrator Costle, dated October 4, 1977.)

However, the continuation of jurisdictional conflicts over drinking water poses potentially significant problems for public water supplies, the bottled water industry and those who utilize drinking water in food processing or packaging. Before allowing the confusion to continue, Congress should consider at least the following costs:

1. With conflicting public policy comes public confusion over what is hazardous and what is safe. The efforts of the Interagency Regulatory Liaison Group to achieve greater uniformity in the regulation of toxic materials is implicit recognition of this fact.

2. Potential unfair discrimination may exist between the public and private sectors insofar as each sells water; that is, public drinking water before it leaves the tap may, in fact, be subject to the Safe Drinking Water Act only; whereas this same drinking water added to food, or used in food processing, or sold in a bottle may face <u>additional</u> regulatory hurdles of the FDA.

3. Research and development may be somewhat impaired due to confusion over what government standards must be met in order to introduce and use a new water treatment chemical or to add water to food or to use water in the packaging of food.

There appears to be no compelling scientific or public interest reason for treating water as a "food" differently from drinking water which is directly consumed from the tap.

<u>Jurisdictional Conflicts Between Federal and Local Governments</u>

The point has been well made by Mr. Ongerth and others that Congress left state and local governments little substantive decision-making authority over safe drinking water. Putting aside the issue of whether such a decision is wise from a public policy or constitutional standpoint, I believe the

Congressional decision to attempt to force the cooperation of local governments has important ramifications for the success of a national safe drinking water program.

With the exception of cigarettes and possibly saccharin, the federal government has consistently taken the position that the consumer is incapable of making an informed judgment to accept the risks of exposure to carcinogenic substances. Such federal preemption of the individual's decision-making authority appears to rest principally on grounds that the weighing of risks and benefits is too complex a task to be trusted to ordinary citizens. (There is, in addition, the question of involuntary exposure to carcinogenic substances, such as might occur through air pollution, which should not be included in the category of risks as to which there is legitimate argument over consumer choice.)

A similar preemption of consumer choice has occurred with respect to the MCL proposed for trihalomethanes and the treatment techniques proposed for synthetic organics under the Safe Drinking Water Act. However, one important additional step has been taken. The federal government has not only deprived the consumer of any choice in the matter of what the MCL or method of treatment should be, if any; it has ordered the consumer to spend his <u>own</u> money to achieve the standards which Congress and the EPA have set. There does not appear to be any precedent for federal environmental action in which consumers, on the one hand, are deprived of any direct decision-making authority in setting restrictions on carcinogenic risk, while on the other hand these same consumers are being ordered to pay the clean-up costs and, finally, are given no financial assistance to do so.

This discussion is preliminary to my principal point, that is, with standards adopted by the federal government there is no appreciable opportunity to assess society's willingness to accept various levels of risk. Unfortunately, the success of a national safe drinking water program which individual consumers are being asked to underwrite is in large part dependent upon achieving public understanding and acceptance. The Safe Drinking Water Act is geared in the worst way to achieve these goals. While sophisticated concepts of carcinogenic risks and economic benefit are required for individual decision making and for acceptance of additional costs of treating drinking water, the very levels of government which most readily elicit public participation, thereby achieving public education and acceptance, have been deprived of decision-making authority. In effect, state and local governments are left the task of convincing the public that higher costs for drinking water are warranted in order to achieve controversial standards in the development of which they played no substantive role. Clearly, there is little incentive for them to do so.

The Congress has thereby largely ignored the opportunity to obtain public acceptance by encouraging public participation in decision making. Education as the first form of regulation has been rejected.

It seems that for the foreseeable future the opportunity is lost to alter the basic jurisdictional structure of the Safe Drinking Water Act. For purposes of congressional oversight hearings, potential amendments could be considered which might mitigate certain of the act's jurisdictional shortcomings. In this respect, modification of the act's variance or exemption provisions is a practical goal.

Currently, a variance is permissible from a prescribed treatment technique only if the raw water source for the drinking water is such that the treatment technique is unnecessary. A variance from an MCL is permissible only if the raw water source is such that despite application of the best treatment technology available, the MCL cannot be met. An exemption from compliance with an MCL or a treatment technique is permissible only if for "compelling reasons" the public water supply cannot comply. In either case, a variance or an exemption provides only temporary relief.

It would seem feasible to grant temporary variances (or permanent exemptions) from prescribed treatment techniques or specific MCLs if a water supply could demonstrate viable alternative methods of achieving the required result. Thus, for example, a community faced with a high nitrate level could propose to publicize the hazard of methemoglobanemia to pregnant women and to deliver nitrate-free bottled water to such at-risk persons, in lieu of removing nitrate from the entire municipal water supply.[4] Similar opportunities would seem to be available to utilize alternative sources of water during temporary periods of high agricultural runoff, or during periods of drought when the absence of dilution may concentrate contaminants.

More flexible standards for variance from or exception to the Safe Drinking Water Act should also take account of the limited resources of local public health departments. State and local governments are perhaps in the best position to determine priorities for spending limited public health dollars. In certain cases, cost-benefit analysis may indicate convincingly that public health is better served not by expending available resources on additional water-treatment technology which may provide only marginal improvement, but rather by spending available funds on some other health measure.

Given the fact that public health and drinking water are traditional concerns of state and local governments, and given the fact that the federal government so far as drinking water is concerned has chosen not to subsidize the financial capabilities of local health departments, such flexibility would seem appropriate.

Finally, beside adding flexibility to the variance and exemption provisions of the act, Congress and the EPA, by

---

[4]Ed. note: This is currently done in parts of the United Kingdom. See Chapter 6 by Atkinson in this volume.

acting to increase the program's credibility, may help to achieve greater public acceptance (and therefore greater effectiveness) of the national safe drinking water program.

In certain instances compliance problems and costs of a water supply will undoubtedly be directly related to upstream point and non-point sources of pollution. The EPA, by undertaking an aggressive regulatory program against such sources could demonstrate its commitment to a basic economic premise, that is, that social costs should be internalized, not borne by others (the downstream users). In this regard Congress may consider amending the Safe Drinking Water Act to require that the EPA pursue administrative and judicial remedies against any upstream point or non-point source found to be causing or substantially contributing to violation of the Safe Drinking Water Act by a downstream public water supply.

A second area in which the credibility of the national safe drinking water program may be enhanced concerns the extent of coverage for the proposed trihalomethane MCL and the activated carbon treatment technique. As proposed, the EPA would initially (and for up to three to six years) exempt between one-third and one-half of the population from these standards. The EPA has stated that the proposed grounds for such an exemption are (1) economic difficulties of communities under 75,000 population in achieving immediate compliance, and (2) the belief that generally the raw water sources of communities under 75,000 population are neither in violation of the proposed trihalomethane standard nor subject to synthetic organics. It may be that the EPA, by proposing to exempt for such a long period such a large percentage of the population, has undermined the credibility of its already controversial regulatory effort. Those who are immediately subject to the regulation will hardly be convinced of the pressing need for them to undertake additional treatment costs, if millions of others similarly situated are left at-risk, at least for several more years. Similarly, those in the population who would remain at-risk, but who are convinced of the scientific validity of the EPA's position on cancer-causing agents, may with some reason accuse EPA of acting arbitrarily in ignoring them.

In fact, there may be a strong correlation between community population under 75,000 and economic inability to finance necessary additional controls for trihalomethanes and synthetic organics or the lack of necessity for so doing. But clearly the EPA has the burden of demonstrating such correlations.

It would seem that the exceptions may be so numerous as to negate the rule. Furthermore, if the economic correlation exists, it is not clear why the EPA considers economic conditions to be such that a grace period of several years in meeting the standard would be significant. Similarly, if there is a correlation between community size of under 75,000 population and raw water sources which are relatively free of trihalomethanes and synthetic organics, it is not clear why the EPA intends eventually to include such communities under the regulation.

In either event, the EPA must undertake open public discussion of any data underlying its proposal to exempt millions of persons from these controversial standards.

Discussion by Leo E. Ganster

I welcome this opportunity to represent the U.S. General Accounting Office (GAO) at this national conference on drinking water. This setting of distinguished experts gives the GAO the good fortune and opportunity to become acquainted with your views and concerns regarding safe drinking water, and will add an important perspective to our work, as we embark on reviews on the new program. I will discuss some of the emerging issues we have identified in work on the Environmental Protection Agency's (EPA) implementation of the Safe Drinking Water Act.

Very briefly, I would like to start with a bit of background about the GAO. The GAO is in the legislative branch of the federal government and reports to the Congress of the United States. Our primary responsibility involves reviewing how federal agencies implement and carry out legislative programs. By federal agency standards, the GAO is a relatively small agency, having a staff of about 5,000. About 3,000 of our staff are auditors, made up primarily of accountants, engineers, economists, and mathematicians. We also have chemists and public health doctors on our technical staff.

At the EPA we have a staff of twenty-two auditors and analysts assigned on a permanent basis. Over the past three years, we issued about forty reports to the Congress on various aspects of the EPA's programs.

Before I begin discussing our current work, it might be helpful to go over the purpose and thoughts behind the Safe Drinking Water Act. Basically, the purpose of the act is to assure that public water supply systems meet national standards that are intended to protect the public health. The Congress passed this legislation because of increasing outbreaks of disease and poisonings that could be attributed to drinking water. In addition, tests showed that the drinking water provided by public water systems had increased amounts of organic and inorganic chemicals in them; many of these chemicals were not being removed during the treatment process.

The GAO's report to the Congress, on November 15, 1973, entitled "Improved Federal and State Programs Needed to Insure the Purity and Safety of Drinking Water in the United States," highlighted this and other problems. We pointed out that the water supply programs in six states were delivering potentially dangerous water to some consumers, particularly in communities of 5,000 or less; that the federal-state administration of the interstate carrier water-supply program was sometimes inadequate; that the federal government did not monitor the quality of water provided to travelers at interstate highway rest areas; and that there was no effective program for ensuring that bottled water was pure, safe, and free of potential health hazards.

The Congress concluded that the existing legislation was inadequate to assure that America's public water supplies were safe to drink. Thus, the Safe Drinking Water Act of 1974 was enacted, marking the first time this country made a national commitment to protect drinking water.

Last year we began our review efforts to evaluate the effectiveness of the new act. This review is still underway and we expect to issue a report to the Congress sometime this summer. Because the Safe Drinking Water Act gave major responsibility to the states to design and operate a drinking water program (the EPA calls this "primacy"), we have directed our first efforts at determining how successful the states have been in assuming primacy; and the implications to the EPA if states did not assume primacy. We addressed these questions: (1) why some states are not assuming primacy; (2) what problems face states which have assumed or are seeking primacy; (3) whether the EPA can effectively carry out enforcement responsibilities in those states that do not assume primacy.

We have been obtaining information at the EPA headquarters in Washington, D.C., and at its regional offices in Boston, Chicago, Philadelphia, Kansas City, Seattle, and San Francisco, and have also visited federal, state, and local agencies in nine states--Pennsylvania, Vermont, Indiana, Connecticut, Oregon, California, Washington, Missouri, and Arizona.

Here are the three major issues noted so far:

1. Not all states are going to assume primacy. Currently, eight states have decided not to seek primacy. More may follow.
2. Many states are encountering problems which hinder their efforts to assume primacy.
3. The EPA will not operate a monitoring and enforcement program that is as stringent as one required by those states that assume primacy.

Let us take a look at each of them.

Some States Will Not Assume Primacy

Eight states have already decided not to assume primacy, and it is possible that more states might follow suit. The EPA is required by the act to operate a monitoring and enforcement program in those states that do not participate.

The act authorized a joint federal-state program of public water-system supervision in which the EPA is responsible for establishing regulations and national drinking water standards which, in turn, are to be adopted and enforced by the states. The states, therefore, were to be the keystone of the program. Although state participation is voluntary, the Congress considered such participation as critical to the successful implementation of the program. The states have until October 1979 to assume primacy.

As of February 1978, the scorecard looks like this. Twenty states and territories had assumed primacy.[5] Another twenty-eight states and territories have not met the primacy requirements but are working toward primacy. The remaining eight states have advised the EPA that they will not assume primacy.

Why have eight states rejected primacy? Well, there are a number of reasons. The primary reason, of course, is a lack of state resources. Some states do not have the funds or manpower to operate a primacy program. Pennsylvania is the best example of this. Some states do not believe that grant funds provided by the EPA under the act are enough to cover the costs of carrying out the federal requirements. Other reasons are that states (1) do not have the necessary legal authority; (2) do not agree with the requirement that states enforce federally set drinking water standards without receiving corresponding financial support; (3) distrust the federal government because of their previous experience, that is, the government imposes a program on the states and provides funds to get the state started, but then discontinues funding the program, leaving the state holding the bag; and (4) do not believe a problem exists, that is, they believe that their drinking water program is already satisfactory for protecting the public health.

---

[5]Ed. note: For purposes of the act, six territories, including the District of Columbia, are considered as "states."

The States' Problems in Assuming Primacy

Some of the twenty-eight states currently seeking primacy are having difficulty obtaining the legislation needed to enforce the federally mandated program. In one state, the legislature has not passed the necessary statutory authority because it feels that the state will not be able to afford to meet the federal requirements anyway. The problem centers around the fact that 87 percent of this state's public water treatment plants need some type of improvement. The state estimates that it will cost $100 million to improve the systems to comply with the requirements of the Safe Drinking Water Act in order to have safe and reliable public water systems. The act does not provide any financial assistance for capital improvements to water systems. Although this state has a small grant program, state health officials believe it will take a minimum of fifteen years to update the deficient systems using the limited grants they have available. Officials in five of the states we visited cited this lack of capital improvement assistance as one of the major shortcomings of the act.

Another problem we noted that could affect the states' incentive to work toward primacy is that adopting the implementing regulations can be a lengthy and time-consuming process. In one state, for example, the legislature enacted the statutory authority in April 1977. Writing the implementing regulations and obtaining the approval of the state board of health then took seven months. The next step is the approval of a state legislative review committee which did not meet until January 1978. So it could take almost a year to get the regulations approved and begin statewide implementation.

The EPA's Enforcement and Monitoring Program

Our current impression is that, in states which refuse to seek or do not get primacy, the EPA's program will not be fully effective. The act assumes that the programs the EPA established in nonprimacy states would meet the same requirements as those that were set for states carrying out their own programs. This has not been the case in two key areas--sanitary surveys and construction plan reviews.

The Congress intended that the EPA's role under the act was to establish policy guidance and regulations, provide technical and financial assistance to the states, conduct research, monitor state programs, and carry out primacy enforcement responsibility on Indian lands. Additionally, it was expected that the EPA's role would be expanded considerably in states which did not meet the primacy requirements, and the EPA would be responsible for the program.

Under general guidance from the EPA headquarters, regional offices developed the primacy programs in the eight nonparticipating states. These offices for the most part established minimal programs. Certain activities which are required of participating states will either not be performed by the EPA or will not be performed as frequently.

For example, one requirement which has not been adequately addressed is sanitary surveys. A sanitary survey is an on-site inspection of a water system for the purpose of evaluating the adequacy of the water source, facilities, equipment, operation, and maintenance for producing and distributing safe drinking water. Sanitary surveys are the one activity that could have the greatest impact on improving the overall quality of drinking water. A comprehensive sanitary survey of a public water system by a qualified person is essential to locate and identify health hazards which might exist in a system. Sanitary surveys should be conducted annually for all public water systems. Most states we visited have established programs to routinely conduct the annual surveys. The EPA, however, intends to conduct sanitary surveys only when violations occur and then only on a priority basis, or when public water systems request exemptions from the regulations or variances from drinking water standards.

A second key element to be addressed is a requirement for review of planning and construction of water systems. Planning and construction reviews involve the review of plans and specifications and inspection of water systems during and after construction in order to certify that they were built according to plans. The objective of the review is to assure that the new or modified system will be capable of producing water which will meet the requirements of the national standards and regulations. The review also includes an approval of the sources of supply. From a long-range standpoint, the plan review program is the key to improving the quality of drinking water delivered by public water systems. An effective planning and construction review program could avoid future problems because new water systems would be properly designed and constructed initially. The EPA's regions, however, are either not going to conduct planning and construction reviews at all or will require only that the construction engineer certify that the system was designed and constructed in accordance with state regulations.

The reason the EPA will not be able to operate an effective monitoring and enforcement program is that old bugaboo--a lack of resources. Frankly, the EPA's regions do not have the staff to operate a program on the same level as a state. In Pennsylvania, for example, state officials estimated that 137 persons were needed to implement a primacy program. The EPA estimated that from twelve to forty persons were needed to operate an effective program in each nonprimacy state, depending on the number of water systems. The EPA's staff for operating Pennsylvania's program totals only ten people. This lack of staff is typical throughout the EPA's regions. Its problems might be lessened to some extent if the Congress looks kindly on the EPA's budget request for fiscal year 1979, for the agency is asking for a substantial increase in resources for the drinking water program.

In conclusion, I want to remind you that the issues we identified are tentative and subject to change, as we obtain more information. However, it does appear that certain legislative and programmatic changes will be needed if the safe drinking water program is to be effective.

Chapter 13

LOCAL GOVERNMENTS AND THE SAFE DRINKING WATER ACT

Tom Moody

As the mayor of Columbus, Ohio, and president of the National League of Cities, an organization representing some 15,000 cities, it is my pleasure to be here today to offer you a perspective on how municipal officials view their responsibilities in providing safe drinking water to their citizens, especially, of course, under the requirements of the Safe Drinking Water Act. But first, let me review the status of our nations' cities and where they have been over the last several years.

The economic recession and inflation over the past few years has been devastating for local governments. Costs continue to rise, and in the past year, inflation has added a 6 percent burden to the cost of every product or service a city provides. America's cities could only afford to increase their budgets 5 percent, and that 1 percent difference between budget increases and the rate of inflation inevitably represents a reduction in services.

Revenues are lagging and some cities have been forced to raise their property tax rates nearly 25 percent to meet their fiscal needs. And of course, while jobs are declining and unemployment continues to be high in many central cities, the tax burdens on our citizens become even more difficult.

The federal government, on the other hand, has continued to increase the number of its programs affecting cities. Today there are some 1,200 of them. A patchwork of federal policies governing housing, transportation, the environment, economics, and taxation continue to increase the financial burden on local budgets. Now, while cities of this country have been at the center of business, cultural, and political life for more than three centuries, and their vitality has been the standard by which much of the progress of our nation has been measured, it is important also to note that much of the inadvertent urban policy created by conflicting federal programs has tended to expend rather than conserve our urban centers.

The Safe Drinking Water Act was enacted at a time when environmental quality of our cities was an extremely important and urgent issue. The dual need to protect and improve the total environment in the midst of an economic recession is a difficult task for local elected officials, and because the quality of our drinking water supplies is linked so closely to our health, welfare, and safety, city officials recognize it is the responsibility of all levels of government to act positively to protect and improve it.

Providing safe drinking water has long been a local government service and responsibility for more than 15,000 public water systems. With the advancement of new technology and research, the Safe Drinking Water Act has now brought federal standards into the picture. It is the consensus of the nation's cities that where federal standards are established, the federal government must give local governments adequate capacity and time to achieve those standards. Local governments must have flexibility to select among alternate means to achieve federal and state mandates and be able to weigh the economic, social, energy, and environmental costs and benefits of alternate strategies with locally determined plans and priorities.

Since the 1974 act contains no financial assistance for cities, many local officials are becoming increasingly uneasy as compliance costs are beginning to rise far above the EPA's original estimates. The increased number of samples and tests to be taken will greatly increase personnel costs, additional equipment will need to be purchased, and major construction of treatment facilities to meet the strict turbidity levels and organic contaminant levels will be required.

The EPA has estimated that the nationwide capital cost to cities serving populations over 100,000 will be approximately $135.6 million for treatment of turbidity. A recent survey of just thirty-one cities over 100,000 population, conducted by the city of Los Angeles, indicates that a total capital cost of $257 million will be needed to comply with the turbidity standards. This cost alone exceeds the EPA's estimate by nearly 90 percent.

Many of the cities responding will need water rate increases from 10 percent to 100 percent. The EPA has stated that "most public drinking water systems, particularly the larger ones already meet the standards," and that the "new program is therefore a form of preventive medicine."

With the many economic, social, and physical ills facing our cities today, I ask, is this really a cost-effective way to meet a national priority? Although the regulations contain variance and exemption procedures for those water supplies that are already safe for the public, the act still requires the spending of substantial tax dollars for marginal improvements in water quality in many cases.

Very few elected officials come into office by promising clean drinking water. That is expected of us. But many of us might leave office soon after we enclose the first noncompliance notification in our water bills to the public. I should stress that local government officials lobbied actively for the passage of the Safe Drinking Water Act, but we are faced with three major problems: First, there is a limit to what the public will accept as reasonable costs for safe drinking water. Those attending this conference, perhaps better than anyone, know that citizens of this country have received generally high-quality water at fairly minimal costs. However, improvements to drinking water facilities must be financed--and many communities are no longer in a political or financial position to continue to issue revenue bonds to support these capital improvements. Citizen reactions to increased utility rates, including water, are becoming increasingly angry, thus making the job of providing clean water much more difficult.

Second, there is, as I mentioned earlier, the problem of cost-effectiveness. The problem confronts not only the drinking water program but all environmental efforts. In Texas, for example, I understand that the state was proposing that local governments go beyond secondary treatment of sewage wastes to a 98 percent removal standard. Cities in that state were able to bite the bullet and convince their citizens of the need for treatment beyond what is required in the federal Water Pollution Control Act. Elected officials made tough decisions and arrived at a 96 percent removal rate. But removing another 2 percent of pollution requires tremendous investments of public funds which may not be worth the cost. The same problem applies in air quality or the drinking water program: How much pollution is tolerable from a public health standpoint, and where, on the graph, do the costs of clean-up outweigh the benefits to the public?

Finally, we must consider the impact of other environmental standards and goals on the particular pollution problem we are addressing. When I spoke earlier about an inadvertent federal urban policy, I meant that to apply to the environmental field as well. We all know the classic problem:

1. We can't burn solid waste because it violates air quality standards. We can't landfill it in many

cases because it ends up in the groundwater supplies; open dumping is prohibited under the Resource Conservation and Recovery Act; and resource recovery itself is often prohibitively expensive.

2. We clean up point-source pollutants from our water produced by both industries and municipalities, and suddenly find that our waters continue to be dirty because of a lack of federal commitment or program expertise in the non-point area.

3. To promote air quality we limit the size of parking facilities for a surburban shopping mall, which creates even greater dispersion of shopping facilities and greater energy waste in the roads built and gasoline used to reach them.

We all recognize that there are no easy answers to these questions. The National League of Cities is grappling with these issues on a day-to-day basis in the Congress, and in ongoing policy-analysis activities, and is trying to promote the notion of urban conservation as a reasonable analytical framework for looking at the problems of our urban areas. The concept of "throwaway cities" must come to an end.

We elected officials are committed to providing safe and healthy communities and to ensuring a safe supply of drinking water. But because of the increase in financial pressures from the vast array of federal programs and the increase in providing municipal services that comply with these programs, cities simply need assistance in their struggle to keep their communities viable and living environments.

Part IV

Panel Discussion of Policy Alternatives

Chapter 14

Panel Discussion

<u>ALLEN HAMMOND</u>: I am going to share what I hope will be a fair, wide-open, free-for-all here this afternoon. We have heard two and a half days of discussion on what the problem is with regard to drinking water quality, and this afternoon we have a panel of experts to address the question, What do we do next? What direction should drinking water policy take at this juncture? First, I would like to introduce the panel to you. First, N. J. Davoust, retired from the Chicago Water Department; second, Robert Harris, staff scientist with the Environmental Defense Fund; third, Peter Hutt, formerly general counsel of the Federal Drug Administration and now with Covington and Burling; fourth, Victor Kimm, director of the Office of Water Supply, of the U.S. Environmental Protection Agency; and, finally, F. Pierce Linaweaver, a private consultant and former director of public works for the City of Baltimore.

<u>N. J. DAVOUST</u>: Let me elaborate a bit on my background to establish the basis from which I speak. I retired last June from my position as the deputy commissioner for the City of Chicago. My entire thirty-eight-year career with the Chicago Water Department was in the field of water treatment--the earlier days were spent under the reknowned John R. Baylis, probably the most intensive investigator who ever worked in

the field of water-treatment. Besides being an almost tireless researcher, Baylis was also an advocate of the highest possible water quality. He has been quoted as saying that if one dollar would provide the necessary chemical treatment to furnish safe water, spend another dollar and make it as palatable and acceptable as possible for the consumers.

"Acceptable" was a word that I was listening for throughout this conference, and several speakers used it. If you look at it from the consumer's point of view, there are a lot of "acceptables" to be considered--for example, acceptable costs, acceptable quality, and so forth. But let us go on to the proposition that it is up to the management of the water utility to make most of the decisions regarding acceptability parameters.

Since I happen to be the first speaker, I have the choice of naming the policy options as I see them. I will list eight options, attaching no priorities, and hope that the following discussion will provide elaboration. I am not an advocate for any one of the options at this time.

First, there will always be the "opposition" point of view, or the option that says the entire concept of quality regulation at the federal level is bad. The active opposition can resist taking action of any kind, delay in the courts, and generally stall, hoping that some form of mid-course corrections to the Safe Drinking Water Act might provide relief.

The second option is one that some will say is traditional for the waterworks industry--the "do-nothing" option. While this might relate to the opposition concept, I see it as being more passive. I will say more later about some of the advantages of this option. Third, and one of the more acceptable options that has already been used by some water utilities, is the revision of the treatment processes. Moving the point of chlorine application has helped reduce the formation of trihalomethanes (THMs) in some of the places it has been tried. Going to combined chlorine by the use of ammonia has been shown to prevent the formation of the THMs--although the proposed regulations want to "outlaw" this treatment. There are other possible revisions, such as the use of chlorine dioxide, or possibly ozone, which should be field-tested on problem waters.

Another option would be to change the source of the system's raw water. Generally, this would be a switch from a surface water to a groundwater source. This, of course is not as simple as it may sound, and there can be many problems.

You could, as a fifth option, consider the concept of dual- or even triple-distribution systems, wherein only a small percentage of the water would receive high-level treatment for potable use. It is easy to say, "Yes, if we were building a new town, we would consider a multiple distribution system," but it is actually a complex subject. If we consider dual distribution, why not multiple waste collection? We would end up with streets full of pipes at tremendous costs, which is not too practical.

Next would be the "under-the-sink" option, the so-called home treatment devices. More on these later.

Another option would be the furnishing of bottled water, possibly in addition to centralized treatment points. I will discuss these choices as they were used last year in Duluth, Minnesota.

Finally, as the eighth option, there is the EPA's treatment recommendation for granular activated carbon (GAC) for controlling the THM and synthetic organics. I have heard many pros and cons on this option during this conference, and I hope that further comment will be generated by this panel.

Let me go back over some of these options. In the process of retiring to northern Minnesota, I went through Duluth frequently at the height of the asbestos controversy. This was, and is, the site of the celebrated Reserve Mining case, wherein the taconite tailings being dumped into Lake Superior are considered to be the source of potentially dangerous asbestos fibers in the drinking water for Duluth and other towns in the area. Litigation was likely to occur and to be well publicized. Judge Miles Lord became very impatient with all parties--the state, the EPA, and the Reserve Mining Company. The lengthy trial aggravated him to the point where he ordered the federal government to provide an alternate supply of drinking water for the affected population. The job wound up in the lap of the Corps of Engineers, and they proceeded to provide alternate water supplies in a short time. They retained a contractor to put up water in half-gallon cartons--the kind normally used for milk. It was a very extensive program; water was dispensed through supermarkets and other business establishments. In addition, there were some twenty-five filtration systems installed in firehouses and other public buildings where consumers could fill their own containers.

In the 403 days that the program was active, the contractor put out 4,733,811 half-gallon cartons of water. While the program was pretty well accepted, the allocation of this amount of water to the estimated 150,000 people involved implies that less than 15 percent were averaging about one quart of water per day.[1] No figures were available on the amount of water dispensed at the centralized filtration points since meters were not installed. The cost of the water program was $508,000 or about 11¢ per half-gallon. Ultimately this cost was passed on to the Reserve Mining Company. It was an emotional issue that managed to flare up in direct relationship to the TV and newspaper publicity about the trial, to unusual turbidity due to lake storms, and so forth. Surprisingly, the older people in the area proved to be the most concerned, despite the EPA's advice that young children were the prime concern. In retrospect, the consumers were relatively blasé about the whole thing.

---

[1]Ed. note: Baumann and Dworkin in chapter 10 discuss other evidence of consumer response to the Duluth water problem.

In the meantime, back at City Hall, the council could not determine a course of action. The selection of a consultant to do a pilot study in conjunction with the EPA was a controversial subject that never did come out of the council. Once again, at the behest of the court, the Corps of Engineers was asked to select the consultant, and the project went ahead with little or no work on the part of the city of Duluth. The study for the pilot plant was completed and work on the design of the filtration plant was started, without Duluth spending any money. Under a greatly accelerated construction program the plant was completed in something less than eighteen months. The city did not want to pick up the $6 million cost of the project, feeling that since things were fine before the taconite tailings problem occurred, the bill should be paid by the state (who awarded the dumping permit), the federal government (who discovered the asbestos fibers), or Reserve Mining (who was the polluter). The City Council finally realized that there were monies available and applied for a demonstration grant. The EPA awarded $4 million and the state of Minnesota some $2 million. I understand that Duluth paid for the plant landscaping, and that was about all. Also, somewhere in that federal grant of $4 million there is an ongoing research grant for $900,000, which provides for three full-time employees in the water department laboratory.

The Duluth experience illustrates both the bottled-water option and the do-nothing option. By stalling, and by relying on the federal courts to provide the impetus, the city was able to furnish its consumers with an alternate supply of drinking water for the interim period that was required to design and construct a treatment facility--and all with someone else's money. The greater majority of waterworks personnel are not in favor of this concept. Water utilities are, I think, capable of paying their own bills, as Henry Graeser has already stated (see chapter 9).

Let us address some of the other options. The so-called home or "point-of-use" treatment devices have possibilities. There has never been objection to supplemental water treatment for water to be used in high-pressure boiler plants, nor against treatment systems that are installed in resort hotels, as for example, in Acapulco, Mexico. As a matter of fact, the most popular bottled water in Chicago is produced by the supplemental treatment of city water. Essentially, water is run through a GAC filter in order to dechlorinate it and to remove any residual taste or odor. It is done quite well and no attempt is made to conceal the water source.

The Chicago laboratories did some evaluation work on the home treatment devices in cooperation with a drug store chain. It was found that most of the units will do a fair job if used strictly in the manner prescribed. The problem seems to be the potential for poor maintenance when the user, untrained and unaware of the consequences of neglect, has to make the decisions to change the filter cartridge or to replace some element of the device.

I was able to find that Water Pic sold some 600,000 of its faucet-mounted GAC filter units last year, and that they expect to sell about 800,000 this year. I was not able to learn how many replacement cartridges were sold over this period.

The EPA is about to start a comprehensive evaluation of the point-of-use devices with results anticipated in about eighteen months. If this investigation is properly conducted, we can expect some controversy, most probably to the good of the customer. I believe I would seriously consider the use of one of the better-designed devices if I were in an area that was receiving water of questionable quality. I would be most careful how the device was used--low flow rate, element changes, and so forth--but would feel more assured of having quality water.

As to the treatment revision, the just published proposed regulation regarding control of organic chemical contaminants in drinking water would prohibit the use of chloramines "as the primary disinfectant in drinking water." I am not sure this is a rule that can be enforced. Springfield, Illinois, has been adding ammonia and chlorine at the raw water intake for over forty years. The treatment was started in order to prevent the formation of chlorophenols (severe taste and odor) and has continued to this day. Bacterial results for the system are excellent, and there is no indication that the management of the Springfield water system is about to change its treatment sequence. In cases like this, with the preponderance of evidence in favor of the chloramine treatment, it is difficult to see how the regulations can be enforced. Does the EPA know how many systems are successfully using chloramines from the head of the process? Without a doubt, if good bacterial control can be demonstrated, the chloramine treatment is a valid option. There seems to be an inference in the proposed treatment techniques that the point of application for chlorine be moved as far back in the process as possible--ideally, after the GAC contact. Depending, of course, on the quality of the raw water, this can lead to some fantastic growths in the settling basins, and even in the filters, and in fact, more so on the GAC itself. I have seen filters that were operated with a low residual chlorine level in the applied water. There were severe growths that acted like clumps in the filter media; it took some super chlorination and extended high-rate backwash to clean up the units. We <u>must</u> predisinfect--I am inclined to say prechlorinate, but I would like to see more done in this country with ozone. Should we consider flash-type super chlorination with dechlorination immediately downstream and prior to the GAC? Don Williams in Bantford, Ontario, used a technique something like this, using ammonia as a dechlorinating agent. It could be a technique worth investigation.

Before I close, I would like to give you a quote from an environmental publication put out by the Region V office of the EPA. This is a public involvement type of magazine which makes a definite play to the environmentally oriented segment of the public. The quote is from Hank Switzer, a Michigan water planner, who is referring to possible new treatment methods for wastewater. He says, "We have been telling the people, 'You've got to do this and you've got to do that, and you have to pay for improved water treatment, and waste disposal, and you've got to pay for better waste water treatment.' What we had better do is to tell them that they have to go out and make more money somewhere. The money has to come from someplace." How true it is.

HAMMOND: We will move on to Robert Harris.

ROBERT HARRIS: Some of you may know that I am now on a leave of absence from the Environmental Defense Fund, working with Dr. Bruce Ames in the Biochemistry Department at Berkeley. When my wife and I arrived last month in Berkeley, we were invited to a dinner party at a friend's house. During the course of the evening, one of the guests started discussing the inefficiency of the postal service, whereupon someone asked, "Well, are there any institutions left in this country that are really efficient? Are we getting what we pay for from any of our public or semi-public institutions?" Another guest answered, "Yes, from the telephone company." Of course, this led to quite a debate, and a vigorous discussion ensued for about twenty minutes. After things had quieted down somewhat, I said, "There is one service that public institutions provide that is very efficient--drinking water."

The immediate response was, "We would have never thought of that," yet everyone seemed to agree with my statement. As one guest said, "We turn on the tap, and we pay the bills, but we don't think very much about either. The water bill is a heck of a lot less than most of the bills we get, and we haven't worried very much about the quality of our drinking water, that is until Dr. Harris came along." Everyone laughed! I recount this story not to be glib, but rather to suggest what I feel is the general mood around the country. Most people accept the municipal drinking water authority for what it is--an efficiently run public service. I also believe that most people accept the relatively low cost of water as one of the last great bargains--sort of like the last nickel candy bar. Like the price of the nickel candy bar, however, the cost of water is about to go up if the EPA has its way and new regulations on carcinogens are promulgated. But here is where the analogy ends. When the nickel candy bar went up to a dime, it also shrank in size, and, many believe, in quality. The disillusionment and resentment with the performance of public services is in large measure based on the widespread perception that as the cost of utility rates and public services has gone up, the quality has gone down. The rising cost of drinking water, however, will be accompanied by substantial increases in quality, both from an esthetic as well as a public health perspective. Unfortunately, over the past several months the public has heard far too much about the rising cost, and far too little about the offsetting benefits.

Despite the disagreement over whether the EPA should regulate carcinogens, the controversy that ensued after the Environmental Defense Fund's 1974 study in New Orleans[2] has been accompanied by a considerable amount of research. The EPA has analyzed the drinking water in over one hundred cities nationwide for the presence of chemical carcinogens, pilot and bench-scale studies have been conducted on the technology for their removal, and, as Dr. Schneiderman has reported (see chapter 3) approximately a dozen epidemiologic investigations have been completed. Although the vast majority of the approximately 700 organic chemicals so far identified in drinking water have yet to be tested for their potential carcinogenic effect, numerous carcinogens and mutagens have been found to be rather ubiquitous contaminants of treated drinking water (table 14-1). Furthermore, the epidemiologic studies represent a reasonably consistent pattern of association between drinking water quality and cancer mortality rates, primarily of the gastrointestinal and urinary-tract organs.

The question for policymakers, of course, is how much information should be required before remedial measures are taken to reduce the public's exposure to carcinogens in drinking water. Unlike the waterborne diseases prevalent during the latter part of the last century, exposure to carcinogens is typically followed by twenty- to thirty-year latency periods before symptoms of the disease are clinically detectable.

---

[2]For the New Orleans study, see Science vol. 193, no. 55, 1976.

Table 14-1. Selected Carcinogens and Mutagens Found in an Eleven-City Survey

Compounds	New Orleans	Miami	Seattle	Ottumwa, Iowa	Philadelphia	Cincinnati	Tucson	N.Y.C.	Lawrence	Grand Forks	Terrebonne Parish
Benzene*	X			X	X	X					
Carbon tetrachloride*	X	X		X	X	X		X	X	X	X
Bis(2-chloroethyl)ether*	X				X						
Chloroform*	X	X	X	X	X	X	X	X	X	X	X
1,2 Dichloroethane*	X	X			X	X					
Dieldrin*	X	X	X	X		X					
DDT, DDE*	X										
Heptachlor*	X										
Hexachlorobenzene*	X										
Hexachlorocyclohexane*						X					
Lindane (γ-BHC)*						X				X	
PCB*						X					
Tetrachloroethylene*	X	X		X	X	X	X	X	X	X	
Trichloroethylene*	X	X		X	X	X			X		
Vinyl chloride*		X			X						
Bromodichloromethane	X	X	X	X	X	X		X	X	X	X
Chlorobenzene	X	X	X	X	X	X		X	X	X	X
Chloromethylether	X										
Dibromochloromethane	X	X	X		X	X	X	X	X	X	X
1,3 Dichlorobenzene	X	X			X	X			X		
Dichloroiodomethane	X	X			X	X		X	X	X	X
Methylene chloride	X	X	X	X	X	X		X	X	X	X
Vinylidene chloride	X	X			X	X			X		

*Carcinogenic compounds.

Therefore, the relatively recent expansion of chemical production (mainly post-1950) is probably not yet reflected in cancer mortality rates. The epidemiologic studies described by Schneiderman probably reflect exposures occurring thirty years or more ago, when the levels of chemical carcinogens (except possibly the trihalomethanes) were undoubtedly lower than they are today. Thus, decision makers are not only forced to make decisions in the face of the uncertain risk that today's levels of carcinogens pose to the public, but they also will be unable to evaluate the effectiveness of remedial measures for possibly several decades in the future.

In addition, there were several points made by Drs. Lijinsky (see chapter 4), and McCann that bear repeating. First is the concept of thresholds; it is really not important to decision makers that theoretical arguments can be made supporting a threshold exposure to chemical carcinogens, or even that a threshold has been demonstrated for one or two carcinogens, which of course it has not. Both Drs. Lijinsky and McCann give examples of chemical carcinogens which have caused cancer in experimental animals at exceedingly low doses. For example, aflatoxin, DES, and dieldrin have all been demonstrated to be carcinogenic at the parts-per-billion or lower level. As was evident from Dr. McCann's talk, however, there is a million-fold range in the potency of chemical carcinogens in animal tests, which means that the risk posed by a chemical carcinogen in drinking water is related not only to the concentration but also to that chemical's carcinogenic potency.[3] Therefore, considering the diversity in the mixture of synthetic chemicals in a typical polluted water supply, decision makers have little choice but to assume that some carcinogenic risk would accompany the use of that water for drinking purposes. The question that remains, of course, is whether the cancer risk is of sufficient magnitude to warrant the expenditure of public monies to lower the risk.

Determining the cancer risk posed by chemicals in drinking water is not a simple matter. Even if those twenty-odd chemicals that have been found to be carcinogenic were the only contaminants present, it would not be appropriate to simply assess the risk from each and then determine the overall risk by assuming that it is additive. Dr. Lijinsky quite properly pointed out that for many mixtures of carcinogens the effect of the mixture is much greater than the sum of the individual effects would be if each acted alone. Dr. Lijinsky mentioned that such synergism has been demonstrated in skin-application studies, for example.

Determining the risk from chemical carcinogens in drinking water by conducting epidemiologic studies also has its limitations. Because of the twenty- to thirty-year latency period for most chemical carcinogens, today's exposure will not be reflected in changes in cancer rates in human populations

---

[3] Ed. note: Joyce McCann did not prepare a written version of her remarks for inclusion in this volume.

for twenty to thirty years hence. And, as Dr. Schneiderman has pointed out, once we have solid epidemiologic information, it is too late, or at least it is too late for those individuals who are part of the study. It is important to remember that epidemiology is a science, and a rather inexact science at that. There is no well defined end-point in an epidemiologic investigation of a cause of human cancer. Conclusions as to cause represent judgments, and those judgments are often influenced by political and socioeconomic biases. For example, Sir Richard Doll, the famous Oxford epidemiologist, recounted in one of his books the story of a single epidemiologic investigation of a certain African tribe that had the peculiar habit of smoking their cigarettes with the lighted portion on the inside of their mouths rather than on the outside. It was also observed that this tribe had an exceptionally high rate of cancer of the roof of the mouth. The epidemiologist concluded that the peculiar smoking habit was the cause, and epidemiologists worldwide accept this single study as proof of causality. In stark contrast to the nearly twenty-five years of investigations of the relationship between smoking and lung cancer that was required before the epidemiologic community admitted to causality, Doll suggested that in the case of the African tribe the acceptance of causality was inversely related to the number of epidemiologists worldwide that smoked their cigarettes with the lighted portion inside their mouths!

Therefore, despite the dozen or so epidemiologic studies that were discussed by Dr. Schneiderman, many more will undoubtedly be required before the epidemiologic community will be willing to accept the causal relationship between drinking water and cancer. But as Dr. Reiches emphasized (see chapter 3), the epidemiologic studies of drinking water and cancer, while not sufficient to prove causality, nonetheless provide, when taken together, consistent support for that hypothesis. The coefficients of the drinking-water variable and the relative risk suggested from these studies, together with the anatomical sites that appear to be influenced by drinking-water contaminants are surprisingly consistent from study to study. Therefore, considering that further delay will result in irreversible exposure of large populations, I have little doubt about the wisdom of the proposed EPA regulations.

Implicit in my support of these regulations is the belief that the benefits outweigh the cost of installing granular activated carbon (GAC) filters or other treatment methods in order to reduce the levels of trihalomethanes and industrial carcinogens. At a conference last year in Cincinnati, Ohio, I made a crude estimate of the potential benefits to be derived from installing GAC filters in Cincinnati. Based on estimates of the direct and indirect quantifiable cost of cancer, I arrived at a cost of approximately $80,000 for every cancer death. Although most economists would value the human life at far greater than $80,000, I nonetheless used this estimate to ask the question, "How much would we have to reduce the cancer rate in Cincinnati before the cost of carbon filtration would be offset by the benefits from reduced cancer deaths?" The answer was approximately 1 percent, which means that the installation of carbon filters would have to reduce the cancer rate by approximately 1 percent or more before the benefits would exceed the cost. Since the epidemiologic studies in Ohio had suggested that as much as 10 percent of the cancer mortality rate in Cincinnati could be due to drinking water from the Ohio River, I concluded that the installation of carbon filters would appear to make good economic and public health sense.

More recently, I have performed a similar exercise with regard to the benefits of installing GAC filters in the New Orleans metropolitan area, which includes approximately one million inhabitants. From a recent study by the National Academy of Sciences, various methods of estimating the value of a human life suggest that the most reasonable estimate of the per capita value associated with preventing a cancer death ranges from $100,000 to $1 million.[4] A range of risk from carcinogens in New Orleans drinking water was estimated by various methods (table 14-2). A method of risk estimation recommended by a recent panel of the National Academy of Sciences was used to estimate the risk of chloroform and dieldrin.[5] This method

---

[4]National Academy of Sciences, <u>Nonfluorinated Halomethanes in the Environment</u> (Washington, D.C., NAS Environmental Studies Board, 1978).

[5]National Academy of Sciences, <u>Contemporary Pest Control Practices and Prospects</u>, vol. 1 (Washington, D.C., NAS, 1975).

Table 14-2. Estimates of Cancer Risk from Drinking Water in New Orleans Area Based on Animal and Human Epidemiologic Studies

Method of risk estimation	Risk estimate (Cancer deaths per million per year)
NAS method for 200 ppb chloroform, lifetime exposure, risk equal to most sensitive animal	20
Method same as above for dieldrin at 70 ppt	20
Assume risk at least 20 percent excess for GI and urinary-tract cancer deaths	70
Assume risk equivalent to N.Y. study; odds ratio of 2.7 for GI and urinary-tract cancer deaths	250
Based on New Orleans studies, drinking water accounts for 10 percent of total cancer mortality rate	200

of risk estimation, it should be noted, results in a higher risk than the method used by the Rohlich report. As can be seen in table 14-2, the risk estimation for 200 parts per billion of chloroform and 70 parts per trillion of dieldrin is the same, that is, approximately twenty cancers per million population per year. This underscores what Dr. McCann emphasized in her discussion; namely, the potency of chemical carcinogens varies enormously, and in this case the potency of dieldrin is approximately 3,000 times the potency of chloroform. Although a risk estimate could be made for several other carcinogens in New Orleans' drinking water, the paucity of carcinogenicity data for the vast majority of the chemicals found in New Orleans' water militates against this approach for risk estimation. A risk estimation for chloroform and dieldrin is presented only for comparative purposes.

Four of the epidemiologic studies discussed by Dr. Schneiderman are useful for estimating the risk in New Orleans. Dr. Alavanja's study in New York State, since it was a carefully conducted case-control study, is probably the most precise study from the viewpoint of risk estimation. If one can assume for the moment that the New Orleans population is similar to his New York populations, then we can assume a relative risk of 2.7 for gastrointestinal and urinary-tract cancers between urban populations receiving chlorinated water and urban populations receiving nonchlorinated water. This would suggest an excess risk of cancer associated with New Orleans drinking water of approximately 250 deaths per million population per year. Based on the three ecologic studies in New Orleans, it would appear that approximately 10 percent of the total cancer rate in New Orleans is associated with the drinking water. This would translate to approximately 200 cancer deaths per million per year.

Using this range-of-risk estimate together with the suggested range of the economic impact of a cancer death, we can compare the impact of carcinogens in New Orleans drinking water with the expected cost from installing and operating GAC filters (table 14-3). Based on Dr. Clark's estimate of the

Table 14-3. Estimated Range of Economic Impact Resulting from Cancer Deaths Due to Carcinogens in New Orleans Drinking Water Compared with Cost of Activated Carbon Treatment

Estimates of excess cancer deaths due to chloroform (200 ppb)	Range of economic impact (million $ per year)
20 cancer deaths per year (NAS method, lifetime exposure, risk equal to most sensitive animal)	2-20
70 cancer deaths per year (assumes 20 percent excess risk of GI and urinary cancers)	7-70
200 cancer deaths per year (assumes 10 percent of total cancer death rate due to drinking water)	20-200
250 cancer deaths per year (assumes relative risk ratio of 2.7 for GI and urinary tract cancers)	25-250
EPA's estimate of the cost of activated carbon filtration	5

upper bound for installing and operating GAC filters for a population of approximately 1 million, it would appear that the New Orleans community would have to pay approximately $5 million per year for such treatment. Assuming that chloroform was the only carcinogen in the New Orleans drinking water, the benefits from complete removal of 200 parts per billion of chloroform would range between $2 and $25 million dollars per year. Therefore, since the cost of GAC treatment is estimated to be approximately $5 million per year, it suggests that it may be possible to justify carbon treatment solely on the basis of removal of chloroform at these levels. When one considers the presence of approximately 100 synthetic organic chemicals in the New Orleans water, of which at least 20 are carcinogenic, it is evident that the benefits would be far greater than those estimated from removal of chloroform alone. In addition, a more realistic estimate of risk based on the epidemiologic studies would suggest the benefits would range between $20 and $250 million per year.

It is also perhaps instructive to consider the limits of sensitivity of epidemiologic investigations. I believe that most epidemiologists would agree that the consistency observed in the twelve studies reviewed by Dr. Schneiderman would suggest that the risk of gastrointestinal and urinary-tract cancer must be at least as large as the limits of sensitivity of epidemiologic methodology. These limits are assumed to be approximately a 20 percent excess of gastrointestinal and urinary-tract cancers. Therefore, the epidemiologic studies are suggesting a lower bound on the economic impact of $7 to $70 million per year. Despite the obvious difficulties in doing a rigorous cost-benefit analysis, what I have presented here should give us a great deal of confidence that the proposed EPA regulations should result in benefits far in excess of the increased cost of treatment.

Another policy issue, of course, relates to the hostility that has developed across the country regarding federal intervention into what has for the past century been a state responsibility. However, in the case of regulations on carcinogens, the EPA, and not the states, is in the best position to conduct the health effects and technology research necessary to develop such regulations. Where the EPA is failing is in not properly educating the public and the state and local officials about the results of their analytical and epidemiologic studies. I am convinced that once this scientific case has been laid before the public and their representatives, the EPA will receive overwhelming support for what they are trying to do.

There is one aspect of the proposed regulations which concerns me, however, as was mentioned by Timothy Harker, the lack of coverage for populations less than 75,000 population may suggest to some that our rural areas are populated by second-class citizens. However, considering the political, economic, and practical problems of implementing these proposed regulations, I believe that there is probably no alternative but to phase in the regulations. However, the EPA should include in the regulations an explicit timetable for phasing in communities having a population less than 75,000. Without such a timetable, I believe there is a substantial question as to whether the existing proposed regulations are in violation of the Safe Drinking Water Act.

Another issue which has been discussed at this conference relates to the possible need for construction grants to communities to build activated carbon facilities. Frankly, I find it hard to believe that the waterworks community would be satisfied with such a construction grant program. Historically, waterworks officials have been proud of the self-sufficiency of municipal water authorities, and to involve the federal government in the financing of improvements would seem to me to be an intrusion that would result in important inefficiencies and further erode the prerogatives of local officials. However, I do believe that a system should be developed whereby upstream polluters pay for the necessary treatment in downstream communities in order to prevent the public's exposure to industrial carcinogens. This could be done through effluent charges and an equitable redistribution system that would help communities finance needed treatment systems. The question has also been raised as to whether these regulations place an unfair burden on low-income groups who can less afford the rising cost of water, however justified it may be. On the other hand, the present situation is far more intolerable; namely, low-income groups who populate urban areas often receive water of lower quality than more affluent suburban areas, and at the same time are less able to afford expensive alternatives such as bottled water and water-treatment devices.

In conclusion, from this conference I sense a great deal of optimism among representatives of both the private and public sectors. Although we are lagging behind the Europeans in building treatment plants capable of removing chemical carcinogens from drinking water, these new EPA regulations appear to represent a justifiable means of beginning to redress this imbalance. Cancer has become the second leading killer in the United States, and it is on the increase. There seems little question that the majority of human cancers is preventable, although it is likely that there is a multitude of causes, none of which, with the exception of cigarette smoking, accounts for a large percentage of the total. Therefore, if we as a society are to develop a truly effective cancer prevention program, we must not ignore what appear to be relatively minor causes.

HAMMOND: Thank you. And now Peter Hutt.

PETER HUTT: I would like to point out at the outset that I was invited here, not as an expert in any way on the Safe Drinking Water Act, but rather to discuss its relationship to the federal Food, Drug, and Cosmetic Act under which the status of water is of great importance. Clifford Russell asked me, a short while ago, to extend my remarks into broader comments on how one goes about establishing safe levels for contaminants under any federal regulatory statute. I will try to cover both of those subjects in my remarks.

You have to start with the basic proposition that water is itself a food. In addition, it becomes an ingredient in most of our food products. Water also comes under the jurisdiction of the Food and Drug Administration (FDA) insofar as it is used in processing or making a cosmetic or a drug.

Thus, the FDA has as big an interest in the issue of water quality and water safety as does the EPA. When water is used in food--and I will concentrate on that, rather than its use in drugs and cosmetics--it has to be analyzed under the Food, Drug, and Cosmetic Act in the same way as any other food. The status of bottled water under the act is no different, conceptually, than the status of peas, carrots, meat, potatoes, or anything else. From a legal standpoint, the status of water as an ingredient in food is no different than if it were an emulsifier, a stabilizer, or a preservative.

One has to look at whether any ingredient is, or contains, any poisonous or deleterious substance, as defined under Section 402(a) of the act. One has to look at whether it meets the food additive requirements: is there prior sanction for it; is it generally safe; or the last alternative, does a food additive regulation authorizing its use already exist?

One also has to look (under Section 406 of the act) to see whether there should be a tolerance, even though the ingredient is a poisonous or deleterious substance, because it is needed in the production of the food and is unavoidable in good manufacturing practice. I am not going into any of that in detail. Those are the basic legal elements of any aspect of the FDA's review of water.

I would like to talk a little, however, about the consequences of those issues. In the past--and this goes back to at least 1950, and probably well before then--the FDA took the position on several food ingredients, that if there were evidence of carcinogenicity in test animals, any food ingredient would automatically be banned as a poisonous or deleterious substance. This was a full eight years before anybody even dreamed of the Delaney Clause. I will not go into any great discussion of the Delaney Clause, which is, and always has been, utterly irrelevant to food safety policy in the United States, and has almost never been invoked except in the rarest of circumstances. If it were repealed tomorrow morning, it would not change any safety decision made by the FDA in the past, or any that might be made in the future. This is because the general safety provisions of the law have been interpreted for many years to have the same impact as the Delaney Clause. Those who fear that the Delaney Clause will somehow be applied to water, simply have their eye on the wrong ball. They do not understand the Food, Drug, and Cosmetic Act and how it has been applied over the years.[6]

The real issue, and the issue that is far more serious and far more difficult, is how, in the future, we are going to create a food safety policy--which includes a water safety policy--that is consistent with new scientific knowledge that we never had in the past, even as recently as five years ago. This new knowledge shows that we have animal carcinogens in

---

[6]See Legislative History and Application of the Anti-Cancer Clauses, Hearings before a House Subcommittee of the Committee on Appropriations, 93 Cong., 2 sess., Pt. 8, (1974) pp. 180-222.

vitually every food that is currently marketed in the United States and in almost the entire water supply.

The problem that we have is that the old FDA approach, of banning all animal carcinogens, is no longer feasible because we now know we have too many of them to ban. Some of the more recent studies include one sponsored by the National Center for Toxicological Research, showing that vitamin D produces cancer in test animals. It is quite clear that we cannot ban eggs, cinnamon, tea, coffee, bacon, vitamin D, and all of these other things that we now know are carcinogenic or which contain constituents that are carcinogenic in test animals. We cannot ban all foods that contain traces of chloroform, nitrosamines, and aflatoxin.

The future FDA approach to food safety is going to have to be very close to the EPA approach for drinking water standards. It is going to be a question of setting tolerances for contaminants at the lowest feasible level. As new technology becomes available, it will force the industry and the American public to pay for still lower traces of contaminants.

I see no possibility for a conflict between the EPA and the FDA on this fundamental issue. I find it unlikely that FDA will, in any way, object to letting EPA have the lead in the drinking water area because FDA has enough problems without worrying about the entire water supply as well.

The real issue today is how both of these agencies can somehow come to grips with the basic question of safety. What is safety? How can one best protect consumer health?

First, let me observe that there is no concept of absolute safety. The expectations that I had when I was at the FDA, that we would somehow ban all unsafe ingredients, are now shown to be ludicrous.

To confront this issue, Congress has included in the Saccharin Study and Labeling Act, passed in 1977, a provision requiring that the National Academy of Sciences (NAS) review and make a report to Congress on exactly this question. On the basis of that report, which is due in some fifteen months, Congress will in turn consider, not just the future of saccharin, but rather the question of the entire future regulation of food safety.

Without anticipating what NAS is going to say, I would like to offer a brief outline of some of my own views on the general conceptual framework for the future food safety policy of this country--and that includes water safety policy.

These are concepts that I laid out in a paper delivered at MIT last September and, more recently, in a paper I delivered last month.[7]

---

[7]Peter Hutt, "The Basis and Purpose of Government Regulation of Adulteration and Misbranding of Food," the Underwood-Prescott Memorial Lecture, presented at the Massachusetts Institute of Technology (September 27, 1977); Peter Hutt, "Public Policy Issues in Regulating Carcinogens in Food," Keynote Address to the Conference on the Scientific Basis for Interpretation of the Delaney Clause, sponsored by the International Academy of Environmental Safety (February 6, 1978).

My first rule is that there is no such thing, quite obviously, as absolute safety. All food components, including water, are capable of producing harm in some people under some circumstances. Some risks must therefore be tolerated by the public if we are to have any food or water at all. And that is not just for the next twenty years. That is for the next twenty centuries.

One of the most important tasks--and this is one that Bob Harris just alluded to--is to review those risks and priorities for dealing with them. To my knowledge, no government agency has thus far successfully done that. The Occupational Safety and Health Administration (OSHA) proposal to establish a classification system for all toxic substances posing a potential occupational carcinogenic risk, published last fall, was the first real attempt at it.[8] I think it was faulty in some important respects, but at least it began to address the issue.

Recently the American Industrial Health Council (AIHC) has submitted comments proposing a modification of the OSHA approach. The AIHC comments would classify chemicals which may present a carcinogenic risk into four categories: Category I would include proved human carcinogens; Category II, proved animal carcinogens; Category III, substances which would require further testing; and Category IV, all other substances. That still is too broad for purposes of setting

[8] Federal Register vol. 42 (October 4, 1977) p. 54148.

priorities, however, so AIHC has proposed that the first two categories be further subdivided into high, medium, and low potency. There is, as Bob Harris said, a very major difference over the potency range of various carcinogens. If the AIHC approach were adopted, we would for the first time--not just for OSHA, but for EPA, FDA, and the Consumer Product Safety Commission (CPSC) as well--begin to set priorities whereby the country would at least attack these chemicals on the basis of their potential risk.

My second rule is that safety must be viewed as a relative concept among competing alternatives. If, for example, there are several similar food components which satisfy a particular need--for example, several emulsifiers--and one poses a significantly greater risk, elemental principles of safety would justify eliminating that one component solely on the ground that there are other acceptable safer alternatives. But implementation of this rule is going to be very difficult for the government. The determination of acceptable alternatives will, in many instances, simply be impossible, or at least will depend on many elusive facts and difficult judgments.

The third rule is the need for a special analysis for those components, like water, which do pose significant safety risks, but for which there are no alternatives. For it is unlikely that we are going to find a substitute for water in the very near future. I would certainly be interested if someone has come up with one during the last two days.

This special analysis I have set out in my papers involves a three-step process. The first step is to determine whether it is feasible from both the technological and economic standpoints to reduce the risk by further processing. This is, of course, the approach of setting standards to reduce contaminants. The second step will be to conduct a risk assessment on the lowest risk that one is ultimately able to reach. The third, and final, step, is to determine appropriate regulatory action, if any, based on that lowest level of risk that one is able to reach. Let me briefly go through each of these steps. In some instances, without question, the risk can be reduced but only at a specified cost. It might be feasible, for example, to reduce or eliminate all aflatoxin in peanuts, corn, and milk, or chloroform in water, but the technology may not be available yet, or the cost of the treatment needed to accomplish these steps might be prohibitive. The cost might simply price the product out of the market. What the government must do--and this will be extraordinarily difficult--is to balance the cost against the reduction in risk, and to come out with a conclusion on how far one can force this process at any given moment. Certainly, no one could oppose reducing any risk, of any kind, if the cost is trivial.

In particular, human exposure to animal carcinogens should be reduced to the lowest feasible level, taking into consideration what I have mentioned, as long as potential risk to humans remains uncertain. And I see virtually no possibility of having resolved in my lifetime the scientific issues of threshold and when animal studies show human risk.

Where the technology is currently unavailable and the cost is substantial, however, we reach the difficult issue of cost-benefit analysis. It is perhaps here, more than anywhere else, that the scientific uncertainty about the existence of the hazard itself, when the sole evidence consists of animal studies, may be of particular importance. Expenditure of major amounts of money may well not always be warranted to achieve a relatively small reduction in what can only be determined to be a theoretical human risk.

In any event, once the risk is reduced as far as is warranted under a cost-benefit analysis, then a risk assessment must be conducted. This would be designed to quantify the risk involved in order to place it in perspective relative to other risks that we all face. Without question, this kind of risk analysis is difficult and uncertain on the basis of our scientific knowledge. One of the mathematical models developed to estimate human risk from animal data, or some variation of those models that we all know about, is going to have to be used for this purpose. This approach, while undoubtedly not precise, nonetheless, in my judgment, represents a significant improvement over the current views of arbitrary safety factors (the old one-in-a-hundred, or a hundred-to-one safety factor) or levels of toxicological insignificance (such as two parts per billion) which have been used for many years by the FDA.

Again, it is important to keep in mind that any risk analysis based on animal studies represents a projection of only a potential, not a proven risk, and is therefore an appropriately conservative method of assuring adequate public protection.

The third step, once the risk is quantified, is to determine an appropriate regulatory response. In the past--and I have participated in this on many occasions--this has been done on an entirely ad hoc basis. In the future I believe that a consistent rationale for all government action with respect to carcinogenic risks can and must be enunciated if the public is to accept the governmental decisions involved.

Conceptually, there are basically three levels of risk. For lack of other names, I call them "high," "moderate," and "low."

For high risks, society has always imposed a ban. Examples of high risks--and frankly, outside the food and drug area there are not very many--that are banned in our society include putting frank poisons in the food supply, attempted suicide (which is a crime virtually everywhere), and, in my home town of Buffalo, New York, trying to go over Niagara Falls in a barrel. In each one of those instances, the risk is extraordinarily high. The only high risks that we might tolerate would be those with an extraordinary potential benefit, such as anticancer drugs that also themselves clearly produce cancer.

For moderate risks, we provide information and appropriate warnings to the public, but we have never banned them. We advise mountain climbers, canoers, swimmers, and even people who are foolish enough to drive on the Indianapolis Speedway, exactly what they are facing. We warn them to take safety precautions. And then we allow them to make their own decision whether to do so. The same thing applies to putting warnings on aerosol-type food containers where death can result when the contents of the container are sniffed. For low risk, we do not usually even provide any kind of warning. Society, basically, is not greatly concerned about low-level risks.

This analysis depends, of course, upon there being an opportunity for free choice by the consumer. There is no free choice when we take the water straight out of the tap. But there is free choice in the water treatment that can be provided by individual food companies. Therefore, there is a possibility of regulating water when used in the food supply and in bottled water, in a different manner. For example, if bottled water were to fall in the moderate-risk category, then one might handle it by labeling rather than by setting standards and banning it.

I do not have a sufficient mathematical background to decide where one draws the line between these three risk levels. Dr. Richard Wilson of Harvard has not only come up with the theoretical basis for such dividing lines, but by analyzing actual situations he believes that it can be shown that there have been fairly clear dividing lines applied by society in the past. He contends that an activity with a risk of 10^{-2} (one in a hundred) or greater has always been banned by society.[9] He contends that activities with a risk of 10^{-5} (one in 100,000) or less have very seldom been banned. He says that between these risk levels are the moderate risks for which society attempts to lower the risk as far as possible but if the risk cannot be lowered sufficiently, society handles it by providing information to the public in the form of labels, warning statements, and educational materials. Thus the consumer is allowed to decide whether the benefit that one perceives in that activity or that product outweighs the risk that has been discerned to be associated with it.

HAMMOND: Our next speaker is Victor Kimm.

VICTOR KIMM: In the last week, I have participated in meetings with the Environmental Quality Committees of the Association of the State Legislatures and with the National League of Cities. In both sessions and during the last few days here, I have heard comments which lead me to believe that the EPA's drinking water program has been perceived in three roles. Some seem to think that we will do the right thing if they just get our attention and nudge us along. There are clearly those who believe that we are devious and running a bureaucratic empire-building campaign of monumental proportions. Finally, there are a fair number of other people who look at what we are doing and say, "Gee, the EPA people are trying to do the right

---

[9]See the comments submitted by Dr. Wilson on the OSHA proposal.

things, but they just don't realize the consequences of their actions."

I think all of these views are erroneous. I think we have a pretty good grasp of the water supply problems facing the nation and are trying to take reasonable steps to deal with them. We make decisions on a day-to-day basis, often with incomplete knowledge, in order to meet the responsibilities established in the Safe Drinking Water Act. As long as our decisions are subjected to a full public airing, I think it is an appropriate way for a federal regulatory program to operate, and in the EPA's drinking water program we are moving the program along.

Today, I would like to concentrate on what I see as the major priorities for our program in the immediate future, after which I will make some specific comments on the points raised over the last two and one-half days at this conference. To begin, I would like to note the EPA's interest in promoting and facilitating meaningful participation by state and local officials, as well as representatives of all groups and individuals interested in the quality of the nation's drinking water, in formulating the regulations and policies guiding the implementation of the Safe Drinking Water Act.

I think the National Drinking Water Advisory Council, with its broad membership, has made a major contribution to opening up the regulatory process for public review and comment.

Similarly, our participation and support of this meeting demonstrates interest in seeing the best available scholarship applied to the problems facing the drinking water program. We are very pleased to have heard a diversity of views and a variety of suggestions on how we should proceed in the future.

With the change in administration, there are two major new initiatives that should be noted. First, there has been a very strong reemphasis on the public health implication of all of the EPA's programs. The priorities for increased resources in the president's fiscal year 1979 budget requests are in the areas of toxic substances, drinking water and hazardous waste control related to the solid waste programs, as well as the control of toxic and hazardous emissions under the air and water pollution control programs. These priorities reflect the agency's perception of where the potential public health benefits are largest.

Second, there is an increased recognition within the EPA of the need to integrate its programs which have typically operated on a single-media basis. So, for example, we now are dealing with a solid waste program, which is an integral part of our water pollution and groundwater protection programs; and with the toxic substances program, which is intermedia in dimension. In practice, this requires much more coordination of the EPA's programs than has existed in the past, and it is a major management challenge to the agency. I think this program integration is critical for public support and understanding in the future.

To pick up on some of the specific issues that have been touched on by the panel members or the previous speakers, I would first like to comment on regulatory philosophy, by asking, "What is safe?"

For many of the contaminants for which we cannot set any specific threshold or no-effects level, regulatory agencies must look at the magnitude of the risks of exposure and the feasibility of reducing human exposure in order to arrive at an acceptable level of risk, even where our scientific capacity to quantify such risks is inadequate. For example, unlike Mr. Hutt, I do not think it is only a question of the potency of the carcinogen, but also involved are questions about the size of the exposed population and the degree to which this is a voluntary or involuntary exposure. All these facets must be added together in order to assess the benefits to be obtained by reducing exposure to the suspected carcinogen.

Similarly, one must look carefully at the feasibility of reducing human exposure to any non-threshold-level toxicants. Typically, this involves the socioeconomic costs of alternative levels of control technology. Finally, it is necessary to balance the anticipated risks against the feasibility of reduced exposure in order to set a reasonable regulatory requirement. This, in fact, is the way that national standards have been set for a long time. In the last five years, the EPA has set standards for many non-threshold-level toxicants by following this approach and generally employing a philosophy of trying to reduce human exposure to the maximum feasible degree.

I suspect that there will be efforts to come up with schemes for ranking and setting priorities for the regulation of such contaminants. I think that is going to be a fairly arbitrary operation under any set of circumstances, and probably more driven by feasibility of control than by anyone's firm belief that an estimated 10^{-6} risk can really be differentiated from a 10^{-7} or 10^{-5} risk. I do think that the tools we now have give us some idea of the order of magnitude of the risks, but I don't think those numbers can be just linearly extrapolated with any degree of confidence.

If we have a jurisdictional problem, it is that both EPA and FDA have more things to do than they have resources with which to accomplish them. In the area of bottled water, the Safe Drinking Water Act quite explicitly says that it is up to the EPA to generate drinking water standards and that the FDA is to produce bottled water standards based on the EPA's action, and to police them as part of its normal activities in the regulation of interstate commerce.

What has become a problem, and a fairly serious one at that, is the question of who should be saying what about those substances generally regarded as direct and indirect additives, such as the paints and water treatment chemicals, as well as water-supply equipment which may add contaminants to drinking water. Interest in this area was significantly increased when the FDA included reference to pipe made of vinyl chloride in its proposed regulations concerning vinyl chloride.

Though several attorneys have talked about potential conflicts between the FDA and EPA and about duplication and overregulation, there is, in fact, little capacity in either agency to do very much under either set of statutes at the moment. What is going on is a personal exchange of views, between Mr. Costle and Dr. Kennedy, to the effect that we need to look at the impact on drinking water of direct and indirect additives. Second, we do need to find some effective way of dealing with that problem, which may or may not mean legislative changes, or new legislation. Before that, however, we need to decide if there is a sufficient risk associated with these materials to warrant the use of regulatory resources. The forum which will probably be used for later phases of that discussion is the Interagency Regulatory Liaison Group.

Similarly, I know of no intention on the FDA's part of asserting authority over community water supplies. The discussions thus far have tended in the direction of relying on the EPA in order to ensure the quality of public water systems and on the FDA when the water enters the food-processing establishment. The FDA is worrying a great deal about the latter.

The question of small systems is one which has been discussed previously. There is, within the Safe Drinking Water Act, an apparent dichotomy. It says, on the one hand, that standards should consider the costs of medium- and large-sized systems. It is widely recognized that there are economies of scale in the treatment of drinking water as in most other treatment processes. Therefore, the per capita cost of compliance--not to mention the administrative feasibility--for small communities which require complex treatment might very well exceed their financial capabilities in the absence of subsidies.

The Congress recognized this when they passed the act. They explicitly stated that they wanted us to get the program going and to use variances or exemptions as a way of extending deadlines so that we could get a better handle on the true magnitude of the problem. Then, presumably, we could come back to them with some suggested solutions.[10]

Last year's amendments to the Safe Drinking Water Act carried that one step further by giving us an explicit requirement to do a major study of the costs of complying with all the regulations, of the various ways in which those costs could be financed within water rate structures, and of the strengths and weaknesses of a variety of subsidy approaches. Presumably one of those would be federal construction grants, such as the program established to assist publicly owned waste-treatment facilities. This study is getting underway and is due in eighteen months. I would expect there would be some discussion this year in the Congress of the need for construction grant programs, but with a $60 billion deficit, I think that this issue would be more likely to become a major public policy option a year from now, by which time we should also have a better understanding of the magnitude of that problem.

---

[10] Ed. note: Some of the relevant language is quoted by Henry Ongerth in chapter 12.

Finally, I would like to return to where I started, and say that I think doing the public's business in a regulatory fashion today has new dimensions to it. It is no longer simply a matter of mandating standards, but of convincing people that such standards do make sense, and are, in fact, attainable.

HAMMOND: Thank you. Finally, Dr. Linaweaver.

F. PIERCE LINAWEAVER: This conference started out with the stated purpose of reviewing the problems of providing safe drinking water. I think it certainly has done that. It has also identified a number of achievements of the current legislation, and produced a lot of suggestions concerning the future regulation of organic contaminants. As I make my remarks, I am going to identify some of our past achievements.

I cannot underestimate the importance of drinking water to our society. It affects everyone. I appreciated Bob Harris's opening remarks in this regard. The new legislation has, therefore, very broad requirements. In particular, the delegation of monitoring and notification procedures to the states is going to cause a lot of fundamental changes in the way in which water professionals approach the management of the water services. This, of course, affects all supplies. Other elements of the legislation, particularly regulations concerning organics, affect most supplies, but many will end up with variances because of having reasonably well-protected water sources at the present time, resulting in low levels of organics and the trihalomethanes.

During these discussions I could not help but look back on the way in which we have advanced technology in our society, particularly in the water industry. For example, at one point we had, for all practical purposes, individual supplies. Another milestone occurred when we learned how to disinfect our water supplies to control the waterborne diseases.

While we were refining all of these techniques and improving our water supplies over a period of many, many decades, wastewater was receiving much less attention and a much lower priority in public budgets. But recently, we finally decided we were going to have higher ambient water quality. We developed a massive effort to clean up the water, represented by Public Law 92-500.

One of the things the Safe Drinking Water Act has done is to bring us back full circle to the supply end. We are looking at the water that is coming out of that tap. I think this is a notable achievement. Our society seems to make progress in quantum jumps followed by drift for many decades with a sense of security that we have taken care of the problem. Then we find an issue such as the relation between cancer and organics in water, and we identify the problem and then address it.

Let me amplify what I mean by addressing the problem for the case of water professionals. I think the water professionals--that is, the various people having significant responsibility in this effort--recognize this problem as a real one. I think they are very willing to innovate and to demonstrate so that we can identify ways, at acceptable costs, to deal with scientifically accepted estimates of risks.

Municipal water supplies are, for all practical purposes in our society, run at the local level. This is an extremely important point, especially because most supplies are part of the municipal government. There are, of course, privately owned utilities, and there are, in some cases, regional authorities separate from municipalities. But the major part of the population of the United States obtains its water service from municipal governments. This means that water supply is often wrapped up in the budgetary process of the local government.

I would also remind you that, by and large, it is a fully self-supporting service. This has been emphasized throughout this conference. There is no subsidizing of water supplies by local general revenues or by the federal government. I think most municipalities are doing a fundamentally good job in spite of their budget difficulties, but I want to make sure that this relationship between the budgetary problems of the cities and the implications of the new drinking water regulations are properly understood. I think this is very important. The fact that water services are independent of the federal subsidies is also something we have to look at carefully. I have heard a number of my colleagues, as well as participants at this conference, call for some kind of federal subsidy for water services. I believe such a program would be a drastic mistake. I think, for example, that the massive federal subsidy program for wastewater treatment facilities has slowed the pace of technological advance at the local level. This is not to suggest that we do not need some element of enforcement and regulation. I am simply referring to the construction grants as part of the program. I hope we do not make the same mistake in water supply.

One of the common activities of conference participants was to put out numbers and then attempt to qualify those numbers appropriately. Piecing together some of the numbers I heard, one might--and I am not saying one should, but one might--come to the following as a conclusion.

It was stated that as many as 200 of the 300,000 annual cancer deaths in the U.S. may be attributable to drinking water. There are a lot of uncertainties and various categories, but those were two specific numbers which were mentioned. It appears that expenditures to reduce that threat may amount to \$300 to \$700 million. For our purposes here, let us use \$500 million. Let us assume that we could identify some ways in which to reduce the number of cancer deaths in increments of 200 deaths per year until we get to 300,000. Similar expenditures for each additional 200 could suggest a total cost of \$750 trillion in order to eliminate 300,000 deaths per year from cancer.

I do not consider that to be a valid conclusion or argument, but, on the basis of the presentations I heard while I was sitting in the audience at this meeting, it seems to me that it is worth saying because it may give you some sort of perspective on the real implications of the new regulations. Our society is concerned about cancer, but I do not think to that extent. People might rather spend that money some other way or much of it not at all.

This is also, I think, true of the professionals in the water industry. Most feel that we are about to spend a lot of money for very little gain. I think that has also been said by a number of the people at this conference. I would identify that as the major shortcoming of the legislation.

It has been suggested by Bob Harris that the introduction to the EPA regulations clearly identifies the need. I disagree. One of the things that concerns me--and I can recall this kind of concern or point coming up many times in discussing my budget with the mayor when I was with the City of Baltimore--is that the regulations say, toward the end of the introduction, that in accordance with requirements of the Safe Drinking Water Act, revised regulations will be forthcoming which will impose additional MCLs on organic chemicals, as well as to require further reductions in the level of THMs as additional toxicological, technological and analytical information becomes available. That is the scientific way of saying that we really do not know everything; that this is the best we know now; and that we are ready to take additional steps as we find them advisable. But when you are at the local level, wrestling with the budget dollar, and you see a statement like that, then you think that while you can design something specific to meet the current regulations, how are you going to design it now when you know they are going to change the regulations?

This is an extremely difficult problem. When I asked our staff in Baltimore for their costs for the removal of organics from the city's water supply, they identified a cost of $30 million for the renovation of two existing plants plus the building of a third plant. And here is another important point: the renovation costs for the existing plants are far greater than those for putting in a new system.

At any rate, when the operators look at these expenditures, based on their knowledge of the water supply, on their understanding of the NAS study, and on the regulations that are being proposed by the EPA, they quickly conclude that they disagree with spending that kind of money in that way. They can then list all the other things which they think would provide better water service and reduce other risks associated with drinking water. I think they question the basis for establishing, for example, the hundred parts per billion for total THMs. The rationale is not clearly enough identified that the professionals can accept it. Maybe the numbers ought to be goals instead of standards until we can demonstrate with a carefully thought-out program within the utility industry just exactly what we can accomplish in this regard.

My earlier comment about the closing of the cycle also reminds me that I think the water professionals in our society are probably ready for a quantum jump in the way we think about our drinking water and the sophistication with which we deliver that particular service. Various other speakers have identified several areas on which this service could be improved by the expenditure of money on things which concern the operator. I think we are ready to do that, but it is going to have to be done in small increments. Water is a relatively small part of expenses for the typical home owner, but keep in mind that he also, in most cases, gets a wastewater bill along with that water bill. The wastewater bill, which not too many years ago was 60 percent of the water bill, is rapidly approaching 100 percent or even 120 percent of the water bill in many utilities.

Based on what has been said at this conference, I think some of the following initiatives ought to be taken: allowing for demonstration of various technologies and identification, from an engineering point of view, of the costs that are involved. It seems to me that we do not have sufficient experience with these systems to engineer something which ought to last many years. We lack operating experience, for example, for some of the systems that are being proposed. We do have a lot of experience with the handling of taste and odor control but not with the removal of organics.

We need to give the operators an opportunity to show what they can do. I think they are ready for the advances in technology that are implied by the new regulations. But I think the timetables are too short. It was stated by one speaker that EPA gives you one year and that by adding four years of variances or exemptions you get a total of five years to comply. I would remind you that most municipalities work on a capital budget which is projected six years in the future. This is the local government's way of setting priorities. But even the five-year timetable is not adequate to get into that process and to go through design and construction and get out the response that I think is desired by the EPA. Let advances in the technology of treatment follow a proper course, which means allow the expenditures to fit the particular problem of various local areas.

I have one closing remark. There certainly has been an impressive array of opinions put forth here. If you cut through all of that, however, the conclusions are not sufficient to warrant the actions which are being proposed by the EPA. I believe we are ready for a significant advance in drinking water quality, but the proposed regulations would amount to mandating suboptimal solutions by administrative fiat rather than finding solutions based on professional and scientific decision making.

HAMMOND: I would like to thank all the panelists for their contributions. I will now ask for questions and comments from the audience. Please state your name and your institution clearly. I also ask you to make your questions or statements brief and not to make speeches. It would also help if you could address your questions to a particular panelist.

UNIDENTIFIED QUESTIONER: Underlying the whole question of assessment, and given that we live with the policy of pushing contaminant levels as low as practical, it seems terribly significant that the mechanisms of action of chronically toxic substances are really not well understood. I have not heard here any discussion about expanding efforts in basic research to get at this. I am wondering how you, Bob Harris, see the tradeoffs between efforts directed at screening and enforcement and research into the mechanisms which might allow us to actually set such things as acceptable levels.

HARRIS: There certainly is a possible tradeoff there, but I think someone earlier in the conference observed that you do not need to know the mechanisms of cancer to prevent the disease. As an analogous example, John Snow did not know the mechanism of cholera, and yet his studies in London led to dramatic improvements in public health.

Although in my view the epidemiologic evidence is reasonably convincing, let me clear up a misunderstanding which I take to be at the root of one of Pierce Linaweaver's remarks. He said that drinking water is responsible for 200 cancer deaths per year. I heard several people previously deny that that could be considered an estimate as well as an accurate risk and say that 200 was derived from experiments on animals involving only chloroform. Indeed, the risk estimate that led to the 200 figure was based on the NCI tests on chloroform. However, if I use the data on chloroform derived by Dr. Roe (Preliminary Report of Long-term Tests of Chloroform in Rats, Mice, and Dogs; unpublished results, 1976), I can increase that risk estimate several fold.

If I take into account the epidemiological studies, I could increase that risk by several more times, since, if the risk were only 200 deaths, you would never see it in an epidemiologic study--and we have eleven or twelve epidemiologic studies which show the same pattern.

Alavanja's studies in New York state suggest that the risk of chlorination is, at the very least, 10 percent of the gastrointestinal and urinary-tract cancer rate. So I think that if anyone is operating under the illusion that 200 cancer deaths is in any sense a best judgment of the risk, he is badly mistaken.

To answer your question more directly, I believe we need to develop better methods of estimating risks, in addition to worrying about mechanisms. It is pitiful to see the small amount of money that is being spent developing ways of testing carcinogens to get a better handle on what the real risk to humans is. And I do not think that you need to know the mechanism in order to do that. I would not, however, divert funds away from regulating the problem because I think we can demonstrate that the costs to clean up organics in New Orleans, Cincinnati, and elsewhere, are offset by the benefits.

KIMM: The 200 figure is, in fact, within the statement of basis and purpose provided as background information. What it says is that if you assume that trihalomethane levels everywhere are similar to those that were found in the 113 cities we monitored, and if you were to run the models out, that would be what they would show. However, further explanations are needed.

As Dr. Sontheimer has shown us, trihalomethanes can be measured and are perhaps no more than 20 or 30 percent of the total chlorinated organics present, which presumably include intermediaries between the humic acid and the end-product substances. You can assume that the intermediaries had similarly adverse health effects.

Similarly, the figure 200 totally disregards any of the other organic compounds that may be present. There is not enough data available to determine what the effect of those other compounds would be. But, I hope that those who are interested in this area will realize these problems. It appears that no one in this world knows precisely what the magnitude of risk is. If we did, we could do a rigorous cost-benefit analysis.

Joyce McCann's slides, you remember, showed that we know almost nothing about the brominated species, but we do know that in all other instances in which we have compared brominated versus chlorinated analogs, the brominated compounds are generally one or two orders of magnitude more important as carcinogens than the chlorinated ones. So it is indeed possible that a few parts per billion of the brominated species may be a great deal more important from the point of view of potency than 200 or 300 parts per billion of chloroform.

TALBOT PAGE (California Institute of Technology): I want to make a comment about epidemiology. Traditionally, epidemiology has been used to try to find two- and threefold differences between something that is treated in some harmful way and something that is controlled. In the epidemiological studies of drinking water we are using more sophisticated techniques and pushing the limits of detectability down. We are talking of differences on the order of 5 to 20 percent. These are much smaller in terms of the background level than the two- or threefold effect.

To think that epidemiology reveals effects on the order of a 0.1 percent difference over background, which is what the 200 deaths amounts to, compared to the 300,000 deaths, is simply unrealistic. If the effects were really that small, one would not get any results with the tests we have now. Yet we have to live with the fact that eleven of twelve studies show some results, so that the range of effects we are talking about has to be significantly greater than that 0.1 percent. They are probably on the order of 5 to 20 percent.

The figure 200 has to do with chloroform only, and we are talking about many other chemicals as well. The chloroform estimate is not an estimate of the total carcinogenic effect of polluted drinking water, but an extrapolation of the effect of a single chemical. The epidemiological studies, which yield estimates on the order of 30,000 deaths per year, try to incorporate the total, aggregate effect of organically polluted drinking water.

C. C. JOHNSON (Malcolm Pirnie, Inc.): I would like to make a couple of comments.

First of all, I would like to commend Mr. Hutt on a lucid, intelligent, and informative statement of the problem. It had nothing to do with who has the bureaucratic responsibility, whether the FDA, EPA, OSHA, or anybody else. It did have to do with how, where, and when we set standards controlling the contaminants that invade our environment. Until we deal with that, and until each of our various agencies approaches this on a uniform basis, we are always going to be in a no-man's land; we will have conflict and ambiguities, and each agency will be saying that the responsibility and authority is out of its territory. We should have had that statement the first day.

Second, I would like to take issue with Mr. Linaweaver. We are both consulting engineers, but his background is public works and mine is public health. From the public health perspective, I do not want to count bodies in the streets before

I take care of the problem. I look for preventive legislation. The indications are that there are some harmful contaminants in the environment and that efforts to reduce the risk they pose are within our economic capability. So the Safe Drinking Water Act says, we ought to remove them from the environment. Not all of our colleagues support this particular viewpoint, but there are many of us--in the consulting field, public health departments, and the water industry--who feel that we ought to produce the best water that we possibly can, and that drinking water should not expose the public to potential danger if prevention is at all possible.

Dr. Sontheimer reported that Germany takes that approach. They do not have standards in our sense. They are continuing to try to improve the quality of the water product. That is what the policy problems ought to be about.

<u>ERNEST MUELLER (Department of Environmental Conservation, Alaska)</u>: I would like to make a couple of comments. I see the discussion between the EPA and the FDA about who handles drinking water as related to the gap that we have in our state. It amounts to this. Who sets the standards for water that is used in the seafood-processing industry but is not drinking water? There are enormous quantities of seawater, for example, that are being used in seafood processing--to move fish around and that sort of thing. Generally, the quality of this water is regulated by the Water Pollution Control Authority. But who sets the standards for how it

should be treated at the intake; how much should be chlorinated, and so on?

The other thing I would like to comment on concerns the potential for a construction grant program. Our state has had a construction grant program operated by the State Water Supply System for about ten years, in which we pay 50 percent of the nonfederal share. This program has been very effective in upgrading some rather terrible supply systems over the last ten years. I think it has been administered rather well. We have not had anywhere near the administrative paperwork and bureaucratic problems with the state program that we have had when working with the EPA.

I am not sure what my view would be, as a representative of a state, on a federal construction program, because I see a lot of strings coming along with it, such as requirements for water meters and water-conservation measures, and a lot of other things that municipal officials would just as soon do without.

I recognize that water systems have historically been reviewed as enterprise activities. I do not think they were originally conceived that way. But in this connection, I would like to pose that seawater problem to you.

<u>HUTT</u>: I will deal with the seawater issue. The Food and Drug Administration currently regulates not only the water used to process seafood, but indeed, the water in which the seafood grows. The National Shellfish Sanitation Program

originated over fifty years ago as part of the Public Health Service, as a cooperative program with the industry and the states. The FDA proposed regulations in the Federal Register to establish this program in regulatory form, controlling precisely the conditions under which oysters, clams, and so forth can and cannot be grown in their natural habitat, as well as setting out good manufacturing practices in the processing of seafood.[11]

These regulations were stayed by congressional action for eighteen months, because the shellfish-growing states were concerned that the FDA was being too stringent, conditioned upon the Department of Commerce conducting studies to determine what appropriate regulatory activity would be.[12] Those studies have now been completed. I have no information about what the FDA may or may not do.

KIMM: I am glad Peter Hutt was here because I would have answered, "God," in terms of setting standards for sea water.

HUTT: It is the same thing.

KIMM: We make no such claims at the EPA.

HUTT: That must be a new policy.

---

[11] Federal Register vol. 40 (June 19, 1975) p. 25916.

[12] Public Law 94-370 para. 16(b), 90 Stat. 1013, 1033 (1976).

<u>KIMM</u>: That is because I still work there, and you do not.

<u>EMIL MRAK (University of California at Davis)</u>: At lunch today there was a discussion of why more consideration is not given to the relation, if any, of water to heart problems.

<u>KIMM</u>: Obviously, this is a high priority among our research activities (which we are currently in the process of reassessing). There are major efforts going on, particularly in the United Kingdom and Canada, and we try to stay abreast of that work. I think it was correctly pointed out on the first day of this conference that the potential health impact from water softness could be much more serious than many of the things we are looking at. This is one area where we will be focusing our efforts.

<u>HARRIS</u>: I might point out that that was a part of the Environmental Defense Fund's concerns in its 1975 lawsuit. The potential problem of hard and soft water, as related to cardiovascular diseases, may be greater than the cancer problem. It has been unexplainably ignored.

There is one thing I would like to add here. Mr. Davoust said something about how he disagreed with restricting chloramines as a possible chlorine substitute. I was under the impression, from some work I did several years ago, that the chloramines are very ineffective in killing viruses.

DAVOUST: Number one, Mr. Harris, I have no knowledge of anybody, including the EPA, ever finding a live virus unit in a source of potable water. We tried in Chicago for a long time, and we finally had to go up the Chicago River (a highly polluted one) before we could find one live virus unit--and then there were some doubts.

I have always been an advocate of free chlorine and I simply cited that option because I know that Springfield, Illinois, uses this method, and the results are good. After all, it is the state capital--the home of the state health department and the Illinois EPA--and I have never heard of any disease outbreaks.

Indianapolis, Indiana, during the warm-water months also goes to a combined chlorine process. Again, they seem to maintain good quality.

I have no brief for the chloramine residual, but am not about to say that the possibility of "less" virus kill would make me think less of the treatment. If the recommended contact time at a given pH and temperature is maintained, I feel pretty well assured that the water will be adequately disinfected. A few years ago it seemed that the virus potential was going to be the primary concern--now we seem to have dropped the subject in favor of the organics and whatever. Will viruses return?

LINAWEAVER: I would like to respond to the comments on my use of the 200-cancer deaths figure. I am glad I got those comments. I said I did not think it was a valid conclusion, but I wanted to make sure that everyone else heard me the same way.

The thrust of my comments about going full cycle in terms of the level of sophistication with which the water professionals approach the quality of water is that we are going to go through a quantum jump. For example, it was not too many years ago that most water operators had never used parts per billion. They were using parts per million for many decades. This is merely one facet of the very significant change in the sophistication with which the whole issue is approached.

One key caution I wish to stress is that, in order to get the advances in technology which I think are possible, you do not want to jump into any kind of subsidy program. I do think that the time limits are too short, given the way the regulations have been drafted. But the mission will be accomplished. There is no doubt in my mind about that. There are spillover effects which all of these things have inherent in them. In this particular case, they are probably going to allow for more successes than in other environmental issues where you push the solution in one place and a problem squeezes out in another.

The notification provisions will increase by orders of magnitude the number of telephone calls that utility operators will receive. There will be announcements on the radio and the public will want to know if its water is safe. But a little explanation from the people responsible for the quality of water in our cities will have people going away happy. They will continue to use the water, as in the case of the water supply in Baltimore. I think that is true in most other cases.

Notification does have a lot of inherent problems in it. There were a number of comments made about the fine-tuning of that notification process. I do not think we need to go into that. I would hope that we would not have scare tactics, but rather constructive dialogue between the public and the water professionals with all the research and regulatory efforts exemplified by the people attending this conference.

DAVOUST: I wish the response from the public was all that great. It has been mentioned quite a few times here that educational programs are necessary. Somewhere in my material for the conference, I have an article that says that HEW Secretary Califano is going to spend something like $6 million on a new anticigarette campaign. Six million dollars is a drop in the bucket compared to national advertising figures, but, even at that, far more than is spent on water programs.

After the publicity about organics in the New Orleans water supply we sat back in Chicago and waited for the reaction from the public and press. I recall only a couple of telephone calls--that was it. Our consumers seem to be completely blasé regarding water quality, though I would like to think that we had a good enough reputation to inspire confidence.

Generally, the larger utilities seem to get poor response on issues like quality parameters--the public is far more sensitive about system outages and things like rate increases.

LINAWEAVER: A quick footnote to that. I think that the increase in response is not going to appear in the larger utilities; and we happen to be talking about two of the largest utilities in the United States. It will be when the smaller utilities notify the customer of problems that I think you will have a significant public response and significant discussion generated. I would agree that in the large water utilities response is disproportionally small.

HARRIS: I think we had better be careful of how we interpret this apparent lack of public response. There is another way of viewing this. Oftentimes it is the best-educated public that responds, and there is a high correlation between education and income. In many cities, in my experience--and I have had no experience in Chicago, the public did respond. They responded by going out and buying bottled water. They responded by buying the home treatment gadgets. When the public believes that it has an affordable alternative and that solves the problem, it will respond in that way. This is because fighting city hall is not a viable alternative for most people. So, I think the lack of phone calls should not be misinterpreted.

<u>KIMM</u>: The experience with public notification is this. We have tracked a thousand or so instances of public notification now. These have primarily involved small communities, and have dealt with the communities' failing to monitor at all. Generally the response has been pretty good and has led to corrective actions. We are beginning to capture some of this information, and we will be trying to make it available to other utilities.

One example that comes to mind is the State of Oklahoma, which had several hundred small systems in its inventory that were not sending in samples, and presumably were not monitoring at all. The state people called a press conference and issued a notice to all of them, and within a few weeks, not only did those systems arrive with samples, but so did several hundred other systems they had not known about.

There is also the incident of a small community in New York State which the state had been hassling for a long time to build a treatment plant. When the city council finally made the decision to go ahead with a plant, its decision was related to an enforced public notification.

In the big cities, when notifications occur the question will be more a matter of how organizations respond. I do not know how that will come out when it actually happens, but if in your city, some group were to pick up the violation to make an issue out of it, then you might get a great deal of response.

HENRY GRAESER (Black and Veatch Consulting Engineers): I agree with public notification and I agree with the way it has happened. The states and the federal government are telling the community that they have to notify the public. I hope I made the point clear in my discussion that I do wish they could exercise discretion, because when you get a bad sample and you go out and do something about it, then it makes us all appear a little bit ridiculous if you come out on the television two or three weeks later and say that you had a bad sample.

I do not want to lose the effectiveness of public notification. That is the reason I was and am so violently opposed to placing the responsibility of public notification on the purveyor. I fully agree with the right of the federal government or the state agency--that is, the regulator--to notify the public when, in his opinion, it is essential to the public health to do it promptly. In the meantime, I would like to be working on the problem.

LEILA BOTTS (Great Lakes Basin Commission): Up until a couple of years ago, I was director of the Lake Michigan Federation, which is a watchdog operation in Chicago. You, Mr. Davoust, may not have been aware of the public concern in Chicago. My organization was contacted by newspapers and by other citizen organizations about whether there was a drinking water problem in Chicago. We went to see if there were.

Either there really is not much of a drinking water problem in Chicago, or the city has done a terrific propaganda job in convincing people that there is not, because we were not able to find evidence of a significant problem in the city itself. The water comes from Lake Michigan, and there is a public perception that the water supply is okay.

If it is not, then you have not done your job of telling the public. If, in fact, the water supply in Chicago is pretty good, that is the reason you were not hearing from people. It was not that people did not care.

DAVOUST: We did receive inquiries from newspapers and TV stations. I was interviewed on TV, and there were several newspaper articles. Fortunately, we had been working closely with the Region V EPA laboratory in Chicago and had a pretty good handle on the organic (THM) levels in our water. Some reporters tried hard to prove we were not aware of the concentrations and had analyses done on their own. Our figures prevailed and we came out looking good.

HAMMOND: I want to thank the panelists, others who participated this afternoon, and those who put together this conference.

CLIFFORD RUSSELL (Resources for the Future): I would just like to close by thanking all of you for your fortitude in sitting through three tough days; and to repeat our thanks to the sponsors who made those days possible. I would especially like to mention two people who made this conference run smoothly and effectively: Margaret Parr and Audrey Lord-Hausman. And finally, I would like to thank the hotel for good food, excellent service, and reliable equipment.

Part V

Lessons from the Drinking Water Conference

Chapter 15

LESSONS FROM THE DRINKING WATER CONFERENCE: ONE MAN'S VIEW

Clifford S. Russell

It would be a lengthy and tedious undertaking, and probably not a particularly useful one, to summarize the papers and discussions from such a wide-ranging conference as this one. A summary would also present the difficulty that, to justify the name, the summarizer would have to take immense care to see that each spokesman for the several sides of the relevant issues was presented at something like equal time. While the product of such an exercise might be useful to that apparently large but mysterious group which thrives on "executive summaries," it would amount to an opportunity lost. One of the reasons for having a conference, instead of merely commissioning a few papers, is to provide a chance for participants to see the authors and spokesmen, to watch them interact, and thus to judge their arguments on the basis of additional information to that available in their prepared papers. The conference audience presumably knows this, and comes, in part at least, because of it. The reader of a proceedings volume is foreclosed from this direct experience but need not be deprived of all its benefits if some effort is made to reflect this extra dimension. This is what I attempt in this concluding essay--a distillation of what I learned in the course of the conference. Therefore, the reader is warned that not every position argued is reported here; and busy executives will be receiving a dose of my interpretation of how the various sides looked and sounded when arrayed together--how they "came off"--in addition to some summarization.

The chapter is divided into four sections. The first deals with what we know about the costs and benefits implied by existing policy as represented by the Safe Drinking Water Act (SDWA) and the implementing regulations being promulgated by the Environmental Protection Agency (EPA). The second discusses the overall mechanism by which the nation is trying to capture the benefits and allocate the costs. The third section attempts to relate that mechanism to other parts of our apparatus for dealing with environmental problems, broadly construed. Finally, the fourth section sets out some suggestions for the revision of current policies which seem to me deserving of serious attention. Not all of these alternatives are equally desirable, but there are links among them such that the least desirable could probably be ignored if the more desirable changes were made.

On Costs and Benefits

During the closing panel session, Pierce Linaweaver cited estimates of the total cost of implementation of the SDWA of around $500 million annually. He then paired this number with an estimate from Carol Tate's discussion of the National Academy Committee Report (National Academy of Sciences, 1977) that, as he understood it, 200 cancer deaths per year were attributable to drinking water. Even assuming that all the deaths were eliminated by adherence to the new regulations, the average cost per death avoided would be enormous, some $2.5 million. On the other hand, Robert Harris, referring to some work he has done in Cincinnati, Ohio, said that if adherence to the new regulations caused even as much as a one percent decrease in the cancer rate, the amount of money saved, based on the average U.S. cost per cancer case (including lost earnings), would be such that at any valuation of human life above $70,000, the SDWA regulations would be economically a good deal.

What can we say about the roots of this rather striking disagreement? Quite a bit, it seems to me, although none of it is especially encouraging for those who feel uncomfortable with public policy which cannot be supported by a straightforward economic efficiency calculation. One problem was, of course, that Linaweaver had misinterpreted the figure on cancer deaths that he used; 200 was the number of deaths tentatively attributed to chloroform alone in drinking water supplies. That figure ignores all other organic chemicals individually and, a fortiori, does not account for interactions among them. But, more fundamentally, the problem is that on the benefit side, we do not begin to know enough to estimate dollar benefits of the SDWA; and on the cost side, though the range of disagreement is much smaller, publicly available estimates also vary widely. Why do we have the grave difficulties with benefits? The catalog of problems has some entries which are familiar from other environmental policy debates. But there are also some unique difficulties. And the overall prognosis for dramatic improvements in our abilities to eliminate them cannot be optimistic.

First, we have only incomplete knowledge of what potentially harmful contaminants are in drinking water. While our knowledge seems to be growing at a rapid pace, it also seems clear that we are going to go on discovering how much we did not know yesterday as we increase the sophistication of our analytical methods tomorrow. If we do not know what is present, we may not be attributing observed morbidity and mortality effects correctly, and, potentially as important, we may not be setting out to reduce the concentrations of the "right" substances.

Our second major difficulty is that, even given complete knowledge of contaminant levels, we would not have the ability to estimate the effects of chronic exposure to those substances. Consider only cancer rates, to single out the disease which dominated discussion at the conference.[1] The difficulties in the way of obtaining the desired knowledge

---

[1]It was pointed out by several speakers that there may be even more important problems in the apparent link between water hardness/softness and cardiovascular disease, but no one came to grips with this question, and certainly no one claimed we know more about it than about cancer. So the remarks in the text probably apply to other potential concerns as well.

are immense. Manageable toxicological experiments require high doses to keep down the number of animals. This dosage leaves the knowledge gained open to objections concerning not only the appropriateness of the animal "models" but also the possibility that very high doses overload some natural defense mechanisms and are thus qualitatively different from the low, chronic doses actually at issue. But beyond that, we have a devilish problem deciding what to feed our animals, for if we take seriously the arguments about synergistic and antagonistic effects of chemicals, we are not faced with the "simple" task of testing 400 or 700 substances, but of exploring an essentially infinite space of combinations of those substances. Thus, as a practical matter, toxicological tests may never be able to tell us quantitatively what the effects of observed contamination of drinking water are, though as William Lijinsky said, it may be that laboratory studies of toxicity are "the best we have now" as a basis for trying to avoid risks.[2]

Epidemiology has its own problems when it comes to estimating effects. It has to work with the results of natural experiments which, while involving the desired human model and the actual chronic, low-dose contamination, also involve

---

[2]There is a strong ray of hope here in the shape of short-term toxicity tests such as that of Bruce Ames. Their development and refinement would certainly provide a powerful new tool for identifying problems, though the conference discussion, particularly the exchanges between Lijinsky and Joyce McMann, leaves us with doubts about the prospects for quantitative results here as well. (McMann's discussion was not available in written form for inclusion in these proceedings.)

the inevitable noise associated with people who change locations over the crucial time periods, who eat tainted foods as well as drink contaminated water, and who change their habits over time; and with environments which also change, sometimes without recorded evidence. (Nancy Reiches cataloged these and other difficulties.) Thus, we turn in vain to epidemiologists for quantification of effects. As Reiches said, "The collective weight of all related (epidemiological) evidence suggests a real problem exists." It is a long step, however, from interpreting a set of a dozen studies with partially conflicting results to saying that X cancers of site Y per year are attributable to the drinking water contamination of city Z.[3]

Thus, we simply do not have now the knowledge to estimate what the effects of drinking water contamination are, and our knowledge and techniques cannot tell us what we would gain by reducing that contamination.[4] This is _not_ to say that there

---

[3]If contaminants in water merely add to the total body burden of contaminants received in other ways as well--as through food--and if the morbidity/mortality effects are non-linear in total body burden, the attribution of cases to water supply has very little meaning. But, we could still hope to estimate the marginal _reduction_ of cases attributable to _cleaning up_ the water supply.

[4]If we could tell this, we would still face grave obstacles to obtaining benefit estimates, for the valuation of benefits which consist of avoiding sickness and pain, or gaining a year or two of average life expectancy, is a perilous business. In the circumstances, the tack taken by Harris in the example above is a sensible one, that is, we can hope to show what less sickness or longer life would have to be worth to justify a policy, the costs of which are fairly straightforwardly estimated. (See below for more on costs.)

is no agreement on the general outline of the problem. There was very little inclination at the conference--and none among the scientific participants--to deny that the small concentrations of known or suspect carcinogens found in water supplies have some effect on cancer rates and life expectancies. But we are left in the uncomfortable situation of having to choose an appropriate insurance policy and premium on the basis of almost no quantitative knowledge of the risk. We do know that if we postpone action while waiting for more convincing evidence, we may be generating many times the currently observable cancer rates. This is because we are now seeing the effects of twenty or thirty years' exposure beginning in the late 1940s, before the production of synthetic organic chemicals began its tremendous growth. Since there has almost certainly been a continuing sharp increase in exposure to the products and residuals of this industry over these years, even if morbidity and mortality are only linearly related to exposure, we can expect nonlinear increases in these measures of effect.[5] On the other hand, it simply does not seem possible to reject out of hand the possibility that the effects of

---

[5]There has been a sharp reduction in the rate of increase of exposure over the past seven or eight years, as environmental regulation has removed some pesticides from use and tightened discharge constraints. But some of the compounds in question are very stable and we may be living for a while on a plateau regarding them.

doing nothing now would be quite small. Here we have the makings of a most difficult and confusing policy debate. The scientists, to whom we are accustomed to turn for answers, do not really have any. They give us impressions of fragmentary evidence, which descriptions tend to get mixed up with policy prescriptions based on personal judgments about the appropriate degree of risk aversion for society.

The debate also involves costs, of course, for if removal of trace contaminants were very cheap, we would be much less concerned about the elusive benefit measurements. The cost side is easier to pin down, or, more accurately, to peg within a range which is much narrower than the range of estimated benefits. Tables 1 and 2, which summarize some results from the conference papers, give some feeling for this assertion.

Accepting, for a moment, the EPA figures provided by Robert Clark and Richard Stevie, we can see from Table 2 that annual costs for a family of four can vary enormously depending on what actions are required and what size plant is appropriate. A large system with a pure, upland source requiring only monitoring and perhaps a change to a new disinfectant might only have to increase the annual bill of this family by less than one dollar. But a small, 1 million gallon per day (MGD), system with turbidity <u>and</u> organic problems might be increasing annual bills by \$114 for the same average family. Estimates of the aggregate cost of SDWA are therefore difficult--requiring case by case estimation of actions required--

Table 15-1. Estimated Unit Costs of Complying with Safe Drinking Water Act

(cents per 1000 gallons)

Activity	System Size (million gallons per day) 1	10	100
Organics Control			
Monitoring[a]	1¢	0.5¢	0.1¢
Avoiding Chlorination[a]	4	1	0.5
Granular Activated Carbon Treatment[a]	70	24	13
Granular Carbon[b] (for comparison)	-[d]	-[d]	(20)
Total	75¢	25.5¢	13.6¢
Turbidity Control			
Clark[a]	35	20	9
Sinnott[c] (for comparison)	(80)	-[d]	(16)

Note: For a system with organics and turbidity problems, cost of compliance with SDWA will be additive. The estimates from Sinnott are included for comparison.

[a]Source: Robert Clark and Richard Stevie, "Meeting the Drinking Water Standards: The Price of Regulation," this volume.

[b]Source: Henry Graeser, "Financing System Changes, this volume.

[c]Source: Walter Sinnott, Discussion of Clark and Stevie paper, this volume.

[d]Not available from this source.

Table 15-2. Costs for a Family of Four of Complying with the Safe Drinking Water Act

(dollars per year per household)[1]

Activity	System Size (millions gallons per day) 1	10	100
Organics Control			
Monitoring[a]	$1.0	$0.5	$0.1
Avoiding Chlorination[a]	4.1	1.0	0.5
Granular Activated Carbon[a]	72.8	24.5	13.3
Granular Carbon[b] (for comparison)	-[d]	-[d]	(20.5)
Subtotal	$77.9	$26.0	$13.9
Turbidity Control			
Clark[a]	35.7	20.4	9.2
Sinnott[c] (for comparison)	(81.6)	-[d]	(16.3)

[1]Estimated on the basis of average daily use of 70 gallons per person, or 102 thousand gallons per year per family of four.

Note: For a system with organics and turbidity problems, cost of compliance with SDWA will be additive. The estimates from Sinnott are included for comparison.

[a]Source: Robert Clark and Richard Stevie, "Meeting the Drinking Water Standards: The Price of Regulation," this volume.

[b]Source: Henry Graeser, "Financing System Changes," this volume.

[c]Source: Walter Sinnott, Discussion of Clark and Stevie paper, this volume.

[d]Not available from this source.

and of very little use in understanding how the act will appear to particular utility managers and customers. It is entirely conceivable, for example, that if we knew the aggregate benefits of the act, we would find that they exceeded the aggregate costs, but that a majority of water system customers were in fact incurring costs greater than their expected benefits.

To make such matters more difficult, even on this side of the equation, we can see from the tables that the EPA cost estimates are not universally accepted as the basis for debate. Sinnott, of Hazen and Sawyer, for example, presented the conference with turbidity control cost estimates which are roughly twice as large as EPA's, and Graeser's cost estimate for carbon adsorption, attributed to Black and Veatch, is about 50 percent larger than EPA's for the 100 MGD system. The reasons for differences on costs include differences in engineering judgment on such items as the appropriate peak-to-average flow rate, required carbon contact time, and required regeneration frequencies for carbon. But the differences also reflect guesses about future system growth which enter through the chosen degree of excess capacity.[6] If one accepts

[6]Where there are economies of scale and expectations of future growth, it is optimal to build ahead of demand. How far it is optimal to go with this, however, depends on the plant scale factor, the interest rate, the allowance made for errors in prediction, and a "political factor" reflecting the ease or difficulty in getting future projects approved. For a discussion of optimal building-ahead policies, see Russell, Arey, and Kates (1970).

the two to one range indicated by the table and applies it across the board, with EPA assumed to be at the low end,[7] we can see that the SDWA bill for a family of four may be anywhere from less than one dollar per year to perhaps $230 per year, even when we talk about systems of at least 1 MGD capacity (around 14,000 customers assuming a 100 percent domestic system with average use of seventy gallons per day per capita). Because economies of scale continue to work in reverse below that size--even allowing for the package plants discussed by Clark and Stevie--very small systems might be facing annual increases in costs per family on the order of $300 or more. Therefore, though we know what we have to do and what it will cost to do it, we will still have considerable differences over specific costs and over the cost-benefit balance in particular instances. Informed customers of different systems, even were they to hold similar judgments about the likely benefits and the appropriate attitudes toward risk, might very well differ on the desirability of the act, either because of differences in their judgments about "best" cost estimates, or because of differences in system size.

---

[7]Because EPA has, to some extent, to "sell" the act to reluctant states as discussed in the next section, it is reasonable to suppose that the agency would take great care not to overestimate costs. My experience with water and air pollution control cost estimates persuades me that this motivation does tend to be reflected in EPA cost estimates, which are often lower than those of other parties.

The discussion of costs at the conference concentrated almost entirely on supply treatment alternatives, applicable to more or less polluted sources, and available now (subject to financing and construction lags). Later in this chapter I mention alternatives applicable at the household level, but even at the system level, total costs will be influenced by arrangements and alternatives not covered in the papers, some of which can only become important over time as system expansion allows choices among alternative sources.

First, there is the modern restatement of the venerable tradition that clean, upland sources are to be preferred to polluted sources lower down in river systems, even if costs of storage and transmission are higher. In the context of concern about synthetic organics this tradition, originally directed at communicable disease problems, takes on new appeal. Whether the answer for a particular place and time is a pumping station on the local estuary or one or more reservoirs in more distant mountains depends on not only a complex of facts about treatment cost and efficacy, and residual risk from contamination, but on social judgments about the importance of the preservation of free-flowing upland streams for aesthetic, wildlife, and recreation reasons. A related question concerns the value of continued protection of reservoir watersheds, with one argument being that since filtration will now be undertaken in any event, there is no longer reason to lock up huge watershed areas under unproductive, and sometimes completely unused forests. This issue was raised in a particularly clear fashion by the plans of several private water companies in Connecticut to sell large parcels of watershed land.[8] Again, questions of treatment effectiveness and cost, residual risk, and the value of undeveloped green space--sometimes open to recreation and sometimes not--must be considered in arriving at a specific answer for a specific place.

Second, it may be contended that costs of compliance with SDWA or any alternative legislation can be significantly reduced by regional water management practices. These practices go beyond interconnection and centralization of storage and treatment to reap economies of scale, and include integration of water supply and wastewater disposal to allow best use of the available natural resource.[9]

Third, the possibility of establishing dual water systems is receiving increasing attention, as this reduces the volume of water which must either be drawn from strictly protected watersheds or subject to extensive and expensive treatment. And while participants tended to dismiss this alternative as

---

[8]New York Times, December 5, 1976, p. 35. A moratorium on such sales, pending study of the issue, has successfully been put into effect by the state legislature.

[9]For a complete and careful discussion of regionalization in England and Wales and its relevance to the U.S. situation, see Daniel A. Okun (1977).

impractical, their comments were based implicitly if not always explicitly on the assumption that "dual system" must always mean two pipes to every home. In fact, there exists a wide range of possible adjustments which conserve high purity water by supplying lower quality water where that will suffice. This may be especially attractive where industrial users are tied into the municipal system so that the problem of providing parallel piping is small relative to quantities of water involved.[10]

On the Mechanism of the Act

The Safe Drinking Water Act establishes a national goal which is to be achieved through national standards and treatment requirements promulgated by a federal agency. But putting these standards into effect at the level of the water system and enforcing them over the long haul are supposed to be the responsibilities of the states and of the water suppliers themselves. Furthermore, the costs of meeting these national standards are to be borne by the customers of the utilities directly. A key instrument of enforcement is the notification requirement, whereby water suppliers which fail

---

[10]Okun, one of the nation's principal advocates of dual systems, has written about their advantages in many places. See, for example, Daniel A. Okun (1975), pp. 1-21; and, in the English context, Okun (1977), pp. 197-205.

to meet the quality or monitoring requirements of the act must so inform their customers. There is also provision for state action to call attention to failures if the local systems do not.

This approach came in for repeated criticism during the conference, particularly, as is hardly surprising, from state and local officials. Mayor Moody, of Columbus, Ohio, President of the National League of Cities, is worth quoting at length in this regard.

> Providing safe drinking water has long been a local government service and responsibility for more than 15,000 public water systems. With the advancement of new technology and research, the Safe Drinking Water Act has now brought federal standards into the picture. It is the consensus of the nation's cities that where federal standards are established, the federal government must give local governments adequate capacity and time to achieve those standards. Local governments must have flexibility to select among alternative means to achieve federal and state mandates and be able to weigh the economic, social, energy, and environmental costs and benefits of alternate strategies with locally determined plans and priorities.
>
> Since the 1974 act contains no financial assistance for cities, many local officials are becoming increasingly uneasy as compliance costs are beginning to rise far above EPA's original estimates.... Many of the cities responding will need water rate increases from 10 to 100 percent. EPA has stated that: "Most public drinking water systems, particularly the larger ones, already meet the standard"; and that the, "new program is therefore a form of preventive medicine." With the many economic, social, and physical ills facing our cities today, I ask, is this really a cost-effective way to meet a national priority? Although the regulations contain variance and exemption procedures for those water supplies that are already safe for the public, the act still requires the spending of substantial tax dollars for marginal water quality improvements in many cases.

> Very few elected officials come into office by promising clean drinking water. That's expected of us. But many of us might leave office soon after we enclose the first noncompliance notification in our water bills to the public.

What was perhaps slightly more surprising was the extent to which those with a stake in the act seemed to recognize the problems created by the chosen mechanism. Robert Harris of EDF and Victor Kimm of EPA, speaking during the final panel session, both called for more education and public participation to create incentives for local utilities and governments to respond as Congress hoped and intended.

For example Harris said:

> Another policy issue, of course, relates to the hostility that has developed across the country regarding federal intervention into what has for the past century been a state responsibility. However, in the case of regulations on carcinogens, the EPA, and not the states, is in the best position to conduct the health effects and technology research necessary to develop such regulations. Where the EPA is failing is in not properly educating the public and the state and local officials about the results of their analytical and epidemiologic studies. I am convinced that once this scientific case has been laid before the public and their representatives, the EPA will receive overwhelming support for what they are trying to do.

And Kimm stated:

> To begin, I would like to note EPA's interest in promoting and facilitating meaningful participation by state and local officials, as well as representatives of all groups and individuals interested in the quality of the nation's drinking water, in formulating the regulations and policies guiding the implementation of the Safe Drinking Water Act....

> I think the National Drinking Water Advisory Council, with its broad membership, has made a major contribution to opening up the regulatory process for public review and comment....
>
> Finally, I would like to return to where I started, and say that I think doing the public's business in a regulatory fashion today has new dimensions to it. It is no longer simply a matter of mandating standards, but of convincing people that such standards do make sense, and are, in fact, attainable.

The problem is a complex one, involving attitudes, institutions, and incentives. There is, first of all, a tradition of local control of the water supply with only minimal interference, from the state. This fact creates a bias against federally imposed standards, however clear their justification. The second problem is one discussed above: the justification, still a matter of some dispute and uncertainty among scientists, is far from self-evident to the customer at the tap. Third, as just observed, even were there an agreed level of benefits from the act, many customers of small or badly placed systems might oppose the local implementation of federal standards because their costs would be very high relative to benefits. Fourth, considering that this requirement for increased spending on what might reasonably be called an environmental issue comes on top of a long period of publicity of other, similar increases, local officials may be forgiven for thinking of straws and camels when they see plans for water treatment plants.

These observations suggest further questions. Why should there be federal involvement at all? How could an act which makes so many state and local people angry pass? What incentives are there in the act that might give EPA sufficient leverage to achieve the act's goals in the face of this dissatisfaction? Because the first two questions open up deep areas of political and economic theory, they cannot be completely dealt with here. As for federal involvement, we can note that the classic justification for federal policies in the environmental field--the existence of interstate pollution effects through air or water borne wastes--is missing here. With few exceptions, we are talking about the supply of water to customers in a single political jurisdiction, by a creature of that jurisdiction--whether public or private--with full control over at least the final treatment of the product, if not the quality of the raw water. Other justifications, however, were expressed or implied at the conference. These included:

- The desirability of uniform national standards, both because citizens and visitors should know that the water is equally safe everywhere, and because no group should be made "second-class citizens" by having a lower quality water supply forced on them;

- The lack of information and technical expertise at the local level, forcing the federal government, with its superior command of these two key resources, to take charge;

- The inefficiency involved in trying to carry the complex and subtle arguments about costs, benefits, and uncertainty inherent in this policy area out to each local jurisdiction or each state government, and to persuade each of them to act.

Reasonable people can differ on the validity and weight to be given such considerations, but we should note the heavy dose of paternalism (a word which was used several times at the conference).

How does such an approach become law? A complete answer is certainly beyond my powers, and in any particular case could probably provide the material for a full-length study of the political process. But one observation which may be relevant came out of the conference. There exists in almost every field a Washington-oriented arm or subgroup. The units of such a group may be law firms, trade associations, retained lobbying firms, corporate offices, or special organizations, such as public interest groups headquartered here. These toilers in the Washington vineyard naturally think of national solutions and have well-developed relations with relevant offices of the federal bureaucracy, with the key staff people on Capitol Hill, and with "the national media" (New York Times, Washington Post, Associated Press, United Press International, and network television, for example). These groups provide a significant fraction of the input from "outside" received by the people who shape national policy. Collectively with their contacts, these groups do much of the defining of national issues and determining of what is on the national agenda. Without questioning the motives of any of the actors involved, one may ask whether such a closed system may not be slightly unstable, in the sense of forwarding an idea whose time has come in the District of Columbia, without checking its timeliness in the hinterlands. Such tension between leadership and representativeness is, of course, endemic in representative democracy. Only referenda decisions on all issues can avoid it, and I think we would all (or nearly all) agree that the cure would itself be a serious disease.

What about incentives? One obvious incentive, common to much of the rest of our environmental legislation, would be a grant program to shift the burden of compliance from water system customers to federal taxpayers. The omission of such a program from the SDWA was apparently intentional, and, according to Philip Cummings of the staff of the Senate Committee on Environment and Public Works (discussant of the Eaton, and coauthors, paper) Congress has no intention of adding a grant program. Apparently, experience with the water pollution control construction grant program has left its mark. (Cummings' remarks were not available for inclusion in the proceedings.)

What possibilities are there if a grant program is ruled out? There is first and foremost the notification requirement, designed to let Congress and EPA speak directly to the water system customers whenever a system is in violation of standards or monitoring requirements, saying: "We have set up standards to protect your health, and your water supplier is failing to meet them." The fact that the actual notice would come from the utility itself seems merely to make the procedure more galling--it certainly does not alter the potential power as seen from the utility manager's point of view. Indeed, this incentive may well be powerful enough to succeed despite the lack of financial incentives and despite local opposition in principle to federal standards. Then again, it may not. It depends on the reactions of customers. The discussion at the conference of evidence on reactions available so far was inconclusive. Kimm said that experience was accumulating which would soon allow some definite lessons to be drawn. But the conference audience had to be content with a few anecdotes showing only that people sometimes get aroused and other times are passive. Duane Baumann's study of the Duluth experience during the Reserve Mining Case (before additional filters were built) suggests that about half the system's customers could not be bothered to react to that well-publicized episode. But if a quarter of a utility's customers _were_ to be aroused by a notification--and, of course, what is important is what the local manager and politicians _expect_ will happen--there would be more than enough reaction. And if notification does not provide sufficient incentive, it is difficult to see how the ends will be achieved. It seems unrealistic to look to such draconian measures as criminal proceedings against local officials. Shutting down an offending system is completely out of the question, implying vast costs and health risks for all concerned.

Finally, it should be added that some curious negative incentives are developing as implementation progresses. First, as pointed out by Leo Ganster of GAO, those states which, for one reason or another, do not choose to seek primary responsibility for implementation and enforcement are being treated to a curious spectacle. The EPA effort, which supplants what they were supposed to do, is to be carried out with about one quarter of the manpower the state would have been required to use, and will, thus, be considerably less complete and careful. However, it appears to Henry Ongerth and others that EPA is being overly strict and onerous in supervising the plans and activities of states which <u>are</u> responding to the wishes of Congress by working to assume "primacy." The natural question becomes, "Why bother?"

On Relations to Other Environmental Policies

The Safe Drinking Water Act is only one room in what is by now a rather large and rambling edifice of environmental policy, and some of the facets of its relationship to other rooms are worth reviewing. In particular, I shall examine very briefly:

- direct physical relations between pollution sources and receptors which connect SDWA and the Federal Water Pollution Control Act (FWPCA) as amended;
- potential jurisdictional overlap between SDWA and the Food, Drug and Cosmetic Act (FDCA);

- inconsistencies in views of risk, balancing of costs and benefits, and burden of proof, between SDWA and several other pieces of legislation with broadly similar goals.

Water Pollution Control

It used to be an article of faith among those of us who worked on the analysis of costs and benefits of water pollution control (when biochemical oxygen demand, dissolved oxygen, and perhaps bacterial contamination were the central problems) that you could never justify high levels of waste reduction at upstream pollution sources on the basis of downstream water supply treatment cost reductions. That treatment was simply too cheap, especially when you took account of the self-purification capacity of the intervening water course (Frankel, 1965).[11] At least among researchers, therefore, the water course itself was the focus of attention on the benefit side. And depending on the current state of the water, its location and history, and the researcher's background, that focus meant concentrating on aesthetic, recreational, or ecological benefits.[12] The recreational and ecological justifications survived the ivory tower stage to end up in the FWPCA amendments of 1972 as "water quality which provides for the protection

---

[11]For estimates of the actual savings to downstream cities (in 1968 dollars) see Clifford Russell (1972).

[12]The famous Delaware River estuary, water quality modeling project, involved generalized recreation benefits and estimates of shad-run survival as balances to waste treatment costs (Kneese and Bower, 1968).

and propagation of fish, shellfish, and wildlife and provides for recreation in and on the water," commonly known as "fishable, swimmable waters," the 1983 goal of those amendments.

It now appears to be time to re-examine the conventional wisdom of the past decade or two. Given the persistent nature of many of the contaminants central to the new drinking water problem (especially the organics), we cannot expect even fairly long stretches of river to offer any significant load reduction possibilities. At the same time, the treatment methods required at source and receptor will be essentially the same, involving granular activated carbon. Therefore, the central tradeoff will, in many cases, be between treatment of small volumes with high concentrations of contaminants at sources, and treatment of large volumes with low initial concentrations at water supply intakes. In other situations the bulk of the organic or other worrisome contaminants may be coming from agricultural or urban/suburban runoff, and here our imperfect knowledge of mechanisms and of costs of discharge reduction will complicate the analysis. But the exercise will be equally valuable and should certainly be a high priority of the section 208, FWPCA amendments, planning agencies for their respective regions. At the same time it is necessary, as C.C. Johnson pointed out in his paper, to be cautious about the apparently attractive alternative of on-land disposal of waste water. Though this alternative is now officially to receive preferential consideration in planning, it is far from clear that it should be favored. Again, lack of understanding of mechanisms makes analysis of cost effectiveness a tenuous exercise, but we must try to understand what levels of pretreatment on the surface prior to spraying on the land, in combination with what types of treatment of groundwater withdrawals prior to rise in water systems, are necessary to comply with the SDWA regulations in particular circumstances. Otherwise, serious inefficiencies (externalities) will be created if the disposer ignores the costs of ground water contamination incurred by the "downstream" supplier.

Safe Drinking Water and Pure Food Drug and Cosmetic Act

The conference discussed several times the potential jurisdictional conflicts between the Food and Drug Administration (FDA) and EPA, created by the unresolved tensions between the Food, Drug, and Cosmetic Act (FDCA) and the SDWA. Timothy Harker, discussing Ongerth's paper on jurisdictional problems; James Turner, discussing Mrak's paper on the food industry's special problems; and Peter Hutt in his panel presentation in the closing session: all gave significant attention to these tensions. Some others, notably C. C. Johnson and Victor Kimm, were skeptical of the importance of the issue, taking the practical line that neither agency had sufficient resources to do its own job and, therefore, had neither the incentive nor the capacity to attempt to interfere with the other. I find their argument unpersuasive and believe the issue identified by the other participants must be taken seriously for two reasons. First, there is a disturbing inconsistency between the FDCA and SDWA in their approaches to the risks of trace contaminants. This is, however, part of a much broader pattern of inconsistency which is the subject of the next subsection. I, therefore, skip over it for now.

The second reason for concern about the FDA/EPA relation in the drinking water field was best summarized by Timothy Harker.

> The statutory overlay with respect to drinking water is important for all forms of drinking water, that is, municipal water supplied by public water systems, bottled water, and drinking water used in the processing or packaging of other foods. While it may be true that FDA has not yet exercised its authority over public water supplies, an important distinction exists between having the authority to regulate (which FDA has) and choosing to exercise that authority. Regulatory priorities change from one administration to the next and may even change more frequently in response to public attention or the pressure of public or special interest groups. Reliable long-range business and public policy cannot be made on the basis of an assumption that because FDA has not yet exercised substantial authority over public drinking water, it will not do so in the future....
>
> For example, a food under the FDCA is subject to the provisions of the Delaney Clause and to FDA's aggressive posture toward the addition of carcinogenic substances to the food supply....
>
> Bottled water is a food subject to the FDCA. It is not clear whether FDA would view the bottling of municipal water which contains trihalomethanes within EPA's standard (proposed at 100 ppb) as prohibited.... Such an interpretation would appear to be impermissible, but confusion will persist...until Congress or the FDA provides a clear direction.

> Another example of the potential conflict caused by overlapping jurisdiction pertains to drinking water used in food packaging or processing. FDA has proposed to ban under section 409 of the FDCA the addition of chloroform to food. Yet EPA will apparently permit 100 ppb of chloroform in drinking water. It is uncertain at this time whether the addition of such drinking water to other foods for processing or packaging would be construed by FDA to constitute a violation of the Delaney Clause.

One can add very little to this summary except to recognize that clearing up this problem will involve the Congress in some very emotional issues, not the least of which is the Delaney Clause. One need not be especially cynical to guess that in this circumstance the practical arguments may carry considerable weight on Capitol Hill. Congress may prefer to encourage continuation of the current spirit of cooperation and hope for the best, rather than to come to grips with decisions on the appropriate level of risk to tolerate in different types of environmental exposures and for different contaminants. Congressman Rogers in his address to the conference explicitly took this line, exhorting the agencies to work together and eschew "interagency chauvinism."

The extent of this problem of inconsistency is made clearer in the next subsection.

Toxic Substances and Human Risk in the Law[13]

It is almost inevitable, and certainly not a matter for censure, that governments will frequently produce inconsistent

[13] This subsection is heavily based on an excellent piece by my colleague, Paul Portney (1978).

legislation. After all, problems are recognized sequentially, and the personnel of government change over time--most importantly, the makeup of the legislature changes with shifts in population, public opinion, and the aging of individual legislators. In addition, even problems which are conceptually similar may look quite different when addressed ten years apart because our knowledge has increased.

Therefore, the summary in Table 3 of the inconsistent provisions on fundamental points of legislation relevant to toxic chemicals should be read as a catalog of opportunities as much as a critique of mistakes.

The fact remains, however, that there appears to be no good reason for retaining this welter of inconsistency, while strong arguments of principle support arriving at consistent positions on the several dimensions of major concern. Thus, in the matter of defining a toxic substance, the applicable legislation sometimes uses the forms "_will_ cause" or "_will_ present" or "_will_ have," and sometimes relaxes the definition to "_may_ cause" or "_may_ present" or "_may_ have." Clearly, in proving that some substance should be subject to a particular restriction or requirement, the difference between those two approaches is enormous. Similarly, the degree of protection required by the several acts varies from "an ample margin of safety to protect the public health," through "that _necessary_ to protect human health and the environment," to "_adequate_ measures to the extent feasible that no employee will suffer material impairment." The differences in implication for specific policy

Table 15-3. Toxic Substances in the Law

Legislation	Definition toxic or hazard	Type of regulation	Degree of protection	Balancing of costs	Burden of proof
Clean Air Act (as amended) (CAA) 1970 1977	"an air pollutant ...which...may cause, or contribute to, an increase in mortality or an increase in serious irreversible, or incapacitating illness." Sec. 112(a)(1)	Emission standards	"...an ample margin of safety to protect the public health..."	No	EPA
Federal Water Pollution Control Act (as amended) (FWPCA) 1972 1977	"...pollutants which will...cause death, disease, behavioral abnormalities, cancer, genetic mutations, physiological malfunctions...or physical deformations." Sec. 502(13)	Effluent standards, ambient standards	"...ample margin of safety."	No	EPA
Occupational Safety and Health Act (OSHA) 1970	Not defined	Exposure standards	"adequately assures to extent feasible that no employee will suffer material impairment of health or functional capacity..." Sec. 6(b)(5)	Yes Sec.6(b)(5)	OSHA
Toxic Substances Control Act (TSCA) 1976	those substances "...presenting an unreasonable risk of injury to health or the environment...." Sec. 6(a)	Pre-market notification and testing; prohibitions on manufacturing, processing, and distribution; information on chemical components must be supplied to EPA	Not specified	Yes Sec.2(b)(5)	Proponent

Legislation	Definition toxic or hazard	Type of regulation	Degree of protection	Balancing of costs	Burden of proof
Federal Food, Drug, and Cosmetic Act (as amended) (FDCA) 1958 1962	Not defined	Labeling; bans on products deemed "unsafe"	"...necessary for the protection of public health ..."	No in case of food additives; yes for drugs and cosmetics	Proponent for drugs and food additives; FDA for cosmetic ingredients
Federal Insecticide, Fungicide, and Rodenticide Act (as amended) (FIFRA) 1972	One which results in "...unreasonable adverse effects on the environment or will involve unreasonable hazard to the survival of a species declared endangered..."	Registration of all pesticides and uses; permits for applicators; cancellation or suspension of specific pesticides or uses	Not specified	Yes Sec. 6(b)(2)	Proponent
Safe Drinking Water Act (SDWA) 1974	"...contaminant(s) which...may have an adverse effect on the health of persons." Sec. 1401(1)(B)	Maximum contaminant standards	"...to the extent feasible...(taking costs into consideration) ..." Sec. 1412(a)(2)	Yes Sec. 1412 (a)(2)	EPA
Resource Conservation and Recovery Act (RCRA) 1976	one which "may cause, or significantly contribute to an increase in mortality or an increase in serious irreversible, or incapacitating reversible, illness; or, pose a...hazard to human health or the environment...." Sec. 1004(5)(A)(B)	Standards for generators, transporters of hazardous waste; permits for treatment, storage or disposal of hazardous waste	"that necessary to protect human health and the environment..." Sec. 3002-3004	No	EPA

Source: Adapted from Paul R. Portney, "Toxic Substance Policy and the Protection of Human Health," in Paul R. Portney, ed., _Current Issues in U.S. Environmental Policy_ (Baltimore, the Johns Hopkins University Press for Resources for the Future, 1978), p. 130.

decisions must be tested through the courts, but it is clear that these phrases can be construed differently. In this regard, notice that the acts also differ in whether or not they specify that a balancing of costs and benefits is to be part of policy formation. And finally, we see that the locus of the burden of proof of the potential harm from the substance is also inconsistent. In some cases, such as Federal Insecticide, Fungicide, and Rodenticide Act (FIFRA) and Toxic Substances Control Act (TSCA), the proponent must show that "unreasonable risks" will not be created by use. In the other cases, the federal agency with responsibility for protection must show that risks will or may be created.

The choice, in any attempt to bring more order to our national policy in this area, is not limited to these extreme positions, but they do have heuristic value. Thus, we can set out a conservative, risk-avoiding position which:

- defines a toxic substance as one which <u>may</u> cause problems;
- provides for an <u>ample margin</u> of safety in seeking to avoid those risks;
- <u>does not permit</u> the balancing of costs and benefits in relation to the risk (or even the consideration of economic feasibility);
- places the burden of proof of safety on the proponent.

Or we can set out an approach, more accepting of risk, which:

- defines a toxic substance as one which <u>will</u> cause problems (does pose a risk);
- provides for <u>adequate</u> protective measures against the risk;

- *requires* the balancing of costs and benefits in relation to the risk;
- places the burden of proof of the existence of the risk on the would-be regulator.

It is interesting to note that in fact not one of the eight acts described in Table 3 fits either extreme. All are compromises of one sort or another. The problem is not with compromise *per se*, it is with the apparently random choice of the elements of each one. We might, for example, consider applying a compromise approach which:

- defines a toxic substance as one which *may* pose risks;
- provides for *adequate* protective actions against the risk;
- *permits* considerations of costs and benefits in relation to the risk;
- places the burden of proof of negligible risk on the proponent of the substance.

But whatever the mix chosen, it does seem important to begin thinking and arguing now about adopting some more unified stance in dealing with threats from toxic substances in the environment. One can hope that any reconsideration of the SDWA will be seized as an opportunity to do so.

On Possibilities for Changes in Safe Drinking Water Policy

The last subsection identified some rather grand issues which ought to be addressed by society, and particularly by its elected representatives. There are, however, several possibilities for more narrowly focused changes to drinking water policies which seem to me, on the basis of the conference papers and discussions, to warrant attention. These include the extent of state and local autonomy, the question of how the bill for our national policy is to be paid, and the possibility that alternatives other than central treatment of all water may be desirable for some areas.

State and Local Autonomy

No single complaint was more often heard at the conference than that of federal government arrogance and inflexibility in the pursuit of what all recognized was a worthy goal. Henry Ongerth's paper is the most complete and telling statement of this view, containing some appalling quotations from EPA comments on the California FY 78 draft grant application. It seems clear that this particular sore can only continue to fester and poison other aspects of federal relations, the more so when it is seen as one in a rather long series of similar pains. And there seems to be particular merit in state complaints about the SDWA, for as I have already discussed, there do not appear to be strong arguments for federal direction here.[14]

---

[14]Indeed, one is tempted to guess that the strongest argument is one which we only use at the peril of the entire system--that local and state governments are unresponsive to perceived problems, or to the desires of their constituencies, or both. There is no way to make an objective judgment about this charge, but to accept it is to say that the federal system is inadequate and should be changed.

There are, on the other hand, very good reasons for expecting local judgments about the appropriate level of drinking water purity to differ across the country. The most straightforward of these, discussed in the first section of this essay, is that treatment costs will vary enormously with system size.[15] Even if everyone had the same (correct) understanding of the risks of exposure and valued those risks in the same way, one would expect that, in general, smaller communities would opt for less pure water than large ones. It is important to stress that there is nothing necessarily *wrong* about such outcome. No one is forced to live in any particular community, and if his or her judgment about drinking water quality differs from the local majority--*and* if this quality seems sufficiently important--there is no legal bar to a vote with the feet.[16] It is as fatuous to say that local choice makes "second-class citizens" out of rural and small town people, as it is to say that this is the effect of

---

[15]Costs of removing some fraction of a particular contaminant, or staying within some required maximum contaminant levels, will also presumably vary with the other characteristics of the raw water, such as turbidity, dissolved solid load, and color. The size variation is, however, enough to make the point.

[16]There may be several reasons making it hard to move out: emotional ties, job-search difficulties, and moving costs, for example, but these apply to people on both sides of every issue decided at the local level and, therefore, if taken to be an argument for uniformity in water quality, also justify uniformity in every other aspect of public life.

the absence of a health food store, jogging track, or public swimming pool. All these items may be linked to longer average life expectancies in the community. They do, on the other hand, cost money, and local opinion may prefer a shorter average life with a bigger school auditorium or better science labs or more police protection. If we set out to give every citizen equal access to everything which might result in longer life, we will, as we are already finding out in the medical care field, be opening up a bottomless pit for resource use (L. Russell, 1977).

Federal intervention is now a fact of life in the drinking water field, and there seems to be no possibility that this situation will be reversed after a review of progress to date. What, then, can be done to mitigate the worst strains. One possibility, dealt with in the next subsection, is a federal grant program to help out with the costs incurred by the SDWA. But there are other measures as well. The law and associated regulations could, for example, be changed to stress the guideline nature of the maximum contaminant level (MCL) and treatment specifications. These guidelines could be widely publicized, and then local communities could be given a chance to decide whether to adopt them or not.[17] Those systems opting for the standards could be given primary

---

[17]Private systems might be subject to rulings of state public health departments on this decision.

responsibility for self-policing, subject to state (or probably federal) audit and random checks. Those systems opting for higher contaminant levels might be required to remind their customers of that decision on a regular basis, to give the customers periodic chances to change the decision, to send out easily read summaries of relevant evidence on costs and damages as it changes, and to monitor actual contaminant levels and be subject to the same audits and checks.[18]

Another sticking point with the current law for water system managers is the notification requirement. No one seems to object to being responsible for rapid notification of emergency conditions, such as an upstream spill or an equipment malfunction. But there were complaints about being required to notify the public of a condition which had already occurred. Tests for certain problems may not be fast enough to allow notification before the "slug" of offending contaminant has passed through the system. There was some feeling that notices after the fact merely create distrust and confusion. Further, local managers would prefer to have all except emergency notices come from higher jurisdictions, state or federal. Clearly one does not want to abandon an important and rather cleverly designed instrument for motivating com-

---

[18]A measure which might have the same effects in practice would be to make the current exemption of communities below 75,000 people at least semipermanent--say ten to fifteen years, by which time continuing research might have changed our estimates of costs and benefits dramatically.

pliance with the goals of the SDWA, but fine tuning might do much to reduce ill will.

Introduction of a Subsidy Program

A common method of making the federal government's will felt at the local level has for decades been to take advantage of the more elastic federal purse by setting up a grant (or low-interest loan) program in which the grants are contingent on compliance with federal requirements. Where there is no local option on whether or not to comply (as with the water pollution control treatment requirements), the grant becomes sugar coating on a pill judged otherwise to be unacceptably bitter. The Safe Drinking Water Act contains no grant provisions (except for state planning grants), and the conference audience and program participants differed sharply on the wisdom and desirability of their addition.

There are good reasons both for Congress to want to avoid another grant program (which would almost certainly be a construction grant program), and for people at the state and local levels to be split over its desirability. Congress, one gathers from the remarks of Philip Cummings, has been impressed--if not stung--by criticisms of the water pollution control construction grant program, especially those criticisms which stress the enormous budget commitments involved and the apparent tendency of the program actually to slow down progress on the waste treatment front. On the other hand, there are equity arguments for making some efforts to equalize the burdens on water users of the requirements of the act, burdens which we have seen will generally be heavier the smaller the water system involved. Big-system managers, as we would expect, are not enthusiastic about grants. They see their cost increases caused by the SDWA as manageable, while the costs (in "hassling") of getting and using a grant as high relative to the unpleasantness of sending out higher bills. It appeared at the conference that the smaller the system, the more likely the management is to see a federal grant program as worth its cost in applications, audits, and so forth.

One fairly obvious alternative is, therefore, to amend the SDWA to include a construction grant program applicable to systems serving fewer than X thousand customers or to cities of less than X thousand people (perhaps X should equal 75 as is implied by the current EPA regulations). This would have the advantage of meeting head-on the most troublesome equity issue raised by the act but would also have several significant disadvantages. First, it would probably be impossible to pass in the form suggested; all systems would have to be given some aid before any could get it. Second, it would introduce the perverse incentives common to all capital grants--incentives to use more capital and less subsequent labor (for operation and maintenance) than would be optimal if market prices influenced the decision. Third, there is some risk that plants will be built but that subsequent operation and maintenance will be substandard (being paid for as they will be out of local money) creating new problems which would be the more serious for the false sense of security. A program which made grants for both construction *and* subsequent running expenses has its own problems, of course, particularly the incentive to waste. Fourth, there is the slowdown effect engendered by limited grant appropriations and the necessity of queuing up to get yours. There is no incentive to spend local money with the prospect of a later grant, since at best that costs extra interest and at worst exposes the community to a change in federal policy.

Is there any program which would solve the equity problem without setting back the causes of efficiency and progress unacceptably? If there is, it is certainly not evident. It might, however, make sense to combine a grant program with the voluntary local compliance policy outlined above. This would be a return to grants as payment for compliance, a transaction with at least the dignity of a quid pro quo. Further, the grant eligibility could be calculated from a table of per-customer values reflecting the economy-of-scale problem. Thus, a tiny system might receive $100 for each of its 100 customers, while New York City might receive $1 for each of its millions. Finally, the grant need not be earmarked for construction as opposed to operation, but only for the water system and its problems in complying with the drinking water standards.

Alternatives to Central System Changes

The above suggestions for change are based on the implicit assumption that the goals of SDWA can only be obtained through changes in the operations of entire public and private water systems, changes which involve making sure that all water supplied by the systems meets the mandated standards. The conference heard some evidence that this assumption deserves critical examination. In particular, two broad alternatives to system-wide changes were discussed. In one, bottled water was the key; in another, at-the-household treatment devices would do the job.[19] In both cases the central rationale rests on the assertion that only about one-half of one percent of all the water supplied by a typical water system is actually used for drinking and cooking.[20] What

---

[19]Another alternative, which some well-known figures advocate, is movement toward dual water supply systems, with separate pipes for drinking water and general purpose water. There was, unfortunately, no advocate of this alternative at the conference to present the cost-effectiveness case for it. Thus, it received only the briefest of mentions in the discussions.

[20]Estimated average community water use, expressed in per capita terms, was 175 gallons in 1975; 0.5 percent of this is 3 1/2 quarts per day. In any event, it is probably true that provision of one gallon per capita per day of potable water would be sufficient for household drinking and cooking.

sense is there, the argument goes, in treating 200 times as much water as is necessary? Who cares what the contaminant levels are in toilet flushing water or in lawn irrigation water?

Asking how these alternatives might actually come to play a part brings us back to issues of free choice, equity, and information adequacy which are far from simple. Before getting into those, however, let me consider the alternatives in slightly more detail. Bottled water could play a part in meeting the goals of the SDWA in any of a number of ways:

- Consumers could simply be advised of the relative purity of their own system and of the available bottled water and advised (or not) to depend on the latter for drinking.
- Consumers could be given vouchers, in return for part of their water bills, redeemable only for bottled water.
- Bottled water could be made available at low cost at a central point by the utility itself.
- Bottled water could be delivered door-to-door to all households or only to those households containing someone at special risk.

Similarly, there are several options for making use of point-of-use treatment devices, ranging from provision and maintenance by the water utility to complete laissez-faire, perhaps with an educational campaign. (The possibilities are also large because any number of specific quality problems and combinations of quality problems can be dealt with by different available devices.) Here, however, the engineering and economic evaluation of a program is made difficult or impossible by our current serious ignorance about the costs, relative levels of effectiveness, and amenability to do-it-yourself servicing of the many alternative pieces of equipment. An EPA study on these questions is currently underway, but no results were presented at the conference. Baumann, in his paper on this general topic repeatedly had to admit frustration in finding out even the most rudimentary facts; and Porth, from the industry, provided none.[21]

One clear possibility for congressional action, is the establishment of a major demonstration and testing program for these approaches to providing high quality drinking water. We heard references at the conference to an unnamed Illinois town that wants to meet its obligations by providing bottled water. Several such experiments should be encouraged and the costs and quality results carefully observed.

---

[21]It has come to my attention since the conference that the National Science Foundation has been funding a study which touches on these questions. It is being performed by Weston Environmental Consultants - Designers, Arun K. Deb, principal investigator, and is titled, "Multiple Water Supply Approach for Urban Water Management." In the annual report of this project, dated 20 March 1978, some calculations are presented on the relative costs of a "dual system" in which all potable water would be delivered in bottles by truck with subpotable water in the mains, and a system with all water delivered by pipe (whether dual or not is unclear from the report). The bottled water alternative is found to be twenty to fifty times more expensive in present-value terms over a twenty-five year planning period. These results may, however, overstate the case, because it is not clear from the report what treatment requirements were placed in the potable water in the piped alternative.

Before leaving this last area of policy innovation, however, let me say a bit more on the problems raised by bottled water and point-of-use devices. First, if one chooses either of these routes it will be essential to recognize that individual choice becomes a paramount factor. You can give a customer bottled water but you can't make him (or her) drink it, to paraphrase a familiar aphorism. You can, as a practical matter, require the installation of particular treatment devices through housing codes. (Though it takes some years to convert the existing stock since enforcement really depends on inspection in connection with sale or remodeling.) But it is essentially impossible to require effective maintenance. Therefore if, as seems implicit in SDWA, there are doubts about the adequacy of information available to individuals about drinking water quality and its consequences, <u>or</u> if it can be argued that since our health care system now makes us all liable to pay the cancer (or other disease) bills of those who choose not to play the game, then these point-of-use options lose much of their appeal. There is, finally, the matter of equity. Unless some subsidy is provided, we would, with most of these alternatives, be creating a situation in which the rich could more easily protect their health than could the poor; and from which the poor could opt out, as they cannot from systemwide improvements.[22] The standard remedies, including tax credits and deductions, have their own problems related to the structure of the income tax.

A Concluding Comment

The policy issues and opportunities relevant to the Safe Drinking Water Act review are not only interesting in themselves. They are representative of important problems facing our society in many areas as it attempts to come to grips with the dangers it has created through efforts to control other problems. Making decisions in face of extreme uncertainty is only the most obvious of the issues. Others include the "correct" balance between paternalistic concern and individual freedom to take risks, and the proper balance between federal, state and local governments in these areas. It is important that the participants in the continuing policy debate keep in mind that much more than science (or engineering or economics) is involved and that the actions taken on this issue have implications for our collective life at least as important as the lengthening of average life expectancy by six days or six months.

---

[22]SDWA requirements also burden the poor disproportionately, to the extent that customers of small systems are on average less well-off than customers of large systems.

References

Frankel, Richard. 1965. "Water Quality Management: Engineering-Economic Factors in Municipal Waste Disposal," Water Resources Research, vol. 1, pp. 173-186.

Kneese, A. V., and B. T. Bower. 1968. Manager Water Quality: Economics, Technology, Institutions (Baltimore, Johns Hopkins University Press for Resources for the Future).

National Academy of Sciences. 1977. "Drinking Water and Health," A report of the Safe Drinking Water Committee, Advisory Center on Toxicology, Assembly of Life Sciences, National Research Council (Washington, D.C.).

Okun, Daniel A. 1975. "A Potential for Water Reuse--Dual Supplies," Proceedings of American Water Works Association Seminar on Reuse, June 8 (Denver, Colorado, American Water Works Association).

______. 1977. Regionalization of Water Management: A Revolution in England and Wales (London, Applied Science Publishers).

Portney, Paul. 1978. "Toxic Substance Policy and the Protection of Human Health," in Paul Portney, ed., Current Issues in U.S. Environmental Policy (Baltimore, Johns Hopkins University Press for Resources for the Future).

Russell, Clifford S., David G. Arey, and Robert W. Kates. 1970. Drought and Water Supply (Baltimore, Johns Hopkins University Press for Resources for the Future).

Russell, Clifford S. 1972. "Municipal Evaluation of Regional Water Quality Management Proposals," in Robert Dorfman, Henry D. Jacoby, and Harold A. Thomas, Jr., eds., Models for Managing Regional Water Quality (Cambridge, Mass., Harvard University Press).

Russell, Louise. 1977. "Medical Care Costs," in Joseph A. Pechman, ed., The 1978 Budget: Setting National Priorities (Washington, D.C., The Brookings Institution).

LIST OF PARTICIPANTS
NATIONAL CONFERENCE ON DRINKING WATER POLICY PROBLEMS
6-8 March 1978

Michael C. Alavanja
School of Health Sciences
City University of New York
118 East 107th Street
New York, N.Y. 10029

John W. Atkinson
North West Water Authority
Dawson House
Great Sankey, Warrington
Lancashire WA5 3LW
England

Duane D. Baumann
Department of Geography
Southern Illinois University
Carbondale, Illinois 62903

Christopher Bellavita
Graduate School of Public Policy
University of California
2607 Hearst Avenue
Berkeley, California 94720

Louis Glenn Bennett
Farmer's Home Administration
14th and Independence
Room 5007, South Building
Washington, D.C. 20250

John Boland
Dept. of Geography and Environmental Engineering
Johns Hopkins University
Baltimore, Maryland 21218

Wilbur D. Campbell
Community and Economic Development Division
U.S. General Accounting Office
441 G Street, N.W.
Washington, D.C. 20548

Robert M. Clark
Municipal Environmental Research Lab
Water Supply Research Division
U.S. Environmental Protection Agency
Cincinnati, Ohio 45268

Philip T. Cummings
Committee on Environment and Public Works
U.S. Senate
Washington, D.C. 20510

N. J. Davoust
Chicago Water Department (Retired)
Eagle Nest's Lake, #4
Star Route, Box 5683
Ely, Minnesota 55731

Daniel M. Dworkin
Department of Geography
Southern Illinois University
Carbondale, Illinois 62903

David J. Eaton
LBJ School
The University of Texas
Austin, Texas 78712

Leo M. Eisel
U.S. Water Resources Council
2120 L Street, N.W.
Washington, D.C. 20036

Leo E. Ganster
U.S. General Accounting Office
441 G Street, N.W.
Washington, D.C. 20548

Henry J. Graeser
Black and Veatch Consulting Engineers
555 Griffin Square
Dallas, Texas 75202

Joseph V. Radziul
Philadelphia Water System
1180 Municipal Building
Philadelphia, Pa. 19107

Allen Hammond
American Association for the Advancement of Science
1515 Massachusetts Ave., N.W.
Washington, D.C. 20005

Timothy L. Harker
Suite 1120
1150 Connecticut Ave., N.W.
Washington, D.C. 20036

Robert Harris
Environmental Defense Fund
1525 Eighteenth Street, N.W.
Washington, D.C. 20036

Charles J. Hitch
Resources for the Future
1755 Massachusetts Ave., N.W.
Washington, D.C. 20036

Maynard M. Hufschmidt
Dept. of City and Regional Planning
University of North Carolina
Chapel Hill, N.C. 27514

Peter B. Hutt
Covington and Burling
888 Sixteenth Street, N.W.
Washington, D.C. 20036

Helen M. Ingram
Resources for the Future
1755 Massachusetts Ave., N.W.
Washington, D.C. 20036

Charles C. Johnson
Malcolm Pirnie, Inc.
8757 Georgia Avenue
Silver Spring, Md. 20910

Victor Kimm
Office of Water Supply
U.S. Environmental Protection Agency
Room 1101, East Tower, WH550
401 M Street, S.W.
Washington, D.C. 20460

William Lijinsky
Frederick Cancer Research Center
P.O. Box B, Building 538
Frederick, Maryland 21701

F. Pierce Linaweaver
Village of Cross Keys, Suite 108
Baltimore, Maryland 21210

Joyce McCann
Biochemistry Department
University of California
Berkeley, California 94720

Alan H. Molof
Polytechnic Institute of New York
333 Jay Street
Brooklyn, New York 11201

Thomas Moody
City of Columbus
90 West Broad Street
Columbus, Ohio 45268

Emil R. Mrak
University of California
602 Cordova Place
Davis, California 95616

Norton Nelson
NYU Medical Center
Dept. of Environmental Medicine
550 First Avenue
New York, New York 10016

Henry J. Ongerth
California Department of
Health
2151 Berkeley Way
Berkeley, California 94704

Harold E. Pearson
Metropolitan Water District
of Southern California
P.O. Box 54153
Los Angeles, California 90054

Donald L. Porth
Culligan International
1 Culligan Parkway
Northbrook, Illinois 60062

John T. Quigley
University of Wisconsin
432 North Lake Street
Madison, Wisconsin 53706

Nancy A. Reiches
Comprehensive Cancer Center
Ohio State University
410 W. Tenth Avenue
Columbus, Ohio 43210

Paul G. Rogers
U.S. House of Representatives
Room 2407, Rayburn Building
Washington, D.C. 20515

Gerard A. Rohlich
Dept. of Civil Engineering
8.6 E. Cockrill Hall
The University of Texas
Austin, Texas 78712

Patricia L. Rosenfield
Resources for the Future
1755 Massachusetts Ave., N.W.
Washington, D.C. 20036

Marvin A. Schneiderman
National Cancer Institute
Landow Building, Room 4C-03
Bethesda, Maryland 20014

Walter B. Sinnott
Hazen and Sawyer Consulting
Engineers
360 Lexington Avenue
New York, New York 10017

Heinrich Sontheimer
Chemical Engineering
Department
University of Karlsruhe
Engler-Bunte Institut
Richard-Willstatter-Allee 5
7500 Karlsruhe
Federal Republic of Germany

Richard G. Stevie
Environmental Research
Center
Water Supply Research Div.
U.S. Environmental Protec-
tion Agency
26 West Clair Street
Cincinnati, Ohio 45268

Carol H. Tate
James M. Montgomery Consult-
ing Engineers
555 East Walnut Street
Pasadena, California 91101

James S. Turner
1625 I Street, N.W.
Washington, D.C. 20036

Charles H. Warren
The President's Council on
Environmental Quality
722 Jackson Place, N.W.
Washington, D.C. 20006

Jacqueline M. Warren
Environmental Defense Fund
1525 Eighteenth Street, N.W.
Washington, D.C. 20036